Qing-Hao Zhang, Jun-Guo Lu, YangQuan Chen
Fractional-Order Singular Systems

Fractional Calculus in Applied Sciences and Engineering

Volume 14

Qing-Hao Zhang, Jun-Guo Lu, YangQuan Chen

Fractional-Order Singular Systems

Robust Control Problems, Admissibility and Stabilization

DE GRUYTER

Mathematics Subject Classification 2020
93A10, 93B17, 93B36, 93B51, 93B52, 93C05, 93C55, 93C73, 93D09, 93D15, 93D20

Authors
Qing-Hao Zhang
Shanghai Jiao Tong University
School of Electronic Information and Electrical Engineering
800 Dongchuan Road
Shanghai
People's Republic of China
zhangqinghao@sjtu.edu.cn

Prof. YangQuan Chen
University of California, Merced
School of Engineering
5200 N. Lake Road
Merced CA 95343
USA
ychen53@ucmerced.edu

Jun-Guo Lu
Shanghai Jiao Tong University
School of Electronic Information and Electrical Engineering
800 Dongchuan Road
Shanghai
People's Republic of China
jglu@sjtu.edu.cn

ISBN 978-3-11-162856-1
e-ISBN (PDF) 978-3-11-162864-6
e-ISBN (EPUB) 978-3-11-162882-0
ISSN 2509-7210

Library of Congress Control Number: 2024951599

Bibliographic information published by the Deutsche Nationalbibliothek
The Deutsche Nationalbibliothek lists this publication in the Deutsche Nationalbibliografie; detailed bibliographic data are available on the Internet at http://dnb.dnb.de.

Cover image: naddi/iStock/thinkstock
Typesetting: VTeX UAB, Lithuania
Printing and binding: CPI books GmbH, Leck

www.degruyter.com
Questions about General Product Safety Regulation:
productsafety@degruyterbrill.com

MIX
Papier | Fördert gute Waldnutzung
FSC® C083411

This book is dedicated to our families, mentors, and students

Preface

Fractional-order singular systems, which combine the advantages of fractional-order and singular systems, are a new research direction in the field of control science and engineering in recent decade. Fractional-order singular systems have more flexibility and applicability in modeling practical physical systems, for example, electrical circuit systems. The reason lies in the fact that the state-space model of a regular fractional-order system requires its state variables to fully characterize the system behavior. However, when modeling some practical systems, the chosen physical variables often do not meet the minimality requirement, meaning that the relationships between these variables are not entirely dynamic but also include static or algebraic relationships known as differential-algebraic equations (DAEs). On the other hand, the real industrial environment is quite harsh, with constant changes in temperature, humidity, and pressure around the systems. Additionally, as time passes, the components of the systems will also age. These factors can lead to uncertainties in the modeling of fractional-order singular systems. Robust control methods are important means to address the above-mentioned model uncertainty issues, aiming to design controllers that can ensure that the closed-loop systems remain admissible and meet the performance criteria despite the presence of uncertainties. Due to the complexity of fractional-order singular systems, there are many difficulties and challenges encountered in robust control research for such systems with uncertainties.

This book explores robust control strategies to manage the inherent uncertainties and maintain the admissibility and performance of fractional-order singular systems. It covers essential topics such as system admissibility, robust stabilization, H_∞ control, positive real control, fault detection, delay systems, and discrete systems. The book consists of eight chapters. This book first gives necessary and sufficient conditions for the admissibility of fractional-order singular systems, which can ensure the regularity, non-impulsiveness, and stability of the systems. By employing these fundamental results the robust admissibility and robust stabilization of fractional-order singular systems are addressed. When the external disturbance is considered, the robust H_∞ controllers and robust positive real controllers of fractional-order singular systems are designed. Based on the proposed bounded real lemma, the issues of fault detection of fractional-order singular systems are solved. At last, this book also explores the robust control problems of fractional-order singular delay systems and fractional-order singular discrete time systems.

To the best of our knowledge, this book is the first monograph that systematically studies the robust control of fractional-order singular systems. By investigating these areas the book provides some insights for scholars specializing in control systems, particularly those focused on robust control, fractional calculus, and singular systems, academics interested in the mathematical foundations and applications of fractional-order systems and singular systems, and graduate students pursuing degrees in control engi-

https://doi.org/10.1515/9783111628646-201

neering, applied mathematics, electrical engineering, mechanical engineering, and related fields.

Finally, but importantly, we would like to gratefully acknowledge Prof. Changpin Li, the Editor-in-chief of the series “Fractional Calculus in Applied Sciences and Engineering” of De Gruyter, for his visionary guidance that brought this book to fruition. Additionally, we appreciate the immense patience and assistance of Dr. Damiano Sacco, Acquisitions Editor Engineering/Computer Science, for his remarkable patience and invaluable help during the copy editing stage. The work was partially supported by the National Natural Science Foundation of China (No. 62303309, No. 62073217, No. 61374030, No. 60974002, No. 62403310) and the China Postdoctoral Science Foundation (No. 2022M722090).

Shanghai, China, 2024 — Qing-Hao Zhang
Shanghai, China, 2024 — Jun-Guo Lu
Merced, CA, USA, 2024 — YangQuan Chen

Contents

Abbreviations and Notations

FOCTS	Fractional-order continuous time system
FODTS	Fractional-order discrete time system
FOS	Fractional-order system
FOSCTS	Fractional-order singular continuous time system
FOS-SD	fractional-order system with state delay
FOSDTS	Fractional-order singular discrete time system
FOSS	Fractional-order singular system
FOSS-SD	Fractional-order singular system with state delay
GKYP	Generalized Kalman–Yakubovich–Popov
IOSDTS	Integer-order singular discrete time system
IOSS	Integer-order singular system
KYP	Kalman–Yakubovich–Popov
LMI	Linear matrix inequality
ZOH	Zero-order holder
$\mathbb{R}$	Set of real numbers
$\mathbb{C}$	Set of complex numbers
I_n	The identity matrix of dimension n
0	The zero matrix of the appropriate dimension
$\otimes$	The Kronecker product
$*$	The convolution operation
M^T	The transpose of matrix M
$\bar{M}$	The conjugate of matrix M
M^*	The conjugate transpose of matrix M
$\lambda_{\max}\{M\}$	The maximum eigenvalue of matrix M
$\star$	The conjugate symmetric component in the matrix
$M < 0$	The negative definite matrix M
$M > 0$	The positive definite matrix M
$\mathrm{Sym}\{M\}$	$M + M^*$

https://doi.org/10.1515/9783111628646-202

1 Introduction and overview

Abstract: In this chapter, we introduce and overview fractional-order singular systems (FOSSs). Firstly, we introduce the research background and significance of FOSSs. Secondly, we review present situation of research on FOSSs. Thirdly, we introduce the concept of fractional-order calculus. Finally, we present a preview of chapters.

1.1 Introduction

Fractional-order calculus has a history of over 300 years, first proposed by L'Hôpital and Leibniz [79]. From a mathematical perspective, it extends the traditional integer-order calculus to arbitrary orders. For a long time after the introduction of fractional-order calculus, most scholars focused only on the mathematical aspects due to limited technological advancement, with little exploration of its engineering applications. In 1974, Oldham and Spanier coauthored a monograph related to fractional-order calculus theory and its applications [75], which detailed the research progress and application prospects of rational fractional-order calculus, promoting the integration of fractional-order calculus theory with practical applications. In 1982, Mandelbrot [61] suggested that fractional-order systems (FOSs), where the order of differentiation is not an integer, are prevalent in nature. In 1999, Podlubny [79] published a classical work on fractional-order calculus, systematically describing fractional-order calculus of arbitrary orders and its applications in system modeling, also providing examples of FOSs in real engineering. Compared to integer-order systems (IOSs), FOSs possess characteristics of both infinite- and finite-dimensional systems [66]. For complex systems with global correlations and historical dependencies, applying FOSs can achieve more accurate models [79]. Since then, FOSs have attracted increasing interest and attention from scholars in various fields.

Singular systems, also known as descriptor systems, generalized state-space systems, implicit systems, and differential-algebraic systems, have been found to have applications in many fields. In 1974, Rosenbrock [83] proposed a singular system model while studying complex electrical network systems. Since then, the singular system model has entered the view of scholars. Among them, Luenberger [57, 58] published multiple papers on singular systems in top control journals such as *IEEE Transactions on Automatic Control* and *Automatica*, researching their related properties. This significantly promoted the theoretical research on singular systems and their applications in physical and engineering modeling. For example, the mobile cleaning robot studied in [18] is a typical case of a singular system. Other examples of applications of singular systems include biological systems [44], robotic systems [85], economics [7], and computers [121], demonstrating the significant practical application value of singular systems.

https://doi.org/10.1515/9783111628646-001

FOSSs combine the advantages of FOSs and singular systems, making them highly significant for research both in engineering and academia. FOSSs offer greater flexibility and applicability in modeling actual physical systems. For instance, the circuit systems mentioned in [112] are modeled by FOSs. However, if the circuit system contains branch nodes with inductors, such as the circuit system in Example 3.5 of [41], then it involves not only the dynamic constraints of fractional-order differential equations but also algebraic constraints. Therefore FOSSs must be used for modeling. This is because the state-space model of a normal FOS requires its state variables to fully describe the system behavior. However, when modeling some real systems, the selected physical variables often do not meet the minimum requirements, meaning that the relationships between these variables are not purely dynamic but also static. In such cases, FOSSs are necessary for description. Besides their engineering applications, FOSSs are also of great academic importance. In the analysis of time-domain solutions, the time-domain solutions of normal FOSs only contain Mittag-Leffler responses, whereas the time-domain solutions of FOSSs include both the Mittag-Leffler responses and impulse terms from infinite subsystems, as well as input derivative terms [38]. In terms of system degrees of freedom, for a system of dimension n, a normal FOS has n degrees of freedom, whereas an FOSS has $\operatorname{rank}(E) = r \leq n$. For system stability, normal FOSs only require stability considerations, whereas FOSSs need to take the admissibility (the regularity, nonimpulsiveness, stability) into account [106, 48, 97]. Regularity ensures that the FOSS has a unique solution. Nonimpulsiveness ensures that there are no impulse terms in the system solutions, as the presence of impulse terms could lead to control saturation and affect system performance. In terms of integrated modeling with other systems, FOSSs are widely applied in combined modeling with fuzzy systems, multiagent systems, and switched systems [118, 117, 78].

As mentioned above, FOSSs are widely applied in engineering fields. However, real industrial environments are often harsh, with constantly changing factors such as temperature, humidity, and pressure. Additionally, system components may degrade over time. These factors lead to uncertainties in the modeling of FOSSs, which can severely affect the system admissibility and performance. Therefore studying FOSSs with uncertainties is of great significance. Robust control is an important approach to address these issues. It aims to design controllers that can ensure the admissibility and performance of the closed-loop systems under the influence of uncertainties. Due to the complexity of FOSSs, research on robust control for such systems with uncertainties faces many difficulties and challenges. To solve these problems, this book focusses on the robust control of FOSSs from different perspectives, including admissibility, robust stabilization, H_∞ control, positive real control, and fault detection of FOSSs, as well as robust control of fractional-order singular systems with state delay (FOSSs-SD) and fractional-order singular discrete-time systems (FOSDTSs).

1.2 Overview of robust control of FOSSs

In this section the research on robust control of FOSSs is reviewed.

1.2.1 Admissibility and robust stabilization of FOSSs

An FOSS consists of a set of fractional-order differential equations describing the dynamic relationships of the state variables and algebraic equations describing the static relationships. For normal FOSs, stability is the primary factor to consider. However, for FOSSs, it is also necessary to consider their regularity, nonimpulsiveness, and admissibility. Normal linear FOSs always have a unique solution, but for FOSSs, the situation is much more complex. They may have a unique solution, an infinite number of solutions, or no solutions. For example, consider the FOSS

$$\begin{bmatrix} 1 & 0 & 0 \\ 0 & 1 & 0 \\ 0 & 0 & 0 \end{bmatrix} D^{\alpha}x(t) = \begin{bmatrix} -1 & 0 & 0 \\ 0 & -1 & 0 \\ 0 & 0 & 0 \end{bmatrix} x(t),$$

where $x(t) = \begin{bmatrix} x_1^T(t) & x_2^T(t) & x_3^T(t) \end{bmatrix}^T$. The substates $x_1(t)$ and $x_2(t)$ are constrained by the dynamic fractional-order differential equations, but the substate $x_3(t)$ has no constraints, meaning that this FOSS has infinitely many solutions. The existence and uniqueness of solutions for FOSSs introduce the concept of regularity, indicating that an FOSS is regular when it has a unique solution. The works [69, 106] studied the regularity conditions of FOSSs in the frequency domain. Based on the regularity conditions, the studies [68, 96] designed output feedback controllers to ensure that the closed-loop systems satisfy the regularity. Unlike normal FOSs, the time-domain solutions of FOSSs are more complex. This is because they not only include the Mittag-Leffler response of finite subsystems but also contain impulse terms from infinite subsystems and input derivative components. Some studies [106, 116] directly presented the time-domain solutions of FOSSs, but they overlooked the system control inputs, which play an important role in the completeness of the time-domain solutions. The work [38] provided the time-domain solutions with control inputs, but the conclusions ignored the impulse terms generated by the initial input terms and their derivatives, which can affect the system impulse control performance. The impulse terms do not affect asymptotic stability, but they can likely cause control saturation, leading to performance degradation. The studies [69, 106] investigated the conditions for the nonimpulsiveness in the frequency domain.

The admissibility of FOSSs describes the regularity, nonimpulsiveness, and stability. Numerous interesting results have emerged thanks to the tireless efforts of scholars. The work [100] provided necessary and sufficient conditions for D-admissibility of FOSSs of order $\alpha \in [1, 2)$, based on the regularity, nonimpulsiveness, and D-stability. Subsequently, the admissibility conditions for FOSSs of order $\alpha \in [1, 2)$ were presented in [69], but the systems were assumed to be regular and impulse-free, thus limiting the applicability of the conclusions. To derive the admissibility conditions for FOSSs with order $\alpha \in (0,1)$, the work [106] provided conditions for systems with order $\alpha \in (0,1)$. However, the conclusions had some undetermined variables and were conservative. Then [105] presented necessary and sufficient conditions for the admissibility with orders $\alpha \in (0,1)$ and $\alpha \in [1,2)$. Following this, other similar forms of necessary and sufficient conditions for the admissibility were proposed, such as those in [64, 63], which provided admissibility conditions for systems with orders $\alpha \in (0,1)$ and $\alpha \in [1,2)$, respectively. To resolve the complex variable issues encountered when solving linear matrix inequalities (LMIs) in [105, 64], the work [116] provided admissibility conditions where all the undetermined variables are real matrices. Building on these admissibility conditions, [48] used matrix variable decoupling techniques and designed a more generalized controller based on an observer, ensuring that the closed-loop FOSS meets the admissibility requirements.

When FOSSs are subject to uncertainties, the robust admissibility and stabilization play a significant part in robust control theory of FOSSs, and some valuable works are proposed. For FOSSs with norm-bounded uncertainties, the robust stabilization conditions were obtained, and state feedback controllers were designed in [116, 68]. If the internal states of FOSSs cannot be measured, the controllers in [116, 68] are not applicable, promoting the design of output feedback controllers. For example, the output feedback controllers for FOSSs were designed to stabilize the systems in [36, 63, 120]. However, the input matrix B^T in [36] and the output matrix C in [63, 120] are required to be full row rank, which have some limitation of applications. Two-step methods of designing the output feedback controllers of FOSSs were put forward in [96, 97], where virtual state feedback controllers are designed firstly, followed by the output feedback controllers. This method is indirect and may lead to no solutions if the parameters in the first step are not determined appropriately, as can be seen in [49]. The output feedback controller synthesis of FOSSs with norm-bounded uncertainties was solved in [62], where the results are rather complicated, and an improper selection of the parameter J may lead to no feasible solutions. As mentioned above, most existing works on the robust admissibility of FOSSs focus on norm-bounded uncertainties [68, 36, 116, 62]. In the case where FOSSs are subjected to polytopic uncertainties, the robust admissibility and stabilization are still open problems. Note that not all the internal states of FOSSs can be sensed, which impels us to utilize output feedback controllers instead of state feedback controllers to stabilize the systems. Thus we expect to obtain robust admissibility conditions for FOSSs with uncertainties and then design the output feedback controllers such that the closed-loop systems are robustly admissible.

1.2.2 Robust H_∞ control of FOSSs

The H_∞ performance of a system reflects its disturbance rejection and robust stability capabilities, playing a critical role in control system analysis and design. H_∞ control was first introduced by Zames [107] in 1981 through the study of system sensitivity. Over time, significant research has led to the development of mature frequency domain solution methods. In 1990, Doyle, Glover, Khargonekar, and Francis [17] formalized the general problem of H_∞ control by solving four Riccati equations, thereby establishing a comprehensive theoretical framework for solving the problem in state-space settings. This was a milestone in the development of H_∞ control. Later, the introduction of LMI techniques overcame the challenge of solving high-dimensional Riccati equations, transforming them into simplified LMI conditions, known as the bounded real lemmas, which ensured effective solutions to the problem [26, 34].

For FOSs, the study of H_∞ performance has become a hot topic. One of the most important research tools in this area is the Kalman–Yakubovich–Popov (KYP) lemma. The KYP lemma effectively establishes the connection between time-domain and frequency-domain methods [12]. Although many studies and applications of the KYP lemma have been conducted, it was not until 2009 that Camlibel and Frasca [10] extended it to singular systems for the first time. In [20, 46] the bounded real lemmas for FOSs under the L_∞ and H_∞ norms were presented based on the KYP lemma, and further, the H_∞ control synthesis problem was solved using the LMI method. In [4] a robust H_∞ controller based on an observer was designed to ensure the robust stability and satisfy H_∞ performance for the closed-loop system. Moreover, the work [3] proposed an algorithm to solve the H_∞ norm for FOSs based on the transfer function maximum singular value level set method. In [45] a general framework for the finite frequency band generalized KYP lemma for FOSSs was established, and the bounded real lemmas under the L_∞ norm for different frequency ranges were provided, along with the corresponding controller to improve the L_∞ performance of FOSSs.

H_∞ control analysis is more complex than L_∞ performance analysis because it not only needs to satisfy H_∞ performance criteria but also must meet system stability conditions. For singular systems, system admissibility conditions must also be satisfied. Moreover, H_∞ control for FOSSs plays an irreplaceable role in the design of filters and observers, making the study of H_∞ control for FOSSs highly significant.

1.2.3 Robust positive real control of FOSSs

Positive realness is a crucial research area in system theory, tracing back to the 1930s when Brune [6] first introduced the concept while studying passive networks. It has significant applications in various fields, such as stability analysis, hyperstability, dissipativity, quadratic optimal control, algebraic Riccati equations, and adaptive control [24, 30]. The problem of positive real control refers to designing an internally stable controller for a given system so that the closed-loop system becomes strictly positive real. Such a controller is called a positive real controller. In a feedback interconnection system, where the forward part is a linear system and the feedback part is a time-invariant or time-varying nonlinear system, if the nonlinear part can be characterized by a Popov integral inequality, then ensuring the extended strict positive realness of the linear system guarantees robust asymptotic stability for the entire feedback system. One major advantage of this approach is that the internal structure of the nonlinear component does not need to be explicitly known. It only needs to adhere to the Popov inequality in its input–output behavior. Thus the strict positive realness of a linear system guarantees robust asymptotic stability for the entire system. Compared to Lyapunov stability theory, which addresses the stability of a system without external input, the stability method based on positive realness deals with the system stability in the presence of external input and provides global stability results, offering greater universality and flexibility.

The methods for studying positive real control can be divided into frequency domain and time domain approaches. Due to the large computational effort required for frequency domain methods, which makes controller design difficult, recent research has focused on time domain methods. These can be further categorized into Riccati equation methods and LMI methods. Since solving LMIs is more convenient compared to Riccati equations, LMI methods have become an effective approach for studying positive real control. In [91], clear definitions for system positive realness, strict positive realness, and extended strict positive realness are provided. Based on this, several research achievements on positive realness for IOSs have emerged. In [99], conditions for norm-bounded uncertain systems to satisfy extended strict positive realness were presented using LMI methods, and state feedback and output feedback positive real controllers were designed. For singular systems, [111] provided sufficient and necessary conditions for singular systems to achieve extended strict positive realness in both continuous and discrete cases. When singular systems have parametric uncertainties, [23] studied the conditions under which discrete uncertain singular systems satisfy extended strict positive realness and designed state feedback control to ensure that the closed-loop system is admissible and strictly positive real.

For FOSs, [123] investigated the positive realness of fractional-order linear time-invariant systems of orders $\alpha \in (0,2)$. Using LMI methods and the KYP lemma, it provides necessary and sufficient conditions for the positive realness of FOSs, as well as sufficient conditions for achieving positive realness with pseudo-state feedback in such systems. Building on the positive realness lemma in [123], [29] explored the learnability of FOSs and analyzed the relationship between learnability and the positive realness lemma for FOSs. To the best of our knowledge, there have been no studies addressing the extended strict positive realness of FOSSs, largely due to technical difficulties. On one hand, when deriving the conditions for extended strict positive realness, it is necessary to consider both the system's strict positive realness and asymptotic stability, i. e., whether the system is analytic when $\mathrm{Re}(s) \geq 0$. For instance, the results in [123] can only ensure the positive realness of FOSs but cannot guarantee their stability simultaneously. On the other hand, whereas stability alone needs to be considered for normal FOSs, FOSSs require additional consideration of regularity and nonimpulsiveness. This ensures the system to have a unique solution without impulsive behavior.

1.2.4 Robust fault detection of FOSSs

Fault diagnosis has emerged as a prominent research focus. Among the numerous approaches, model-based fault detection (FD) techniques have been explored extensively and are recognized as one of the most effective methods. Most existing FD strategies rely on robust fault detection observers, which generate a residual signal and evaluate it through a threshold-based residual evaluation function. One of the key challenges in this area is how to distinguish between faults and disturbances. To address this issue, the H_- index has been introduced, measuring fault sensitivity performance, and the H_-/H_∞ FD observer has been established as one of the most effective tools to simultaneously ensure disturbance attenuation and fault sensitivity performance.

However, as indicated by [95], full-frequency H_-/H_∞ FD observers are primarily valid for faults manifesting directly in measurement equations. In real-world applications, faults and disturbances often occur in specific frequency ranges. For instance, incipient faults may be confined to the low-frequency domain [9, 94], as is the case with actuator failures in flight control systems [47]. Consequently, the design of FD observers over finite-frequency domains has gained traction. Various studies have explored this, including methods for linear discrete-time systems [115], T-S fuzzy singular multiple time-delay systems [108], and LPV singular systems [95]. However, these methods are generally restricted to integer-order systems, whereas fractional-order systems (FOSs) remain underexplored. A fractional-order singular system FD observer over finite frequencies would offer more flexible and robust fault detection, allowing differentiation between faults and disturbances.

Research on FOSSs remains limited. Although [81] had addressed sensor fault estimation in FOSSs, their work spanned the entire frequency domain and is not suitable for actuator faults. Furthermore, the finite-frequency H_- index and its corresponding lemmas for FOSSs have not been formally established, posing challenges in addressing disturbances impacting sensors. Thus designing an H_-/H_∞ FD observer for FOSSs over finite-frequency ranges is still an open and valuable problem in the field.

1.2.5 Robust control of FOSSs-SD

In the actual systems, the transmission of signal and energy always takes some time, which may lead to the instability and performance degradation of systems [60, 11]. Therefore the influence of state delay should be considered in the analysis and design of FOSs. To study the stability of fractional-order systems with state delay (FOSs-SD), some important research results in the frequency domain and time domain have been put forward. In the frequency domain the FOSs-SD were studied in [15], and several stability conditions were proposed by introducing the characteristic equation for such systems. To study the exact stability boundary of fractional-order multidelay systems, the characteristic equations of these systems were analyzed in [31, 77], and some different stability analysis methods were given. The method of finite spectrum assignment was given in [52], based on which the robust stability conditions for FOSs-SD were established. To study the stability of disproportionate FOSs-SD, the Nyquist theorem was used to give sufficient conditions for the stability of disproportionate FOSs-SD, and the relationship between the fractional order and time-delay coefficient was provided in [114]. In the time domain the stability problems of FOSs-SD were solved in [2] based on the Lyapunov method, but the required matrix variables must be diagonal, increasing the conservatism of the results. To reduce the conservatism, the indirect Lyapunov method and the diffusion expression method were used to study the stability of FOSs-SD. For example, the stability and stabilization conditions of FOSs-SD were proposed in [5] by applying the Lyapunov–Krasovskii approach and the diffusive representation. Different from the methods presented in [5], the small gain theorem was used, and the delay term was treated as a disturbance term, and then stability and stabilization conditions of FOSs-SD were obtained in [37] by using the fractional bounded real lemma.

Similarly, the delay term affects the stability and admissibility of fractional-order singular systems with state delay (FOSSs-SD). To study the admissibility of FOSSs-SD, some scholars have proposed interesting results. The FOSSs-SD defined by the Riemann–Liouville fractional derivative were studied in [53], and sufficient conditions for the admissibility of the system were provided by using the Lyapunov method. However, the existence of matrix equation in [53] may lead to no solutions. In [98] the admissibility conditions of FOSSs-SD defined by the Caputo fractional-order derivative were given in LMI form by using the Lyapunov–Krasovskii method. However, note that the Caputo fractional-order derivative is different from the integer-order derivative. The method of directly adopting the integer-order Lyapunov–Krasovskii function for the Caputo fractional derivative must be revised. To study the admissibility of FOSSs-SD without using the Lyapunov method, the analysis of Mittag-Leffler function and the scaling of norm inequality were employed in [42], and sufficient admissibility conditions were obtained, but this method cannot be directly used to design the controllers.

1.2.6 Robust control of FOSDTSs

For fractional-order singular continuous-time systems (FOSCTSs), they are composed of fractional-order differential equations and algebraic equations. The admissibility of FOSCTSs, including regularity, nonimpulsiveness, and stability, needs to be taken into account. A significant number of research results on different aspects of FOSCTSs have been proposed in monographs and papers; see, for example, [106, 68, 36, 116, 97, 63] and references therein. For fractional-order singular discrete-time systems (FOSDTSs), however, they are composed of a set of fractional-order difference and algebraic equations. The admissibility of FOSDTSs includes the regularity, causality, and stability. The regularity guarantees the existence and uniqueness of solutions, and the causality ensures that the state at any time is determined by the initial condition and previous inputs. Note that singular discrete systems are not always causal and noncausality indeed exists in real systems [14]. For FOSDTSs, it is necessary to analyze the regularity and causality, which are as important as stability.

The admissibility of FOSCTSs has been widely studied, but the admissibility of FOSDTSs has only been preliminarily researched. In [41] a new class of FOSDTSs was introduced, and the solutions were proposed by using the Weierstrass regular pencil decomposition. The logistic map and Kalman filter of FOSDTSs were investigated in [73] and [74], respectively, where the regularity and causality of FOSDTSs were also studied. For the stability of FOSDTSs, the Lyapunov method was applied, and a sufficient condition for the asymptotical stability was given in [40], but the results are not tractable for controller design. To overcome this problem, the stability conditions for FOSDTSs in terms of matrix inequalities were proposed in [54].

The aforementioned works have some issues to be resolved. For example, the FOSDTSs in [41] were assumed to be regular, and the causality was not considered. In [73, 74] the regularity and causality were analyzed in the frequency domain. It will be more direct and vivid to study them in time domain instead of frequency domain. For the stability of FOSDTSs, only few works have been proposed, and there is a lack of consideration for causality when analyzing the stability conditions [40, 54]. Therefore the admissibility of FOSDTSs is still waiting for further study. In addition, FOSDTSs are inevitably influenced by uncertainties, for example, interval uncertainties. It is necessary to solve the controller synthesis problem for FOSDTSs with interval uncertainties.

1.3 Fractional-order calculus

Currently, there is no unified definition of fractional-order derivatives. In this section, we introduce three common definitions of fractional-order derivatives.

Definition 1.1 (Grünwald–Letnikov (GL) fractional-order derivative [79]). For a real number $\alpha > 0$, the α-order fractional-order derivative of a function f defined on $[0, t]$ is

$$ {}_{GL}D_t^{\alpha} f(t) = \lim_{h \to 0} h^{-\alpha} \sum_{r=0}^{\lfloor \frac{t}{h} \rfloor} (-1)^r \binom{\alpha}{r} f(t - rh), $$

where $\lfloor \frac{t}{h} \rfloor$ represents the floor function of $\frac{t}{h}$.

Definition 1.2 (Riemann–Liouville (RL) fractional-order derivative [79]). For a real number $\alpha > 0$, the α-order fractional-order derivative of an integrable function f defined on $[0, t]$ is

$$ {}_{RL}D_t^{\alpha} f(t) = \frac{1}{\Gamma(k - \alpha)} \frac{d^k}{dt^k} \int_0^t (t - \tau)^{k-\alpha-1} f(\tau) d\tau, $$

where k is an integer satisfying $k-1 < \alpha < k$, and Γ denotes the gamma function defined as $\Gamma(z) = \int_0^{\infty} e^{-t}t^{z-1}dt$.

Definition 1.3 (Caputo fractional-order derivative [79]). For a real number $\alpha > 0$, the α-order fractional-order derivative of a k-times differentiable function f defined on $[0, t]$ is

$$ {}_C D_t^{\alpha} f(t) = \frac{1}{\Gamma(k-\alpha)} \int_0^t (t-\tau)^{k-\alpha-1} f^{(k)}(\tau) d\tau, $$

where k is the integer satisfying $k-1 < \alpha < k$, and $f^{(k)}$ denotes the kth-order derivative of f.

Remark 1.1. One key distinction among these definitions lies in the difference in their initial values of the Laplace transform. The Caputo fractional-order derivative has a significant advantage, as its initial conditions have clear physical interpretations. The Laplace transform of the Caputo fractional-order derivative is

$$ L\{{}_C D_t^{\alpha} f(t)\} = s^{\alpha} F(s) - \sum_{m=0}^{k-1} s^{\alpha-m-1} f^{(m)}(0), $$

which shows that the initial values of the Laplace transform of the Caputo fractional-order derivative are with the same explicit physical meanings as those of the integer-order derivative [79]. Additionally, the Caputo fractional-order derivative of a constant is zero, that is, if $f(t)$ is a constant, then

$$ {}_C D_t^{\alpha} f(t) = 0. $$

Therefore the Caputo fractional-order derivative is widely used in control system analysis and design. Throughout this book, we adopt the Caputo fractional-order derivative, denoted as D^{α} for simplicity.

1.4 Preview of chapters

In Chapter 1, a brief introduction of FOSSs is given, which consists of the research background and significance. Then the literature review of the robust control of FOSSs is provided. The definition of fractional-order calculus is reviewed, which is one of the very important concepts in this book. The preview of chapters is also given, which can help the readers better understand the context of this book.

In Chapter 2, the regularity, nonimpulsiveness, stability, and admissibility of FOSSs are considered. First, the existence and uniqueness of time-domain solutions of the systems are analyzed by using the Kronecker equivalent standard form, and then a necessary and sufficient condition for the regularity is proposed. Second, the explicit time-domain solutions of the systems are presented, which are the basis of necessary and sufficient conditions for the nonimpulsiveness and stability. Third, several versions of novel necessary and sufficient conditions for the admissibility of the systems are proposed.

In Chapter 3, the problems of robust admissibility and stabilization for uncertain FOSSs are considered. The structured and polytopic uncertainty cases are studied. The robust admissibility conditions of FOSSs with structured uncertainties are given by using robust nonsingularity analysis of uncertain matrices, error-free determinant transformation, and μ-analysis methods. The robust admissibility conditions of FOSSs with polytopic uncertainties are given by using the proposed decoupled methods and LMIs, which are easier to design closed-loop controllers. For uncertain FOSSs with $1 < \alpha < 2$ and $0 < \alpha < 1$, the output feedback controllers are designed.

In Chapter 4, the robust H_∞ control problems of FOSSs are addressed. Firstly, based on the LMI method and the generalized Kalman–Yakubovich–Popov (GKYP) lemma, the bounded real lemmas of FOSs are extended to the bounded real lemmas of FOSSs, resulting in more generalized results. The state feedback H_∞ controllers are designed for FOSSs with norm-bounded uncertainties, so that the closed-loop systems satisfy both robust admissibility and H_∞ performance.

In Chapter 5, the robust positive real control problems of FOSSs are studied. Firstly, based on the LMI method and KYP lemma, the extended strictly positive real conditions for nominal FOSs are given, which can ensure the asymptotic stability of the systems. On this basis, the positive real lemmas for nominal FOSSs are given, which can ensure that the nominal FOSSs satisfy both admissibility and extended strictly positive realness. Using the positive real lemmas, state feedback controllers are designed for FOSSss with norm-bounded uncertainties, ensuring that the closed-loop systems satisfy both robust admissibility and extended strictly positive realness.

In Chapter 6, the robust fault detection observer design problems in finite-frequency domains for FOSSs are investigated. An H_-/H_∞ fault detection observer for FOSSs is designed to meet the system admissibility, disturbance robustness, and fault sensitivity indices in finite-frequency domains simultaneously. Four theorems and four corollaries on the system admissibility and the performance index H_-/H_∞ in different frequency ranges are given, and then sufficient conditions are obtained in terms of LMIs.

In Chapter 7, the robust control problems of FOSSs-SD are explored. Firstly, the regularity and nonimpulsiveness conditions and the solutions of nominal FOSSs-SD are proposed. Secondly, the admissibility conditions for nominal FOSSs-SD are given based on the small gain theorem and the bounded real lemmas of FOSSs. Then the methods of designing the state feedback controllers such that the uncertain FOSSs-SD are robustly admissible are obtained. Based on the above results, the robust H_∞ controllers and robust positive real controllers are designed.

In Chapter 8, the admissibility and robust stabilization of FOSDTSs with interval uncertainties are investigated. Firstly, based on the analysis of the regularity, causality, and stability, novel admissibility conditions for nominal FOSDTSs are derived including a necessary and sufficient condition in terms of spectral radius and a sufficient condition in terms of nonstrict LMIs. To eliminate the coupling terms and propose strict LMI results, another novel admissibility condition is obtained, which is more tractable and reliable with the available LMI software solver and more suitable for the controller design compared with the existing results. Secondly, the state feedback controller synthesis for the FOSDTSs with interval uncertainties is addressed, and the state feedback controller is designed.

1.5 Chapter summary

In this chapter a brief introduction of FOSSs has been given, which consists of the research background and significance. Then the literature review of the robust control of FOSSs has been provided. The definition of fractional-order calculus has been stated and reviewed, which is important in this book. The preview of chapters has also been given, which can help the readers better understand the context of this book.

2 Admissibility

Abstract: In this chapter, we consider the regularity, nonimpulsiveness, stability, and admissibility of fractional-order singular systems (FOSSs) with fractional-order $\alpha \in (0, 2)$. Firstly, we analyze the existence and uniqueness of time domain solutions of the systems by using the Kronecker equivalent standard form, and then we propose a necessary and sufficient condition for the regularity. Secondly, we present the explicit time domain solutions of the systems, which are the basis of necessary and sufficient conditions for the nonimpulsiveness and stability. Thirdly, we propose novel necessary and sufficient conditions for the admissibility of the systems.

2.1 Introduction

With the advancement of fractional-order systems (FOSs), FOSSs have been sporadically researched in papers scattered throughout the literature. FOSSs are composed of fractional-order differential equations and algebraic equations. For normal FOSs, the stability is the main issue for consideration. However, for FOSSs, the regularity, nonimpulsiveness, and admissibility also need to be studied. Note that the dynamic orders of FOSSs are less than the dimensions of the systems, which may cause the systems to have no solution or have multiple solutions. Furthermore, similarly to integer-order singular systems (IOSSs), the state response of FOSSs may contain impulsive terms, which can cause control actuator saturation problems [18]. For FOSSs, the importance of regularity, nonimpulsiveness, and admissibility is explained as follows. The regularity guarantees the existence and uniqueness of a solution of FOSSs. The nonimpulsiveness ensures that the state responses of FOSSs have no impulsive terms. The regularity and nonimpulsiveness analysis of FOSSs should be performed to obtain the regularity and nonimpulsiveness conditions of the systems [70, 106]. The admissibility describes the regularity, nonimpulsiveness, and stability of FOSSs, which can be used for robust admissibility analysis and controller design [62].

With the above motivation, the FOSSs of fractional order $\alpha \in (0, 2)$ will be considered in this chapter, and the main contributions are as follows.

(1) A necessary and sufficient condition for the regularity is proposed by analyzing the structure, existence, and uniqueness of solutions of FOSSs based on the Kronecker equivalent standard form. The condition is suitable for not only linear FOSSs but also nonlinear FOSSs, which is more general than the existing results [69, 106].

(2) The time domain solutions of finite subsystems and infinite subsystems of FOSSs are derived by the Laplace transform. It should be emphasized that in this chapter the time domain solutions of infinite subsystems include the impulse terms excited by the initial values of input and its derivatives, which are not considered in the existing references [69, 106, 38, 116].

https://doi.org/10.1515/9783111628646-002

(3) A necessary and sufficient condition for determining whether FOSSs are impulse-free is proposed based on the time domain solutions.
(4) A necessary and sufficient condition for the stability of FOSSs is given by checking whether the finite poles are within the stable area or not. The condition is derived on the basis of the time domain solutions of FOSSs.
(5) Several novel necessary and sufficient admissibility conditions of FOSSs are proposed in terms of linear matrix inequalities (LMIs), which cover the admissibility conditions of IOSSs and have some advantages compared with the existing results.

2.2 Regularity analysis

Consider the following FOSS

$$ED^{\alpha}x(t) = Ax(t) + Bu(t), \tag{2.1}$$

where $x(t) \in \mathbb{R}^n$ and $u(t)$ are the state vector and control input, respectively, D^{α} represents the Caputo fractional-order derivative, $\alpha \in (0, 2)$ is the commensurate order of FOSSs, and $E, A \in \mathbb{R}^{n\times n}$ with $\operatorname{rank}(E) = r \le n$.

To propose the main results in this chapter, we need the following definitions and lemmas.

Lemma 2.1 ([27]). *For arbitrary matrices $A, E \in \mathbb{R}^{m\times n}$, there exist invertible matrices Ψ, Φ satisfying*

$$\begin{aligned}
\Phi A\Psi &= \operatorname{diag}(\mathbf{0}, J_1, J_2, \ldots, J_p, J_1', J_2', \ldots, J_q', A_1, I_g),\\
\Phi E\Psi &= \operatorname{diag}(\mathbf{0}, L_1, L_2, \ldots, L_p, L_1', L_2', \ldots, L_q', I_r, N),
\end{aligned}$$

where

$$A_1 \in \mathbb{R}^{r\times r}, \quad \mathbf{0} \in \mathbb{R}^{m_0\times n_0},$$

$$L_i = \begin{bmatrix} 1 & 0 & & & \\ & 1 & 0 & & \\ & & \ddots & \ddots & \\ & & & 1 & 0 \end{bmatrix}, \quad J_i = \begin{bmatrix} 0 & 1 & & & \\ & 0 & 1 & & \\ & & \ddots & \ddots & \\ & & & 0 & 1 \end{bmatrix},$$

$$L_j' = \begin{bmatrix} 1 & & & \\ 0 & 1 & & \\ & 0 & \ddots & \\ & & \ddots & 1 \\ & & & 0 \end{bmatrix}, \quad J_j' = \begin{bmatrix} 0 & & & \\ 1 & 0 & & \\ & 1 & \ddots & \\ & & \ddots & 0 \\ & & & 1 \end{bmatrix},$$

$$
\begin{aligned}
&J_i \in \mathbb{R}^{m_i\times(m_i+1)}, \quad J_j' \in \mathbb{R}^{(n_j+1)\times n_j},\\
&L_i \in \mathbb{R}^{m_i\times(m_i+1)}, \quad L_j' \in \mathbb{R}^{(n_j+1)\times n_j},\\
&N = \operatorname{diag}(N_{k_1}, N_{k_2}, \dots, N_{k_l}) \in \mathbb{R}^{g\times g},
\end{aligned}
$$

$$
N_{k_s} = \begin{bmatrix} 0 & 1 & & \\ & 0 & \ddots & \\ & & \ddots & 1 \\ & & & 0 \end{bmatrix} \in \mathbb{R}^{k_s\times k_s}, \quad \sum_{s=1}^{l} k_s = g,
$$

$$
g + r + \sum_{j=1}^{q}(n_j+1) + \sum_{i=1}^{p} m_i + m_0 = m,
$$

$$
g + r + \sum_{i=1}^{p}(m_i+1) + \sum_{j=1}^{q} n_j + n_0 = n,
$$

$$
j = 1,2,\dots,q, \quad s = 1,2,\dots,l, \quad i = 1,2,\dots,p.
$$

Lemma 2.2 ([79]). *Consider the following initial-value problem of fractional-order differential equation:*

$$
p_n(t)y(t) + \sum_{j=1}^{n-1} p_j(t)D^{\sigma_{n-j}}y(t) + D^{\sigma_n}y(t) = f(t), \tag{2.2}
$$

$$
\left[D^{\sigma_k-1}y(t)\right]_{t=0} = b_k, \tag{2.3}
$$

where

$$
\begin{aligned}
D^{\sigma_k} &= D^{\alpha_k}D^{\alpha_{k-1}}\cdots D^{\alpha_1},\\
D^{\sigma_k-1} &= D^{\alpha_k-1}D^{\alpha_{k-1}}\cdots D^{\alpha_1},\\
\sigma_k &= \sum_{i=1}^{k}\alpha_i, \quad 0<\alpha_i<1,\\
0<t<T<\infty, &\quad \int_0^T |f(t)|dt < \infty,\\
k &= 1,2,\dots,n.
\end{aligned}
$$

The existence and uniqueness of solutions with respect to the initial-value problem (2.2)–(2.3) *are described as follows.*

If $p_j(t)$ $(0<t<T<\infty)$ are continuous functions and f satisfies $\int_0^T |f(t)|dt < \infty$, the fractional-order differential equations (2.2)–(2.3) *have a unique solution.*

Definition 2.1 ([69]). For a sufficiently differentiable function u and an initial condition, FOSS (2.1) is said to be regular if the system has a unique solution.

Suppose that $f(t) = Bu(t)$ and $f(t)$ is sufficiently differentiable. Then system (2.1) can be rewritten as

$$ED^{\alpha}x(t) = Ax(t) + f(t), \quad \alpha \in (0, 2). \tag{2.4}$$

The following regularity condition for FOSSs can be proposed based on the Kronecker equivalent standard form [18].

Theorem 2.1. *The FOSS (2.4) is regular if and only if there exist two invertible matrices Φ and Ψ satisfying*

$$\Phi A \Psi = \operatorname{diag}(A_1, I_{n-r}),$$
$$\Phi E \Psi = \operatorname{diag}(I_r, N),$$

where $A_1 \in \mathbb{R}^{r\times r}$, and $N \in \mathbb{R}^{(n-r)\times(n-r)}$ is nilpotent.

Proof. Similarly to the case of IOSSs [18], we provide a proof for the integrity of this chapter. From Lemma 2.1 we can obtain the following fractional-order equations by using the coordinate transformation $x(t) = \Psi\tilde{x}(t), \tilde{f}(t) = \Phi f(t)$ by left multiplying system (2.4) with an invertible matrix Φ:

$$\mathbf{0}D^{\alpha}x_{n_0}(t) = f_{m_0}(t), \tag{2.5}$$
$$L_i D^{\alpha}x_{m_i+1}(t) = J_i x_{m_i+1}(t) + f_{m_i}(t), \tag{2.6}$$
$$L_j' D^{\alpha}x_{n_j}(t) = J_j' x_{n_j}(t) + f_{n_j+1}(t), \tag{2.7}$$
$$N_{k_s} D^{\alpha}x_{k_s}(t) = x_{k_s}(t) + f_{k_s}(t), \tag{2.8}$$
$$D^{\alpha}x_r(t) = A_1 x_r(t) + f_r(t), \tag{2.9}$$

where $j = 1, 2, \ldots, q$, $i = 1, 2, \ldots, p$, $s = 1, 2, \ldots, l$, $x_k(t) \in \mathbb{R}^k$, $f_k(t) \in \mathbb{R}^k$,

$$\tilde{x}^T(t) = [x_{n_0}^T, x_{m_1+1}^T, \ldots, x_{m_p+1}^T, x_{n_1}^T, \ldots, x_{n_q}^T, x_{k_1}^T, \ldots, x_{k_l}^T, x_r^T],$$
$$\tilde{f}^T(t) = [f_{m_0}^T, f_{m_1}^T, \ldots, f_{m_p}^T, f_{n_1+1}^T, \ldots, f_{n_q+1}^T, f_{k_1}^T, \ldots, f_{k_l}^T, f_r^T],$$

and L_i, J_i, L_j', J_j' are similarly defined in Lemma 2.1. Since a nonsingular transformation does not change the properties of the solutions of the system, the existence and uniqueness of fractional-order equations (2.5)–(2.9) are equivalent to those of system (2.4). We are ready to continue in the following five steps.

(1) For equation (2.5), if $f_{m_0}(t) = 0$, then the equation has an infinite number of solutions. If $f_{m_0}(t) \neq 0$, then the equation has no solution.

(2) For equation (2.6) and certain i, the dimension of this system is $m_i \times (m_i + 1)$. Represent the state vector and the input vector as $w_\varphi(t)$ ($\varphi = 1, 2, \ldots, m_i + 1$) and $f_\varphi(t)$ ($\varphi = 1, 2, \ldots, m_i$), respectively. Then equations (2.6) can be expressed as

$$D^{\alpha}w_1(t) = w_2(t) + f_1(t),$$

$$D^\alpha w_2(t) = w_3(t) + f_2(t),$$
$$\vdots$$
$$D^\alpha w_{m_i}(t) = w_{m_i+1}(t) + f_{m_i}(t).$$

By Lemma 2.2, $w_1(t), w_2(t), \ldots, w_{m_i}(t)$ can be solved if a specific $w_{m_i+1}(t)$ is given. Thus uncertain $w_{m_i+1}(t)$ will lead to an infinite number of solutions of the above equation.

(3) For equation (2.7) and certain j, the dimension of this system is $(n_j + 1) \times n_j$. Represent the state and input vectors as $w_\varphi(t)$ $(\varphi = 1, 2, \ldots, n_j)$ and $f_\varphi(t)$ $(\varphi = 1, 2, \ldots, n_j + 1)$, respectively. Then equations (2.7) can be expressed as

$$\begin{aligned}
D^\alpha w_1(t) &= f_1(t),\\
D^\alpha w_2(t) &= w_1(t) + f_2(t),\\
&\vdots\\
D^\alpha w_{n_j}(t) &= w_{n_j-1}(t) + f_{n_j}(t),\\
0 &= w_{n_j}(t) + f_{n_j+1}(t).
\end{aligned}$$

By Lemma 2.2, $w_1(t), w_2(t), \ldots, w_{n_j}(t)$ can be uniquely solved according to the previous n_j equations, but the last equation adds an algebraic constraint condition, which indicates that the $n_j + 1$ equations will have no solution if $w_{n_j}(t) \neq -f_{n_j+1}(t)$.

(4) For equation (2.8) and certain s, the dimension of this system is $k_s \times k_s$. Represent the state and input vectors as $w_\varphi(t)$ $(\varphi = 1, 2, \ldots, k_s)$ and $f_\varphi(t)$ $(\varphi = 1, 2, \ldots, k_s)$, respectively. Then equations (2.8) can be expressed as

$$\begin{aligned}
D^\alpha w_2(t) &= w_1(t) + f_1(t),\\
D^\alpha w_3(t) &= w_2(t) + f_2(t),\\
&\vdots\\
D^\alpha w_{k_s}(t) &= w_{k_s-1}(t) + f_{k_s-1}(t),\\
0 &= w_{k_s}(t) + f_{k_s}(t).
\end{aligned}$$

By Lemma 2.2 equation (2.8) has a solution, and the solution is unique because $w_1(t)$, $w_2(t), \ldots, w_{k_s}(t)$ can be uniquely solved if $f_\varphi(t)$ $(\varphi = 1, 2, \ldots, k_s)$ are sufficiently differentiable.

(5) As f is a sufficiently differentiable function, equation (2.9) has a unique solution.

Thus system (2.4) has a unique solution if and only if there exist two invertible matrices Φ and Ψ satisfying

$$\begin{aligned}
\Phi A \Psi &= \operatorname{diag}(A_1, I_{n-r}),\\
\Phi E \Psi &= \operatorname{diag}(I_r, N),
\end{aligned}$$

where $N \in \mathbb{R}^{(n-r)\times(n-r)}$ is nilpotent, and $A_1 \in \mathbb{R}^{r\times r}$. This completes the proof. □

Remark 2.1. The existence, uniqueness, and structure of the solutions of FOSSs discussed in Theorem 2.1 are based on the Kronecker equivalent standard form, which is different from the regularity analysis in [69, 106], where the regularity is studied in frequency domain. Furthermore, the regularity conditions proposed in [69, 106] are only adapted to linear FOSSs. For system (2.4), if f is an external nonlinear sufficiently differentiable signal, then Theorem 2.1 still holds, indicating that the regularity condition presented in this chapter can be used for both linear and nonlinear FOSSs.

2.3 Impulse analysis

In this section, we obtain the time domain solutions of FOSSs by using the Laplace transform and also establish the nonimpulsiveness conditions.

Lemma 2.3 ([38]). *Let $u \in L_1(0,T)$. Then system $D^\alpha x(t) = Ax(t) + Bu(t)$ has a unique solution in $(0,T)$,*

$$x(t) = E_{\alpha,1}(At^\alpha)x(0) + t^{\alpha-1}E_{\alpha,\alpha}(At^\alpha) * Bu(t),$$

where $E_{\alpha,\beta}(z)$ are the two-parameter Mittag-Leffler functions.

Definition 2.2 ([106]). The generalized eigenvectors v satisfying $Ev = 0$ are defined as
(1) The infinite eigenvector of order $k > 1$ satisfying $Ev_i^k = Av_i^{k-1}$, and
(2) The infinite eigenvector of order 1 satisfying $Ev_i^1 = 0$.

Lemma 2.4 ([106]). *The FOSS $ED^\alpha x(t) = Ax(t)$ is with nonimpulsiveness if and only if there exists no infinite eigenvector of order two.*

For the case $\alpha = 1$, the results are given in [18]. For the cases $\alpha \in (0,1)$ and $\alpha \in (1,2)$, the results are given as follows.

Theorem 2.2. *When $T \geq 0$, the time domain solutions of the FOSS (2.1) with $\alpha \in (0,1)$ are*

$$x(t) = \Psi \begin{bmatrix} x_1(t) \\ x_2(t) \end{bmatrix}, \tag{2.10}$$

$$x_1(t) = E_{\alpha,1}(A_1 t^\alpha)x_1(t_0) + (t^{\alpha-1}E_{\alpha,\alpha}(A_1 t^\alpha)) * B_1 u(t),$$

$$x_2(t) = -\sum_{\rho=1}^{\eta} N^\rho(\delta^{(\rho\alpha-1)}(t)x_2(t_0)) - \sum_{\rho=0}^{\eta-1} N^\rho B_2\left(D^{\rho\alpha}u(t) + \sum_{j=0}^{m-1} u_0^{(j)}\delta^{(\rho\alpha-1-j)}(t)\right),$$

where $x_1(t) \in \mathbb{R}^r$ and $x_2(t) \in \mathbb{R}^{n-r}$, and the initial conditions are $x_1(t_0) = x_1(0)$ and $x_2(t_0) = x_2(0)$, $N \in \mathbb{R}^{(n-r)\times(n-r)}$ is nilpotent, η is the nilpotent index, $m-1 < \rho\alpha \leq m \in \mathbb{Z}^+$, $0 < \alpha < 1$, Ψ satisfies Theorem 2.1, u is a piecewise η times continuously differentiable function with initial condition $u_0^j = u^{(j)}(0)$.

If the initial condition is of the form

$$x(t_{0_+}) = \Psi \begin{bmatrix} I \\ 0 \end{bmatrix} x_1(t_0) - \Psi \begin{bmatrix} 0 \\ I \end{bmatrix} \sum_{\rho=0}^{\eta-1} M^{-1} N^{\rho} B_2 \left(D^{\rho\alpha} u(t_{0_+}) + \sum_{j=0}^{m-1} u_0^{(0)} \delta^{(\rho\alpha-1-j)}(t_{0_+}) \right),$$

then solution (2.10) *of system* (2.1) *is unique for* $t > t_0$, *where* $M = \sum_{\rho=1}^{\eta} N^{\rho} \delta^{(\rho\alpha-1)}(t_{0_+}) + I$.

Proof. From the regularity of the system, there exist two invertible matrices Ψ and Φ such that system (2.1) is equivalent to

$$D^{\alpha} x_1(t) = A_1 x_1(t) + B_1 u(t), \tag{2.11}$$

$$N D^{\alpha} x_2(t) = x_2(t) + B_2 u(t), \tag{2.12}$$

where $\Phi B = \begin{bmatrix} B_1 \\ B_2 \end{bmatrix}$. Systems (2.11) and (2.12) are called finite and infinite subsystems, respectively.

System (2.11) is a normal FOS, and by Lemma 2.3 for piecewise continuously differentiable input u, the time domain solutions of subsystem (2.11) are

$$x_1(t) = E_{\alpha,1}(A_1 t^{\alpha}) x_1(t_0) + (t^{\alpha-1} E_{\alpha,\alpha}(A_1 t^{\alpha}))^* B_1 u(t).$$

Using the Laplace transform on system (2.12), $(sN - I)^{-1}$ being equal to $-\sum_{\rho=0}^{\eta-1} (sN)^{\rho}$, where η is the nilpotent index of the nilpotent matrix N, we obtain

$$\begin{aligned} X_2(s) &= (s^{\alpha} N - I)^{-1} (s^{\alpha-1} N x_2(t_0) + B_2 U(s)) \\ &= -\sum_{\rho=1}^{\eta} s^{(\rho\alpha-1)} N^{\rho} x_2(t_0) - \sum_{\rho=0}^{\eta-1} s^{\rho\alpha} N^{\rho} B_2 U(s), \end{aligned}$$

where X_2 and U are the Laplace transforms of x_2 and u, respectively. Then using the inverse Laplace transform on both sides, we get

$$\begin{aligned} x_2(t) = &-\sum_{\rho=1}^{\eta} N^{\rho} (\delta^{(\rho\alpha-1)}(t) x_2(t_0)) \\ &- \sum_{\rho=0}^{\eta-1} N^{\rho} B_2 \left(D^{\rho\alpha} u(t) + \sum_{j=0}^{m-1} u_0^{(j)} \delta^{(\rho\alpha-j-1)}(t) \right). \end{aligned} \tag{2.13}$$

The time domain solutions of system (2.1) are obtained.

To avoid the discontinuous behavior of the FOSS at $t = 0$, the admissible initial conditions have to be considered.

As $t \to t_{0_+}$, equality (2.13) can be transformed into

$$x_2(0_+) = -\sum_{\rho=1}^{\eta} N^{\rho} (\delta^{(\rho\alpha-1)}(t_{0_+}) x_2(t_{0_+}))$$

$$-\sum_{\rho=0}^{\eta-1} N^{\rho}B_2\left(D^{\rho\alpha}u(t_{0_+})+\sum_{j=0}^{m-1}u_0^{(j)}\delta^{(\rho\alpha-1-j)}(t_{0_+})\right),$$

which is equivalent to

$$\left(I+\sum_{\rho=1}^{\eta} N^{\rho}\delta^{(\rho\alpha-1)}(t_{0_+})\right)x_2(t_{0_+})$$
$$=-\sum_{\rho=0}^{\eta-1} N^{\rho}B_2\left(D^{\rho\alpha}u(t_{0_+})+\sum_{j=0}^{m-1}u_0^{(j)}\delta^{(\rho\alpha-1-j)}(t_{0_+})\right).$$

Denote $M=(I+\sum_{\rho=1}^{\eta}N^{\rho}\delta^{(\rho\alpha-1)}(t_{0_+}))$. As $\rho\alpha-1\notin\mathbb{N}^+$, $\delta^{(\rho\alpha-1)}(t_{0_+})$ is neither zero nor infinite but a very large number. Since N is nilpotent, M is a nonsingular matrix, and

$$x_2(t_0)=-M^{-1}\sum_{\rho=0}^{\eta-1} N^{\rho}B_2\left(D^{\rho\alpha}u(t_{0_+})+\sum_{j=0}^{m-1}u_0^{(j)}\delta^{(\rho\alpha-1-j)}(t_{0_+})\right).$$

This completes the proof. □

Remark 2.2. By Theorem 2.2, the time domain solution of FOSSs consists of two parts x_1 and x_2, which are the time domain solutions of the finite and infinite subsystems, respectively; x_1 consists of Mittag-Leffler functions, which are the same as the time domain solutions of normal FOSs, and x_2 is composed of the input and impulse functions. From time domain solutions (2.10) the impulsive behavior of system (2.1) can be analyzed. Since the substate x_1 is continuous, only x_2 needs to be considered.

(1) For $t=t_0=0$, not to lose the generality, let $x_2(t_0)\neq 0$, $x_2(t_0)\notin\ker(N)$ and $u(t_0)=0$. Then

$$x_2(t)=-\sum_{\rho=1}^{\eta} N^{\rho}(\delta^{(\rho\alpha-1)}(t)x_2(t_0)),$$

where $\delta^{(\rho\alpha-1)}(t)\to\infty$ as $t\to t_0=0$. Therefore $x_2(t_0)$ will result in impulse terms.

(2) For $t>t_0=0$, we have

$$x_2(t)=-\sum_{\rho=1}^{\eta}(\delta^{(\rho\alpha-1)}(t)N^{\rho}x_2(t_0))$$
$$-\sum_{\rho=0}^{\eta-1} N^{\rho}B_2\left(D^{\rho\alpha}u(t)+\sum_{j=0}^{m-1}u_0^{(j)}\delta^{(\rho\alpha-1-j)}(t)\right). \tag{2.14}$$

Since $\rho\alpha-1$, $\rho\alpha-j-1$ are not positive integers, and then $\delta^{(\rho\alpha-1)}(t)$ and $\delta^{(\rho\alpha-j-1)(t)}$ are not equal to zero and will result in impulse terms.

According to the final value theorem of FOSs [76], $\delta^{(\rho\alpha-1)}(t) \to 0$ and $\delta^{(\rho\alpha-j-1)}(t) \to 0$ as $t \to \infty$, which indicates that $\delta^{(\rho\alpha-1)}(t)$ and $\delta^{(\rho\alpha-j-1)(t)}$ in (2.14) have no influence on the asymptotical stability of (2.14).

The time domain solutions of FOSS (2.1) with $\alpha \in (1,2)$ are given as follows.

Theorem 2.3. *For $t \geq 0$ and FOSS (2.1) with $\alpha \in (1,2)$, the time domain solutions are*

$$\begin{aligned}
x(t) &= \Psi \begin{bmatrix} x_1(t) \\ x_2(t) \end{bmatrix}, \\
x_1(t) &= E_{\alpha,1}(A_1 t^\alpha)x_{10} + tE_{\alpha,2}(A_1 t^\alpha)x_{10}^{(1)} + t^{\alpha-1}E_{\alpha,\alpha}(A_1 t^\alpha) * B_1 u(t), \\
x_2(t) &= -\sum_{\rho=1}^{\eta} N^\rho (\delta^{(-1+\rho\alpha)}(t)x_{20} + \delta^{(-2+\rho\alpha)}(t)x_{20}^{(1)}) \\
&\quad - \sum_{\rho=0}^{\eta-1} N^\rho B_2 \left(D^{\rho\alpha} u(t) + \sum_{j=0}^{m-1} u_0^{(j)} \delta^{(-j-1+\rho\alpha)}(t) \right),
\end{aligned} \tag{2.15}$$

where $N \in \mathbb{R}^{(n-r)\times(n-r)}$ is nilpotent with nilpotent index η, the initial conditions are $x_{10} = x_1(0)$ and $x_{20} = x_2(0)$, $x_{10}^{(1)}$ and $x_{20}^{(1)}$ are the first derivatives of $x_1(t)$ and $x_2(t)$ at $t = 0$, respectively, $m \in \mathbb{Z}^+$ is determined by ρ, α is such that $m-1 < \rho\alpha \leq m$, u is a piecewise continuous function sufficiently differentiable η times, the initial value of u is $u^{(j)}(0) = u_0^j$, Ψ satisfies Theorem 2.1, and $x_1(t)$ and $x_2(t)$ are subsystems, each possibly having multiple states.

Given the initial conditions

$$\begin{aligned}
x(0_+) &= \Psi \begin{bmatrix} x_1(0_+) \\ x_2(0_+) \end{bmatrix}, \\
x_1(0_+) &= x_{10}, \\
x_2(0_+) &= -M^{-1} \sum_{\rho=1}^{\eta} N^\rho \delta^{(\rho\alpha-2)}(0_+)x_{20}^{(1)} - M^{-1} \sum_{\rho=0}^{\eta-1} N^\rho B_2 \left(D^{\rho\alpha} u(0_+) + \sum_{j=0}^{m-1} u_0^{(j)} \delta^{(-j-1+\rho\alpha)}(0_+) \right),
\end{aligned} \tag{2.16}$$

the time domain solution (2.15) of system (2.1) is unique, and $M = I + \sum_{\rho=1}^{\eta} N^\rho \delta^{(\rho\alpha-1)}(0_+)$, where $t = 0_+$ means that $t > 0$ tends to the initial time 0.

Proof. As the system is regular, two invertible matrices Ψ and Φ can be chosen such that system (2.1) can be transformed into

$$D^\alpha x_1(t) = A_1 x_1(t) + B_1 u(t), \tag{2.17}$$

$$N D^\alpha x_2(t) = x_2(t) + B_2 u(t), \tag{2.18}$$

with $\Phi B = \left[\begin{smallmatrix} B_1 \\ B_2 \end{smallmatrix}\right]$. Systems (2.17) and (2.18) are called finite and infinite subsystems, respectively. The following proof is divided into three steps. Step 1 is to derive the time domain solutions of system (2.17). Step 2 is to derive the time domain solutions of system (2.18). Step 3 is to derive the admissible initial conditions.

Step 1: The time domain solutions of system (2.17) can be found in [39]. For the integrity of this chapter, Step 1 also gives a complete proof. Let $X_1(s) = \mathcal{L}[x_1(t)]$ and $U(s) = \mathcal{L}[u(t)]$. Then performing the Laplace transform on both sides of system (2.17), we have

$$s^{\alpha}X_1(s) - s^{\alpha-1}x_{10} - s^{\alpha-2}x_{10}^{(1)} = A_1X_1(s) + B_1U(s),$$

which can be transformed into

$$\begin{aligned} X_1(s) &= (s^{\alpha}I - A_1)^{-1}(s^{\alpha-1}x_{10} + s^{\alpha-2}x_{10}^{(1)} + B_1U(s)) \\ &= \sum_{k=1}^{\infty}\frac{A_1^k x_{10}}{s^{k\alpha+1}} + \sum_{k=1}^{\infty}\frac{A_1^k x_{10}^{(1)}}{s^{k\alpha+2}} + \sum_{k=1}^{\infty}\frac{A_1^k}{s^{k\alpha+\alpha}}B_1U(s). \end{aligned} \tag{2.19}$$

For equality (2.19), perform the inverse Laplace transform. Then

$$\begin{aligned} x_1(t) &= \sum_{k=1}^{\infty}\frac{(t^{\alpha}A_1)^k}{\Gamma(k\alpha+1)}x_{10} + t\sum_{\rho=1}^{\infty}\frac{(t^{\alpha}A_1)^k}{\Gamma(k\alpha+2)}x_{10}^{(1)} + t^{\alpha-1}\sum_{k=1}^{\infty}\frac{(t^{\alpha}A_1)^k}{\Gamma(k\alpha+\alpha)} * B_1u(t) \\ &= E_{\alpha,1}(A_1t^{\alpha})x_{10} + tE_{\alpha,2}(A_1t^{\alpha})x_{10}^{(1)} + t^{\alpha-1}E_{\alpha,\alpha}(A_1t^{\alpha}) * B_1u(t). \end{aligned}$$

Step 2: For system (2.18), note that $(sN - I)^{-1} = -\sum_{\rho=0}^{\eta-1}s^{\rho}N^{\rho}$ and $X_2(s) = \mathcal{L}[x_2(t)]$. Then we obtain

$$X_2(s) = -\sum_{\rho=1}^{\eta}N^{\rho}(s^{\rho\alpha-1}x_{20} + s^{\rho\alpha-2}x_{20}^{(1)}) - \sum_{\rho=0}^{\eta-1}s^{\rho\alpha}N^{\rho}B_2U(s). \tag{2.20}$$

For equality (2.20), the inverse Laplace transform gives

$$\begin{aligned} x_2(t) = &-\sum_{\rho=1}^{\eta}N^{\rho}(\delta^{(-1+\rho\alpha)}(t)x_{20} + \delta^{(-2+\rho\alpha)}(t)x_{20}^{(1)}) \\ &-\sum_{\rho=0}^{\eta-1}N^{\rho}B_2\left(D^{\rho\alpha}u(t) + \sum_{j=0}^{m-1}u_0^{(j)}\delta^{(-j-1+\rho\alpha)}(t)\right). \end{aligned} \tag{2.21}$$

Thus the solutions of (2.15) are obtained.

Step 3: The admissible initial conditions of FOSSs can ensure the continuous behavior of systems at $t = 0$. Note that system (2.17) is a normal FOS and there exists no discontinuous behavior in system (2.17). Thus the admissible initial condition of system (2.17) is $x_1(0_+) = x_{10}$, and then only the admissible initial conditions of system (2.18) need to be taken into account.

When $t = 0_+$, equality (2.21) can be transformed into

$$x_2(0_+) = -\sum_{\rho=1}^{\eta} N^\rho(\delta^{(\rho\alpha-2)}(0_+)x_{20}^{(1)} + \delta^{(\rho\alpha-1)}(0_+)x_{20})$$
$$-\sum_{\rho=0}^{\eta-1} N^\rho B_2\left(D^{\rho\alpha}u(0_+) + \sum_{j=0}^{m-1} u_0^{(j)}\delta^{(-j-1+\rho\alpha)}(0_+)\right).$$

To ensure that $x_2(t)$ is continuous at $t = 0_+$, we must have $x_2(0_+) = x_2(0) = x_{20}$. Thus we can have

$$\left(I + \sum_{\rho=1}^{\eta} N^\rho \delta^{(\rho\alpha-1)}(0_+)\right)x_2(0_+)$$
$$= -\sum_{\rho=1}^{\eta} N^\rho \delta^{(\rho\alpha-2)}(0_+)x_{20}^{(1)}$$
$$-\sum_{\rho=0}^{\eta-1} N^\rho B_2\left(D^{\rho\alpha}u(0_+) + \sum_{j=0}^{m-1} u_0^{(j)}\delta^{(-j-1+\rho\alpha)}(0_+)\right).$$

Denote $M = I + \sum_{\rho=1}^{\eta} N^\rho \delta^{(-1+\rho\alpha)}(0_+)$. Note that $-1 + \rho\alpha \notin \mathbb{N}^+$. Consequently, $\delta^{(-1+\rho\alpha)}(0_+)$ is a very large number. Since N is nilpotent, M is nonsingular. Then

$$x_2(0_+) = -M^{-1}\sum_{\rho=1}^{\eta} N^\rho \delta^{(\rho\alpha-2)}(0_+)x_{20}^{(1)}$$
$$-M^{-1}\sum_{\rho=0}^{\eta-1} N^\rho B_2\left(D^{\rho\alpha}u(0_+) + \sum_{j=0}^{m-1} u_0^{(j)}\delta^{(-j-1+\rho\alpha)}(0_+)\right).$$

The proof is completed. □

Remark 2.3. The time domain solutions of FOSSs are given by (2.15) in Theorem 2.3, which include $x_1(t)$ and $x_2(t)$; $x_1(t)$ is essentially the solution of FOSs, which is continuous, and $x_2(t)$ contains the impulse function and may cause the impulsive behavior. Thus we analyze $x_2(t)$ below.

(1) $t = 0$. Without loss of generality, suppose that $u(t) = 0$, $x_2(0) \neq 0$, and $x_2(0) \notin \ker(N)$. Then

$$x_2(t) = -\sum_{\rho=1}^{\eta} N^\rho(\delta^{(-2+\rho\alpha)}(t)x_{20}^{(1)} + \delta^{(-1+\rho\alpha)}(t)x_{20}).$$

Since $\delta^{(-1+\rho\alpha)}(t)$ and $\delta^{(-2+\rho\alpha)}(t) \to \infty$ for $t = 0$, x_{20} and $x_{20}^{(1)}$ may cause impulsive behavior.

(2) $t > 0$. Then

$$x_2(t) = -\sum_{\rho=1}^{\eta} N^{\rho}\left(\delta^{(-2+\rho\alpha)}(t)x_{20}^{(1)} + \delta^{(-1+\rho\alpha)}(t)x_{20}\right) - \sum_{\rho=0}^{\eta-1} N^{\rho}B_2\left(D^{\rho\alpha}u(t) + \sum_{j=0}^{m-1} u_0^{(j)}\delta^{(-j-1+\rho\alpha)}(t)\right). \tag{2.22}$$

Note that $\rho, j \in \mathbb{Z}$ and $\alpha \in (1,2)$, which indicates that $-1+\rho\alpha$, $2+\rho\alpha$, and $j-1+\rho\alpha$ cannot always be positive integers. Thus $\delta^{(-1+\rho\alpha)}(t)$, $\delta^{(-2+\rho\alpha)}(t)$, and $\delta^{(-j-1+\rho\alpha)}(t)$ cannot always be zero.

Remark 2.4. Note that the time domain solutions of IOSSs and FOSSs are different because of the properties of the Γ function. The time responses of $\delta^{(b)}(t)$ are related to whether b is a positive integer as $\delta^{(b)}(t) = \frac{t^{-b-1}}{\Gamma(-b)}$, $t > 0$, i. e.,

$$\delta^{(b)}(t) = \begin{cases} \frac{t^{-b-1}}{\Gamma(-b)}, & b \text{ is not a positive integer,} \\ 0, & b \text{ is a positive integer,} \end{cases} \qquad t > 0.$$

For example, when b is taken as 1 and 1.5, respectively, the time responses of $\delta^{(b)}(t)$ are shown in Figure 2.1, where Figure 2.1(a) and Figure 2.1(b) are the time responses of $\delta^{(b)}(t)$ with $b = 1$ and $b = 1.5$, respectively. In Figure 2.1(a) and Figure 2.1(b), $\delta^{(b)}(t) \to \infty$ as $t \to 0$.

Remark 2.5. Based on the final value theorem of FOSs [76], as $t \to \infty$, we can derive that $\delta^{(b)}(t) \to 0$, indicating that $\delta^{(b)}(t)$ do not influence stability of the infinite subsystem (2.12) and (2.18). This will be helpful for analyzing the stability of FOSSs.

The following nonimpulsiveness condition of FOSSs can be obtained based on Theorems 2.2 and 2.3 and Remarks 2.2 and 2.5.

Theorem 2.4. *Let N be the nilpotent matrix defined in Theorem 2.1. Then FOSS (2.1) with $\alpha \in (0,2)$ is with nonimpulsiveness if and only if $N = 0$.*

Proof. From Theorems 2.2 and 2.3 and Remarks 2.2 and 2.5 we can see that the terms containing N always cause impulsive behavior in time domain solutions. The proof is completed. □

Remark 2.6. The same nonimpulsiveness conditions in Theorem 2.4 are also proposed in [69, 106]. The proof of the nonimpulsiveness conditions proposed in this chapter has the following advantages.

1. The proof of the nonimpulsiveness conditions proposed in [69] is based on the analysis of the degree of the polynomial in frequency domain. The proof of the nonimpulsiveness conditions in this chapter is based on the time domain solutions of FOSSs, which is more direct and intuitive.

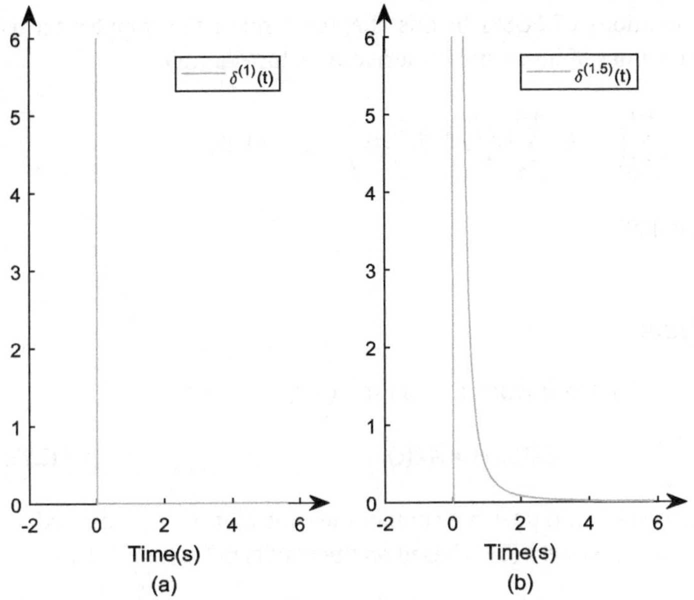

Figure 2.1: The time responses of $\delta^{(b)}(t)$ with $b = 1, 1.5$.

2. The proof of the nonimpulsiveness conditions proposed in [106] is based on the time domain solutions of FOSSs, but the time domain solutions proposed in [106] neglected the control input, which would also result in impulse terms. Thus the proof of the nonimpulsiveness conditions proposed in [106] is short of integrity.

Remark 2.7. The necessary and sufficient condition for the nonimpulsiveness of FOSSs in Theorem 2.4 is based on Theorems 2.2 and 2.3, which deduces the time domain solutions of the finite and infinite subsystems of FOSSs by the Laplace inverse transformation. It should be emphasized that the time domain solutions of the infinite subsystems in Theorems 2.2 and 2.3 contain the impulse terms caused by the initial values of input and its derivatives, which are the significant properties of FOSSs and not taken into consideration in [69, 106, 38, 116].

Remark 2.8. Compared with the results in [38], the structure, existence, uniqueness, and description of the time domain solutions of FOSSs in this chapter are complete and accurate. The reasons are listed as follows.

1. The structure, existence, and uniqueness of the time domain solutions of FOSSs are discussed in Theorem 2.1 based on the Kronecker equivalent standard form, which completely explains the meaning of the regularity. In [38] the investigated FOSSs are directly assumed to be regular without analyzing the structure, existence, and uniqueness of the time domain solutions.

2. The time domain solutions of FOSSs in this chapter contain the impulse terms caused by the initial values of input and its derivatives, for example,

$$-\sum_{\rho=0}^{\eta-1}\left(N^{\rho}B_2\sum_{j=0}^{m-1}u_0^{(j)}\delta^{(\rho\alpha-j-1)}(t)\right),\quad \alpha\in(0,1),$$

which are omitted in [38].

2.4 Stability analysis

Consider the following FOSS of the fractional-order $\alpha \in (0,2)$:

$$ED^{\alpha}x(t) = Ax(t), \tag{2.23}$$

which is the case where system (2.1) does not contain the control input. In this section, we give a stability condition of system (2.23) based on the results in Sections 2.2 and 2.3.

Lemma 2.5 ([67, 65]). *Denote by* spec(A) *the set of all eigenvalues of a matrix A. Then the FOS* $D^{\alpha}x(t) = Ax(t)$, $\alpha \in (0,2)$, *is asymptotically stable if and only if* $|\arg(\mathrm{spec}(A))| > \alpha\pi/2$.

Lemma 2.6 ([21, 84]). *Let* $A \in \mathbb{R}^{n\times n}$, $\theta = (1-\alpha)\pi/2$, *and* $\mathbf{r} = e^{i\theta}$. *The FOS* $D^{\alpha}x(t) = Ax(t)$, $0 < \alpha < 1$, *is asymptotically stable if and only if there exists a matrix* $X \in \mathbb{C}^{n\times n}$, $X = X^* > 0$, *satisfying*

$$\mathrm{Sym}\{A(\mathbf{r}X + \bar{\mathbf{r}}\bar{X})\} < 0.$$

Lemma 2.7 ([21, 84]). *Let* $A \in \mathbb{R}^{n\times n}$, $\theta = (2-\alpha)\pi/2$, *and* $1 < \alpha < 2$. *The FOS* $D^{\alpha}x(t) = Ax(t)$ *is asymptotically stable if and only if there exists* $X \in \mathbb{R}^{n\times n}$, $X = X^T > 0$, *satisfying*

$$\begin{bmatrix}(A^TX + XA)\sin\theta & (XA - A^TX)\cos\theta\\ \star & (A^TX + XA)\sin\theta\end{bmatrix} < 0,$$

i. e.,

$$\mathrm{Sym}\left\{\begin{bmatrix}\sin\theta & -\cos\theta\\ \cos\theta & \sin\theta\end{bmatrix}\otimes(A^TX)\right\} < 0.$$

Theorem 2.5. *Let* spec(E,A) *be the set of all eigenvalues of system* (2.23). *Then the regular FOSS* (2.23) *is asymptotically stable if and only if*

$$|\arg(\mathrm{spec}(E,A))| > \frac{\alpha}{2}\pi.$$

Proof. By the regularity of system (2.23), there exist two invertible matrices Φ and Ψ satisfying

$$\Phi A\Psi = \mathrm{diag}(A_1, I_{n-r}),$$
$$\Phi E\Psi = \mathrm{diag}(I_r, N),$$

and then system (2.23) can be rewritten as

$$D^{\alpha}x_1(t) = A_1x_1(t), \tag{2.24}$$
$$ND^{\alpha}x_2(t) = x_2(t). \tag{2.25}$$

By Theorem 2.2, the time domain solutions of system (2.23) with $\alpha \in (0,1)$ are

$$x_1(t) = E_{\alpha,1}(A_1t^{\alpha})x_1(t_0),$$
$$x_2(t) = -\sum_{\rho=1}^{\eta} N^{\rho}(\delta^{(\rho\alpha-1)}(t)x_2(t_0)).$$

By Theorem 2.3, the time domain solutions of system (2.23) with $\alpha \in (1,2)$ are

$$x_1(t) = E_{\alpha,1}(A_1t^{\alpha})x_{10} + tE_{\alpha,2}(A_1t^{\alpha})x_{10}^{(1)},$$
$$x_2(t) = -\sum_{\rho=1}^{\eta} N^{\rho}(\delta^{(-2+\rho\alpha)}(t)x_{20}^{(1)} + \delta^{(-1+\rho\alpha)}(t)x_{20}).$$

By Lemma 2.5, finite subsystems (2.24) are asymptotically stable if and only if $|\arg(\mathrm{spec}(A_1))| > \alpha\frac{\pi}{2}$.

According to Remark 2.5, the infinite subsystem (2.25) is ultimately asymptotically stable because the substate $x_2(t) \to 0$ for $t \to +\infty$. Thus we desire to prove that $|\arg(\mathrm{spec}(E,A))| > \frac{\alpha}{2}\pi$ is equivalent to $|\arg(\mathrm{spec}(A_1))| > \frac{\alpha}{2}\pi$.

The fact that N is nilpotent implies that $\det(sN - I_{n-r}) = (-1)^{n-r}$, which also implies $\mathrm{spec}(N, I_{n-r}) = \emptyset$. Note that $\mathrm{spec}(E,A)$ is equivalent to $\mathrm{spec}(\Phi E\Psi, \Phi A\Psi)$, i. e., $\mathrm{spec}(\mathrm{diag}(I_r,N), \mathrm{diag}(A_1, I_{n-r}))$ and $\mathrm{spec}(\mathrm{diag}(I_r,N), \mathrm{diag}(A_1, I_{n-r})) = \mathrm{spec}(I_r, A_1) \cup \mathrm{spec}(N, I_{n-r})$. Therefore $\mathrm{spec}(E,A)$ is equal to $\mathrm{spec}(A_1)$. This completes the proof. □

Remark 2.9. A necessary and sufficient condition for the stability of FOSSs in Theorem 2.5 is proposed based on the time domain solutions in Theorems 2.2 and 2.3. Theorem 2.5 in this chapter is less conservative than the stability conditions proposed in [69]. The reasons lie in that the proof of the stability proposed in [69] is based on the assumption that the investigated FOSSs are regular and impulse-free. From the proof of Theorem 2.5 in this chapter we can see that the impulse terms of the time domain solutions do not affect the stability of FOSSs and the proof of the stability in this chapter is only based on the assumption that the investigated FOSSs are regular.

Remark 2.10. According to [18, 101], the asymptotical stability of IOSSs involves checking whether the finite poles are located in the left half-plane, as shown in Figure 2.2. According to Theorem 2.5, the asymptotical stability of FOSSs involves checking whether the finite poles are located in the stable areas. For fractional orders $\alpha \in (0,1)$ and $\alpha \in$

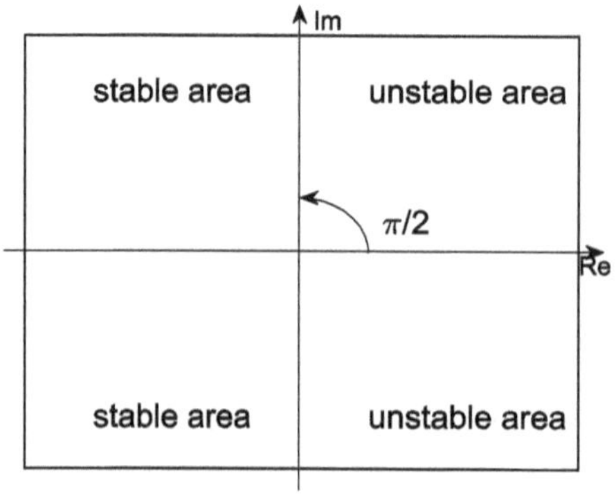

Figure 2.2: The stable areas of IOSSs with the integer order $\alpha = 1$.

(1, 2), the stable areas of FOSSs are shown in Figures 2.3 and 2.4, respectively, which are different from that of IOSSs.

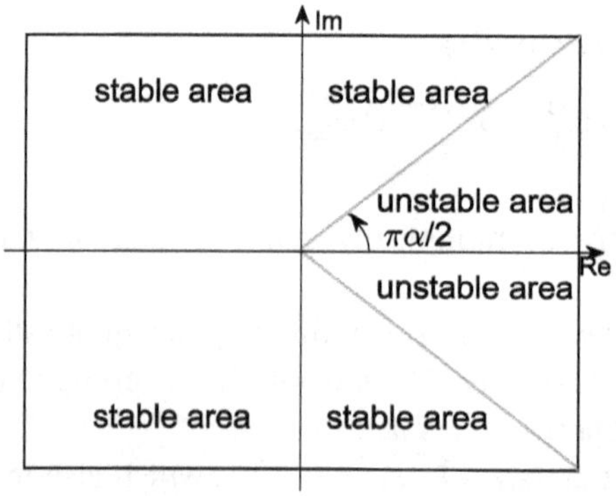

Figure 2.3: The stable areas of system (2.23) of fractional order $\alpha \in (0, 1)$.

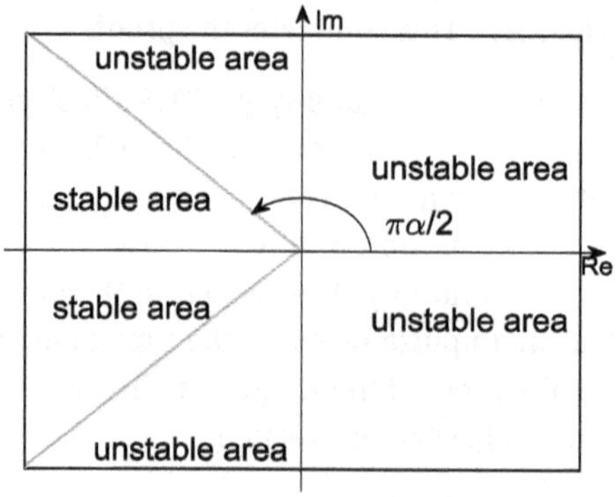

Figure 2.4: The stable areas of FOSSs of fractional order $\alpha \in (1, 2)$.

2.5 Admissibility analysis

In this section, we derive several versions of novel admissibility conditions of FOSSs and give numerical examples to illustrate the effectiveness of the derived theoretical results.

Definition 2.3 (Definition of admissibility [106]). The FOSS $ED^{\alpha}x(t) = Ax(t)$ is said to be admissible if it is regular, impulse-free, and stable.

Lemma 2.8 ([106, 116]). *When the regularity of the FOSS $ED^{\alpha}x(t) = Ax(t)$ is not known, there exist the following matrix decompositions:*

$$MEN = \begin{bmatrix} I_r & 0_{r\times(n-r)} \\ 0_{(n-r)\times r} & 0_{n-r} \end{bmatrix}, \quad MAN = \begin{bmatrix} A_1 & A_2 \\ A_3 & A_4 \end{bmatrix}, \tag{2.26}$$

where M and N are nonsingular matrices.

Denote

$$E^{\dagger} = N\begin{bmatrix} 0_r & 0_{r\times(n-r)} \\ 0_{(n-r)\times r} & I_{n-r} \end{bmatrix}M, \quad E^{\ddagger} = M^{-1}\begin{bmatrix} 0_r & 0_{r\times(n-r)} \\ 0_{(n-r)\times r} & I_{n-r} \end{bmatrix}M, \tag{2.27}$$

which will be used in the main results.

Lemma 2.9 ([116]). *Consider the FOSS $ED^{\alpha}x(t) = Ax(t)$ with the decomposition of* (2.26). *Then*

(1) *The system is impulse-free if and only if A_4 is nonsingular.*
(2) *The system is admissible if and only if $A_1 - A_2A_4^{-1}A_3$ is stable.*

Lemma 2.10 ([101]). *Let Ω be a real square matrix. Denote $\beta(E,A) = \max(\mathrm{Re}(\lambda))$, $\lambda \in \{s \mid \det(sE-A) = 0\}$. For notational simplicity, represent $\beta(I,A)$ by $\beta(A)$ if $E = I$. The quantities $\beta(E,A)$ and $\beta(A)$ are called the generalized spectral abscissa and normal spectral abscissa, respectively. Then the matrix measure $\mu(\Omega)$ is defined as*

$$\mu(\Omega) = \lim_{\theta\to 0^+} \frac{\|I + \theta\Omega\| - 1}{\theta}$$

and has the following properties:

(1) $-\|\Omega\| \le \beta(\Omega) \le \mu(\Omega) \le \|\Omega\|$.
(2) $\mu(\Omega) = \frac{1}{2}\lambda_{\max}(\Omega + \Omega^T) = \frac{1}{2}\beta(\Omega + \Omega^T)$.

Lemma 2.11 ([101]). *Let Π_1, Π_2, Π_3, and Π_4 be real matrices of appropriate dimensions, and let $\Lambda = \left[\begin{smallmatrix} \Pi_1 & \Pi_2 \\ \Pi_3 & \Pi_4 \end{smallmatrix}\right]$. If $\Lambda+\Lambda^T > 0$, then Π_4 is invertible, and $\Pi_1-\Pi_2\Pi_4^{-1}\Pi_3+\Pi_1^T-\Pi_3^T\Pi_4^{-T}\Pi_2^T > 0$.*

Lemma 2.12 ([68]). *For a matrix $\Pi_1 \in \mathbb{C}^{n\times n}$, the following two conditions are equivalent:*

(1) *Π_1 is nonsingular.*
(2) *There exists a nonsingular matrix $\Pi_2 \in \mathbb{C}^{n\times n}$ such that $\mathrm{Sym}\{\Pi_1\Pi_2\} < 0$.*

Now we are in a position to present necessary and sufficient conditions for the FOSS (2.23).

2.5.1 Case $\alpha \in (0, 1)$

Theorem 2.6. *The FOSS* (2.23) *with* $\alpha \in (0,1)$ *is admissible if and only if there exists a matrix* $X \in \mathbb{C}^{n\times n}$ *satisfying*

$$E^T X = X^* E \geq 0, \tag{2.28a}$$

$$\text{Sym}\{A^T(\mathbf{r}X + \bar{\mathbf{r}}\bar{X})\} < 0, \tag{2.28b}$$

where $\mathbf{r} = e^{i\theta}$, $\theta = (1-\alpha)\frac{\pi}{2}$.

Proof. Sufficiency. From Lemma 2.8 we can see that for FOSS (2.23), we can find two invertible matrices $\hat{M}$ and $\hat{N}$ satisfying

$$\hat{M}A\hat{N} = \begin{bmatrix} A_{11} & A_{12} \\ A_{21} & A_{22} \end{bmatrix}, \quad \hat{M}E\hat{N} = \text{diag}(I_r, 0). \tag{2.29}$$

Let

$$P = \hat{M}^{-T}X\hat{N} = \begin{bmatrix} P_{11} & P_{12} \\ P_{21} & P_{22} \end{bmatrix}. \tag{2.30}$$

If $\hat{N}^T$ and $\hat{N}$ are premultiplied and postmultiplied on both sides of (2.28a), respectively, we have

$$\begin{aligned} \hat{N}^T E^T X \hat{N} &= \hat{N}^T E^T \hat{M}^T \hat{M}^{-T} X \hat{N} \\ &= \begin{bmatrix} P_{11} & P_{12} \\ 0 & 0 \end{bmatrix} \geq 0, \end{aligned} \tag{2.31}$$

$$\begin{aligned} \hat{N}^T X^* E \hat{N} &= \hat{N}^T X^* \hat{M}^{-1} \hat{M} E \hat{N} \\ &= \begin{bmatrix} P_{11}^* & 0 \\ P_{12}^* & 0 \end{bmatrix} \geq 0, \end{aligned} \tag{2.32}$$

which implies that $P_{11} = P_{11}^* \geq 0$ and $P_{12} = 0$. It follows from inequality (2.28b) and Lemma 2.10 that

$$\begin{aligned} \beta(A^T(\mathbf{r}X + \bar{\mathbf{r}}\bar{X})) &\leq \mu(A^T(\mathbf{r}X + \bar{\mathbf{r}}\bar{X})) \\ &= \frac{1}{2}\lambda_{\max}(\text{Sym}\{A^T(\mathbf{r}X + \bar{\mathbf{r}}\bar{X})\}) < 0, \end{aligned} \tag{2.33}$$

indicating that $A^T(\mathbf{r}X + \bar{\mathbf{r}}\bar{X})$ is nonsingular. Thus $\mathbf{r}X + \bar{\mathbf{r}}\bar{X}$ is nonsingular, which implies that X is nonsingular. By (2.30), as $\hat{M}$ and $\hat{N}$ are nonsingular, we have $P_{11} \neq 0$. This, together with $P_{11} \geq 0$, results in $P_{11} > 0$. If $\hat{N}^T$ and $\hat{N}$ are premultiplied and postmultiplied on (2.28b), respectively, we have

$$\begin{aligned}
&\hat{N}^T \operatorname{Sym}\{A^T(\mathbf{r}X + \bar{\mathbf{r}}\bar{X})\}\hat{N} \\
&= \operatorname{Sym}\{\hat{N}^T A^T \hat{M}^T(\mathbf{r}P + \bar{\mathbf{r}}\bar{P})\} \\
&= \operatorname{Sym}\left\{\begin{bmatrix} A_{11}^T & A_{12}^T \\ A_{21}^T & A_{22}^T \end{bmatrix}\begin{bmatrix} \mathbf{r}P_{11} + \bar{\mathbf{r}}\bar{P}_{11} & 0 \\ \mathbf{r}P_{21} + \bar{\mathbf{r}}\bar{P}_{21} & \mathbf{r}P_{22} + \bar{\mathbf{r}}\bar{P}_{22} \end{bmatrix}\right\},
\end{aligned}$$

i. e.,

$$\operatorname{Sym}\left\{\begin{bmatrix} A_{11}^T(\mathbf{r}P_{11} + \bar{\mathbf{r}}\bar{P}_{11}) + A_{12}^T(\mathbf{r}P_{21} + \bar{\mathbf{r}}\bar{P}_{21}) & A_{12}^T(\mathbf{r}P_{22} + \bar{\mathbf{r}}\bar{P}_{22}) \\ A_{21}^T(\mathbf{r}P_{11} + \bar{\mathbf{r}}\bar{P}_{11}) + A_{22}^T(\mathbf{r}P_{21} + \bar{\mathbf{r}}\bar{P}_{21}) & A_{22}^T(\mathbf{r}P_{22} + \bar{\mathbf{r}}\bar{P}_{22}) \end{bmatrix}\right\} < 0, \tag{2.34}$$

which implies that

$$\operatorname{Sym}\{A_{22}^T(\mathbf{r}P_{22} + \bar{\mathbf{r}}\bar{P}_{22})\} < 0. \tag{2.35}$$

From Lemma 2.10 it follows that

$$\begin{aligned}
\beta(A_{22}^T(\mathbf{r}P_{22} + \bar{\mathbf{r}}\bar{P}_{22})) &\leq \mu(A_{22}^T(\mathbf{r}P_{22} + \bar{\mathbf{r}}\bar{P}_{22})) \\
&= \frac{1}{2}\lambda_{\max}(\operatorname{Sym}\{A_{22}^T(\mathbf{r}P_{22} + \bar{\mathbf{r}}\bar{P}_{22})\}) < 0,
\end{aligned} \tag{2.36}$$

which shows that $A_{22}^T(\mathbf{r}P_{22} + \bar{\mathbf{r}}\bar{P}_{22})$ is nonsingular, and then A_{22} is nonsingular too. By Lemma 2.9, FOSS (2.23) is regular and impulse-free. From Lemma 2.11 and inequality (2.34), it follows that

$$\operatorname{Sym}\{(A_{11} - A_{21}A_{22}^{-1}A_{12})^T(\mathbf{r}P_{11} + \bar{\mathbf{r}}\bar{P}_{11})\} < 0, \tag{2.37}$$

where $P_{11} = P_{11}^* > 0$. By Lemma 2.7, $A_{11} - A_{21}A_{22}^{-1}A_{12}$ is stable. By Lemma 2.9, FOSS (2.23) is admissible.

Necessity. Suppose that FOSS (2.23) is admissible. By Definition 2.3, FOSS (2.23) is regular, impulse-free, and stable. From Theorems 2.1 and 2.4, it follows that invertible matrices $\check{M}$ and $\check{N}$ can be chosen to satisfy

$$\check{M}A\check{N} = \begin{bmatrix} A_1 & 0 \\ 0 & I_{n-r} \end{bmatrix}, \quad \check{M}E\check{N} = \begin{bmatrix} I_r & 0 \\ 0 & 0 \end{bmatrix}. \tag{2.38}$$

By Theorem 2.5, the stability of FOSS (2.23) is equivalent to the stability of the system $D^{\alpha}x_1(t) = A_1x_1(t)$. By Lemma 2.7, there exists $Y_{11} = Y_{11}^* > 0$ satisfying

$$\operatorname{Sym}\{A_1^T(\mathbf{r}Y_{11} + \bar{\mathbf{r}}\bar{Y}_{11})\} < 0. \tag{2.39}$$

Let

$$X = \breve{M}^T \begin{bmatrix} Y_{11} & 0 \\ 0 & -I \end{bmatrix} \breve{N}^{-1}. \tag{2.40}$$

Then from (2.38) and (2.39), it follows that the matrix X in (2.40) satisfies (2.28). This ends the proof. □

Theorem 2.7. *The FOSS* (2.23) *with* $\alpha \in (0,1)$ *is admissible if and only if there exist matrices* $X = X^* > 0$, $X \in \mathbb{C}^{n\times n}$, *and* $S = S^T$, $S \in \mathbb{R}^{n\times n}$, *satisfying*

$$\mathrm{Sym}\{A^T(\mathbf{r}X + \bar{\mathbf{r}}\bar{X})E\} + A^T SA < 0, \tag{2.41a}$$

$$E^T SE \geq 0, \tag{2.41b}$$

where $\mathbf{r} = e^{i\theta}$, $\theta = (1-\alpha)\frac{\pi}{2}$.

Proof. Sufficiency. The proof of nonimpulsiveness is by contradiction. Suppose that FOSS (2.23) is impulsive. By Lemma 2.4, we can choose a second-order infinite eigenvector $v^2 \in \mathbb{R}^n$ satisfying $Ev^1 = 0$, $Ev^2 = Av^1$. If $(v^1)^*$ and v^1 are premultiplied and postmultiplied on (2.41a), respectively, then the inequality can be rewritten as

$$\begin{aligned} &(v^1)^*[\mathrm{Sym}\{A^T(\mathbf{r}X + \bar{\mathbf{r}}\bar{X})E\} + A^T SA](v^1) \\ &= (v^1)^* A^T SAv^1 \\ &= (v^2)^* E^T SEv^2 < 0, \end{aligned}$$

which does not hold because $E^T SE \geq 0$, so the previous assumption is false. Then we can show that FOSS (2.23) is impulse-free. By Theorems 2.1 and 2.4, FOSS (2.23) is consequently regular. Then there exist invertible matrices $\tilde{M}$ and $\tilde{N}$ satisfying

$$\tilde{M}A\tilde{N} = \begin{bmatrix} A_1 & 0 \\ 0 & I_{n-r} \end{bmatrix}, \quad \tilde{M}E\tilde{N} = \begin{bmatrix} I_r & 0 \\ 0 & 0 \end{bmatrix}. \tag{2.42}$$

Note that $Y = \tilde{M}^{-T}X\tilde{M}^{-1} = \left[\begin{smallmatrix} Y_{11} & Y_{12} \\ Y_{12}^* & Y_{22} \end{smallmatrix}\right]$. It is obvious that $Y = Y^* > 0$. Also, note that $\widehat{S} = \tilde{M}^{-T}S\tilde{M}^{-1} = \left[\begin{smallmatrix} \widehat{S}_{11} & \widehat{S}_{12} \\ \widehat{S}_{12}^* & \widehat{S}_{22} \end{smallmatrix}\right]$. If $\tilde{N}^T$ and $\tilde{N}$ are premultiplied and postmultiplied on (2.41b), respectively, the inequality can be rewritten as

$$\begin{aligned} \tilde{N}^T E^T SE\tilde{N} &= \tilde{N}^T E^T \tilde{M}^T \widehat{S}\tilde{M}E\tilde{N} \\ &= \begin{bmatrix} I_r & 0 \\ 0 & 0 \end{bmatrix} \begin{bmatrix} \widehat{S}_{11} & \widehat{S}_{12} \\ \widehat{S}_{12}^* & \widehat{S}_{22} \end{bmatrix} \begin{bmatrix} I_r & 0 \\ 0 & 0 \end{bmatrix} \\ &= \begin{bmatrix} \widehat{S}_{11} & 0 \\ 0 & 0 \end{bmatrix} \geq 0, \end{aligned}$$

which implies that $\widehat{S}_{11} \geq 0$. If $\tilde{N}^T$ and $\tilde{N}$ are premultiplied and postmultiplied on (2.41a) respectively, then the inequality can be rewritten as

$$\begin{aligned}
&\tilde{N}^T[\mathrm{Sym}\{A^T(\mathbf{r}X+\bar{\mathbf{r}}\bar{X})E\}+A^TSA]\tilde{N}\\
&\quad=\mathrm{Sym}\{\tilde{N}^TA^T\tilde{M}^T(\mathbf{r}Y+\bar{\mathbf{r}}\bar{Y})\tilde{M}E\tilde{N}\}+\tilde{N}^TA^T\tilde{M}^T\widehat{S}\tilde{M}A\tilde{N}\\
&\quad=\mathrm{Sym}\left\{\begin{bmatrix}A_1^T & 0\\ 0 & I_{n-r}\end{bmatrix}\begin{bmatrix}(\mathbf{r}Y_{11}+\bar{\mathbf{r}}\bar{Y}_{11}) & (\mathbf{r}Y_{12}+\bar{\mathbf{r}}\bar{Y}_{12})\\ \star & (\mathbf{r}Y_{22}+\bar{\mathbf{r}}\bar{Y}_{22})\end{bmatrix}\bullet\begin{bmatrix}I_r & 0\\ 0 & 0\end{bmatrix}\right\}\\
&\qquad+\tilde{N}^TA^T\tilde{M}^T\begin{bmatrix}\widehat{S}_{11} & \widehat{S}_{12}\\ \widehat{S}_{12}^* & \widehat{S}_{22}.\end{bmatrix}\tilde{M}A\tilde{N}<0,
\end{aligned}$$

i. e.,

$$\begin{bmatrix}\mathrm{Sym}\{A_1^T(\mathbf{r}Y_{11}+\bar{\mathbf{r}}\bar{Y}_{11})\}+A_1^T\widehat{S}_{11}A_1 & \mathbf{r}Y_{12}+\bar{\mathbf{r}}\bar{Y}_{12}+A_1^T\widehat{S}_{12}\\ \star & \widehat{S}_{22}\end{bmatrix}<0, \tag{2.43}$$

which indicates that

$$\mathrm{Sym}\{A_1^T(\mathbf{r}Y_{11}+\bar{\mathbf{r}}\bar{Y}_{11})\}+A_1^T\widehat{S}_{11}A_1<0. \tag{2.44}$$

Since $\widehat{S}_{11}\geq 0$, we have $\mathrm{Sym}\{A_1^T(\mathbf{r}Y_{11}+\bar{\mathbf{r}}\bar{Y}_{11})\}<0$. Since A_1 is asymptotically stable by Lemma 2.7, A is also asymptotically stable. Therefore the FOSS (2.23) is admissible.

Necessity. Suppose that FOSS (2.23) is admissible. By Definition 2.3, FOSS (2.23) is regular, impulse-free, and stable. From Theorems 2.1 and 2.4 it follows that invertible matrices $\check{M}$ and $\check{N}$ can be chosen to satisfy

$$\check{M}A\check{N}=\begin{bmatrix}A_1 & 0\\ 0 & I_{n-r}\end{bmatrix},\quad \check{M}E\check{N}=\begin{bmatrix}I_r & 0\\ 0 & 0\end{bmatrix}. \tag{2.45}$$

From the proof of Theorem 2.5, it follows that the stability of FOSS (2.23) is equivalent to the stability of the system $D^\alpha x_1(t)=A_1x_1(t)$. By Lemma 2.7, there exists $Y_{11}=Y_{11}^*>0$ satisfying

$$\mathrm{Sym}\{A_1^T(\mathbf{r}Y_{11}+\bar{\mathbf{r}}\bar{Y}_{11})\}<0. \tag{2.46}$$

There exits a small enough number $\varepsilon>0$ such that

$$\mathrm{Sym}\{A_1^T(\mathbf{r}Y_{11}+\bar{\mathbf{r}}\bar{Y}_{11})\}+A_1^T\varepsilon\,\mathrm{Sym}\{\mathbf{r}Y_{11}+\bar{\mathbf{r}}\bar{Y}_{11}\}A_1<0. \tag{2.47}$$

Note that $X=\check{M}^TY\check{M}=\check{M}^T\left[\begin{smallmatrix}Y_{11} & 0\\ 0 & I_{n-r}\end{smallmatrix}\right]\check{M}$ and $S=\check{M}^T\widehat{S}\check{M}=\check{M}^T\left[\begin{smallmatrix}\widehat{S}_{11} & 0\\ 0 & \widehat{S}_{22}\end{smallmatrix}\right]\check{M}$, where $\widehat{S}_{11}=\varepsilon\,\mathrm{Sym}\{\mathbf{r}Y_{11}+\bar{\mathbf{r}}\bar{Y}_{11}\}>0$, and $\widehat{S}_{22}$ is any negative definite matrix. Then (2.47) can be transformed into

$$\begin{bmatrix}\mathrm{Sym}\{A_1^T(\mathbf{r}Y_{11}+\bar{\mathbf{r}}\bar{Y}_{11})\} & 0\\ 0 & 0\end{bmatrix}+\begin{bmatrix}A_1^T\varepsilon\,\mathrm{Sym}\{\mathbf{r}Y_{11}+\bar{\mathbf{r}}\bar{Y}_{11}\}A_1 & 0\\ 0 & \widehat{S}_{22}\end{bmatrix}<0, \tag{2.48}$$

which is equivalent to

$$\mathrm{Sym}\{\check{N}^T A^T \check{M}^T (\mathbf{r}Y + \bar{\mathbf{r}}\bar{Y})\check{M}E\check{N}\} + \check{N}^T A^T \check{M}^T \widehat{S}\check{M}A\check{N} < 0, \tag{2.49}$$

i. e.,

$$\mathrm{Sym}\{\check{N}^T A^T (\mathbf{r}X + \bar{\mathbf{r}}\bar{X})E\check{N}\} + \check{N}^T A^T SA\check{N} < 0.$$

Then

$$\mathrm{Sym}\{A^T (\mathbf{r}X + \bar{\mathbf{r}}\bar{X})E\} + A^T SA < 0, \tag{2.50}$$

Note that $\widehat{S}_{11} > 0$ and thus

$$\begin{bmatrix} \widehat{S}_{11} & 0 \\ 0 & 0 \end{bmatrix} \geq 0, \tag{2.51}$$

which is equivalent to

$$\check{N}^T E^T \check{M}^T \widehat{S}\check{M}E\check{N} \geq 0, \tag{2.52}$$

i. e.,

$$\check{N}^T E^T SE\check{N} \geq 0. \tag{2.53}$$

From inequality (2.53), it follows that

$$E^T SE \geq 0. \tag{2.54}$$

Inequalities (2.50) and (2.54) prove the necessity. This completes the proof. □

Theorem 2.8. *The FOSS* (2.23) *with* $\alpha \in (0, 1)$ *is admissible if and only if there exist matrices* $X \in \mathbb{C}^{n\times n}$, $Y \in \mathbb{C}^{n\times n}$, *and* $Z \in \mathbb{C}^{n\times n}$ *satisfying*

$$\mathrm{Sym}\{A((e^{j\theta}X + e^{-j\theta}\bar{X})E^T + E^{\dagger}Y)\} < 0, \tag{2.55a}$$

$$-EXE^T - \mathrm{Sym}\{E^{\ddagger}Z\} < 0, \tag{2.55b}$$

where $\theta = (1-\alpha)\pi/2$, *and* $E^{\dagger}$ *and* $E^{\ddagger}$ *are given in* (2.27).

Proof. Sufficiency. Denote

$$\begin{gathered} MAN = \begin{bmatrix} A_1 & A_2 \\ A_3 & A_4 \end{bmatrix}, \quad N^{-1}XN^{-T} = \begin{bmatrix} X_1 & X_2 \\ X_3 & X_4 \end{bmatrix}, \\ N^{-1}E^{\dagger}M^{-1} = \begin{bmatrix} 0_r & 0_{r\times(n-r)} \\ 0_{(n-r)\times r} & I_{n-r} \end{bmatrix}, \quad MYM^T = \begin{bmatrix} Y_1 & Y_2 \\ Y_3 & Y_4 \end{bmatrix}. \end{gathered} \tag{2.56}$$

Since E and A in this chapter are real matrices, we can obtain two real transformation matrices M and N by using the methods in [116, Appendix B]. Then we have $N^{-1}\bar{X}N^{-T} = \left[\begin{smallmatrix}\bar{X}_1 & \bar{X}_2\\ \bar{X}_3 & \bar{X}_4\end{smallmatrix}\right]$. By premultiplying and postmultiplying (2.55a) by M and M^T, we have

$$\mathrm{Sym}\{MAN(N^{-1}(e^{j\theta}X + e^{-j\theta}\bar{X})N^{-T}N^TE^TM^T + N^{-1}E^{\dagger}M^{-1}MYM^T)\} < 0. \tag{2.57}$$

Substituting (2.56) into (2.57) yields

$$\mathrm{Sym}\left\{\begin{bmatrix} A_1(e^{j\theta}X_1 + e^{-j\theta}\bar{X}_1) + A_2((e^{j\theta}X_3 + e^{-j\theta}\bar{X}_3) + Y_3) & A_2Y_4 \\ A_3(e^{j\theta}X_1 + e^{-j\theta}\bar{X}_1) + A_4((e^{j\theta}X_3 + e^{-j\theta}\bar{X}_3) + Y_3) & A_4Y_4 \end{bmatrix}\right\} < 0, \tag{2.58}$$

indicating that

$$\mathrm{Sym}\{A_4Y_4\} < 0. \tag{2.59}$$

From Lemma 2.12, it follows that A_4 is nonsingular. Then the FOSS is impulse-free by Lemma 2.9. Denoting

$$\hat{M} = \begin{bmatrix} I_r & -A_2A_4^{-1} \\ 0_{(n-r)\times r} & A_4^{-1} \end{bmatrix}, \quad \hat{N} = \begin{bmatrix} I_r & 0_{r\times(n-r)} \\ -A_4^{-1}A_3 & I_{n-r} \end{bmatrix}, \tag{2.60}$$

we have

$$\hat{M}^{-1} = \begin{bmatrix} I_r & A_2 \\ 0_{(n-r)\times r} & A_4 \end{bmatrix}, \quad \hat{N}^{-1} = \begin{bmatrix} I_r & 0_{r\times(n-r)} \\ A_4^{-1}A_3 & I_{n-r} \end{bmatrix}. \tag{2.61}$$

Combining (2.60) and (2.61) gives

$$\begin{aligned}
\hat{M}MAN\hat{N} &= \begin{bmatrix} A_1 - A_2A_4^{-1}A_3 & 0_{r\times(n-r)} \\ 0_{(n-r)\times r} & I_{n-r} \end{bmatrix}, \\
\hat{M}MEN\hat{N} &= \begin{bmatrix} I_r & 0_{r\times(n-r)} \\ 0_{(n-r)\times r} & 0_{n-r} \end{bmatrix}, \\
\hat{N}^{-1}N^{-1}E^{\dagger}M^{-1}\hat{M}^{-1} &= \begin{bmatrix} 0_r & 0_{r\times(n-r)} \\ 0_{(n-r)\times r} & A_4 \end{bmatrix}, \\
\hat{M}MYM^T\hat{M}^T &= \begin{bmatrix} \Psi_1 & \Psi_2 \\ \Psi_3 & \Psi_4 \end{bmatrix}, \\
\hat{N}^{-1}N^{-1}XN^{-T}\hat{N}^{-T} &= \begin{bmatrix} X_1 & \Psi_5 \\ \Psi_6 & \Psi_7 \end{bmatrix},
\end{aligned} \tag{2.62}$$

where Ψ_i $(i = 1, 2, \ldots, 7)$ are matrices unrelated to the proof process. The matrices $\hat{M}$ and $\hat{N}$ are both real. Denote $MZM^T = \left[\begin{smallmatrix}Z_1 & Z_2\\ Z_3 & Z_4\end{smallmatrix}\right]$. By premultiplying and postmultiplying (2.55b) by M and M^T, we have

$$MENN^{-1}XN^{-T}N^TE^TM^T + \mathrm{Sym}\{ME^{\ddagger}ZM^T\} > 0, \tag{2.63}$$

indicating that

$$\begin{bmatrix} X_1 & Z_3^* \\ Z_3 & Z_4 + Z_4^* \end{bmatrix} > 0, \tag{2.64}$$

i. e., $X_1 > 0$. By premultiplying and postmultiplying (2.57) by $\hat{M}$ and $\hat{M}^T$ it is clear that

$$\mathrm{Sym}\left\{\begin{bmatrix} (A_1 - A_2A_4^{-1}A_3)(e^{j\theta}X_1 + e^{-j\theta}\bar{X}_1) & 0_{r\times(n-r)} \\ e^{j\theta}\Psi_6 + e^{-j\theta}\bar{\Psi}_6 + A_4\Psi_3 & A_4\Psi_4 \end{bmatrix}\right\} < 0, \tag{2.65}$$

from which it follows that $\mathrm{Sym}\{(A_1 - A_2A_4^{-1}A_3)(e^{j\theta}X_1 + e^{-j\theta}\bar{X}_1)\} < 0, X_1 > 0$. By Lemmas 2.6 and 2.9, the FOSS is admissible.

Necessity. Suppose that the FOSS is admissible. For the matrix decomposition in (2.26), applying Lemmas 2.9, 2.6, and 2.12 leads to the fact that there exist $P_1 > 0$ and a nonsingular matrix P_2 such that

$$\mathrm{Sym}\{(A_1 - A_2A_4^{-1}A_3)(e^{j\theta}P_1 + e^{-j\theta}\bar{P}_1)\} < 0, \tag{2.66}$$

$$\mathrm{Sym}\{A_4P_2\} < 0. \tag{2.67}$$

There exist a small enough real constant $\kappa > 0$ and matrices P_3 and P_4 satisfying

$$\begin{aligned}&\mathrm{Sym}\left\{\begin{bmatrix} (A_1 - A_2A_4^{-1}A_3)(e^{j\theta}P_1 + e^{-j\theta}\bar{P}_1) & 0_{r\times(n-r)} \\ 0_{(n-r)\times r} & A_4P_2 \end{bmatrix}\right\} \\ &+ \kappa\,\mathrm{Sym}\left\{\begin{bmatrix} 0_r & 0_{r\times(n-r)} \\ e^{j\theta}P_3 + e^{-j\theta}\bar{P}_3 + A_4P_4 & 0_{n-r} \end{bmatrix}\right\} < 0.\end{aligned} \tag{2.68}$$

The admissibility of FOSS requires A_4 to be nonsingular. Let

$$Y = M^{-1}\hat{M}^{-1}\begin{bmatrix} P_5 & P_6 \\ \kappa P_4 & P_2 \end{bmatrix}\hat{M}^{-T}M^{-T}, \quad X = N\hat{N}\begin{bmatrix} P_1 & P_7 \\ \kappa P_3 & P_8 \end{bmatrix}\hat{N}^TN^T, \tag{2.69}$$

where $\hat{M}$ and $\hat{N}$ are defined in (2.60), and P_5, P_6, P_7, P_8 are matrices of appropriate dimensions. Then

$$\begin{aligned}&\hat{M}M\,\mathrm{Sym}\{A((e^{j\theta}X + e^{-j\theta}\bar{X})E^T + E^{\dagger}Y)\}M^T\hat{M}^T \\ &= \mathrm{Sym}\left\{\begin{bmatrix} (A_1 - A_2A_4^{-1}A_3)(e^{j\theta}P_1 + e^{-j\theta}\bar{P}_1) & 0_{r\times(n-r)} \\ 0_{(n-r)\times r} & A_4P_2 \end{bmatrix}\right\} \\ &+ \kappa\,\mathrm{Sym}\left\{\begin{bmatrix} 0_r & 0_{r\times(n-r)} \\ e^{j\theta}P_3 + e^{-j\theta}\bar{P}_3 + A_4P_4 & 0_{n-r} \end{bmatrix}\right\} < 0,\end{aligned} \tag{2.70}$$

indicating that inequality (2.55a) holds. Note that $P_1 > 0$. Therefore there exist a small enough constant $\kappa_2 > 0$ and matrices $P_9 > 0$ and P_{10} satisfying

$$-\begin{bmatrix} P_1 & 0_{r\times(n-r)} \\ 0_{(n-r)\times r} & P_9 + P_9^* \end{bmatrix} + \kappa_2 \begin{bmatrix} 0_r & P_{10}^* \\ P_{10} & 0_{n-r} \end{bmatrix} < 0. \tag{2.71}$$

Let

$$Z = M^{-1} \begin{bmatrix} P_{11} & P_{12} \\ -\kappa_2 P_{10} & P_9 \end{bmatrix} M^{-T}, \tag{2.72}$$

where P_{11} and P_{12} are matrices of appropriate dimensions. Then

$$\begin{aligned} & - M(EXE^T + \mathrm{Sym}\{E^{\ddagger} Z\})M^T \\ & = -\begin{bmatrix} P_1 & 0_{r\times(n-r)} \\ 0_{(n-r)\times r} & P_9 + P_9^* \end{bmatrix} + \kappa_2 \begin{bmatrix} 0_r & P_{10}^* \\ P_{10} & 0_{n-r} \end{bmatrix} < 0, \end{aligned} \tag{2.73}$$

indicating that inequality (2.55b) holds. The proof is completed. □

Remark 2.11. Theorems 2.6, 2.7, and 2.8 give three necessary and sufficient conditions for the admissibility of FOSSs with $\alpha \in (0, 1)$ in the form of LMIs with equality constraints, nonstrict LMIs, and strict LMIs, respectively. For LMIs with equality constraints, nonstrict LMIs, and strict LMIs problems, SciLAB can be used directly [71].

Remark 2.12. Compared with the results in [69, 88, 106, 116], the novelty and advantages of the admissibility conditions, i. e., Theorems 2.6, 2.7, and 2.8, presented in this chapter are as follows.

1. Compared with the results in [69], Theorems 2.6, 2.7, and 2.8 are the admissibility criteria based on Definition 2.3. The results in [69] are essentially the stability criteria of FOSSs because the investigated FOSSs are supposed to be regular and impulse-free.
2. Compared with the results in [88], Theorems 2.6, 2.7, and 2.8 can be applied more concisely and conveniently. When applying the admissibility conditions proposed in [88], E and A are assumed to have the forms $E = \begin{bmatrix} I & 0 \\ 0 & 0 \end{bmatrix}$ and $A = \begin{bmatrix} A_1 & A_2 \\ A_3 & A_4 \end{bmatrix}$, which have strong limitations. When applying Theorems 2.6, 2.7, and 2.8, LMIs can be solved directly without limitations on E and A.
3. Compared with the results in [106, 116], Theorems 2.6, 2.7, and 2.8 involve less variables to be solved. The admissibility conditions proposed in [106] and Theorem 2.3 in [116] need five and four variables to be solved, respectively, whereas Theorems 2.6, 2.7, and 2.8 only need one, two, and three variables to be solved, respectively.

Now we provide the following illustrative example.

Example 2.1. Consider the FOSS $ED^{\alpha}x(t) = Ax(t)$ with

$$E = \begin{bmatrix} 1 & 0 & 0 \\ 0 & 2 & 0 \\ 0 & 0 & 0 \end{bmatrix}, \quad A = \begin{bmatrix} -10 & 5 & 6 \\ 2 & -5 & -1 \\ -9 & 4 & 8 \end{bmatrix}, \quad \alpha = 0.8. \tag{2.74}$$

The purpose of this example is to determine whether the FOSS with parameters (2.74) is admissible.

Using Theorem 2.6, we can obtain the following feasible solution:

$$X = \begin{bmatrix} 214.3466 & 46.4865 - 12.4935i & 0 \\ 23.2432 + 6.2468i & 53.7086 & 0 \\ -222.1094 & -28.6232 & -27.2741 \end{bmatrix}.$$

By Theorem 2.6 the FOSS (2.74) is admissible.

The eigenvalues of the FOSS with parameters (2.74) are −3.8107 and −1.6893, which lie in the stable areas shown in Figure 2.5. With the initial condition $x(0) = \begin{bmatrix} 1 & -2 & -1 \end{bmatrix}^T$, the time responses of the FOSS (2.74) are shown in Figure 2.6, where we can see that all the states converge to 0, indicating that the system is admissible.

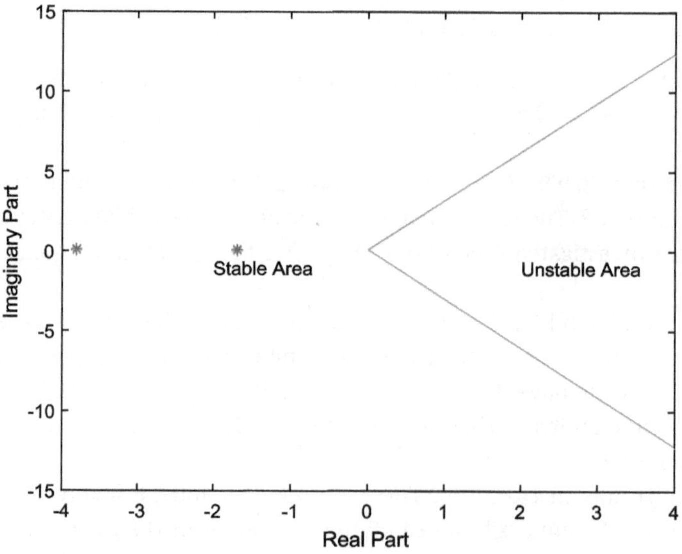

Figure 2.5: The location of eigenvalues of the FOSS with parameters (2.74) in Example 2.1.

The validity of Theorems 2.7 and 2.8 can be verified in the same way as in Example 2.1.

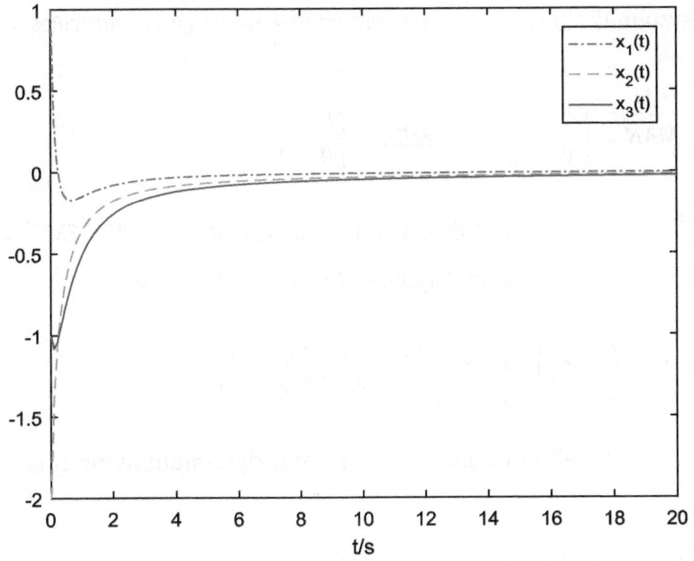

Figure 2.6: The time responses of the FOSS (2.74) in Example 2.1 by means of the predictor–corrector approach [16].

2.5.2 Case $\alpha \in (1, 2)$

Theorem 2.9. *The FOSS* (2.23) *with* $\alpha \in (1, 2)$ *is admissible if and only if there exist matrices* $X \in \mathbb{R}^{n\times n}$ *and* $S \in \mathbb{R}^{n\times n}$, $X = X^T > 0$, $S = S^T$, *satisfying*

$$\mathrm{Sym}\{\Theta \otimes (A^T XE)\} + \Omega \otimes (A^T SA) < 0, \tag{2.75a}$$

$$E^T SE \geq 0, \tag{2.75b}$$

where $\theta = (2 - \alpha)\pi/2$, $\Theta = \left[\begin{smallmatrix} \sin\theta & -\cos\theta \\ \cos\theta & \sin\theta \end{smallmatrix}\right]$, *and* $\Omega > 0$ *is a matrix satisfying* $\Omega \in \mathbb{C}^{2\times 2}$.

Proof. Sufficiency. The proof of nonimpulsiveness is by contradiction. Assume that system (2.23) is impulsive. By Lemma 2.4, we may choose an infinite eigenvector $v^2 \in \mathbb{R}^n$ of order two satisfying $Ev^1 = 0$, $Ev^2 = Av^1$. If $(I_2 \otimes v^1)^*$ and $I_2 \otimes v^1$ are left-multiplied and right-multiplied on (2.75a), respectively, then

$$\begin{aligned} &(I_2 \otimes v^1)^* (\mathrm{Sym}\{\Theta \otimes (A^T XE)\} + \Omega \otimes (A^T SA))(I_2 \otimes v^1) \\ &= \Omega \otimes ((v^1)^* A^T SAv^1) \\ &= \Omega \otimes ((v^2)^* E^T SEv^2) < 0, \end{aligned}$$

which is contradictory to the fact that $E^T SE \geq 0$ and $\Omega > 0$. As a consequence, the assumption is not true, which implies that the FOSS (2.23) is with nonimpulsiveness. By

Theorems 2.4 and 2.1, system (2.23) is regular. We can choose nonsingular matrices M and N satisfying

$$MAN = \begin{bmatrix} A_1 & 0 \\ 0 & I_{n-r} \end{bmatrix}, \quad MEN = \begin{bmatrix} I_r & 0 \\ 0 & 0 \end{bmatrix}.$$

Let $\Lambda = M^{-T}XM^{-1} = \left[\begin{smallmatrix} \Lambda_{11} & \Lambda_{12} \\ \Lambda_{12}^T & \Lambda_{22} \end{smallmatrix}\right]$. It is obvious that $\Lambda = \Lambda^T > 0$. Note that $\widehat{S} = M^{-T}SM^{-1} = \left[\begin{smallmatrix} \widehat{S}_{11} & \widehat{S}_{12} \\ \widehat{S}_{12}^T & \widehat{S}_{22} \end{smallmatrix}\right]$. By left-multiplying N^T and right-multiplying N on (2.75b) we have

$$N^TE^TSEN = \begin{bmatrix} I_r & 0 \\ 0 & 0 \end{bmatrix} \begin{bmatrix} \widehat{S}_{11} & \widehat{S}_{12} \\ \widehat{S}_{12}^T & \widehat{S}_{22} \end{bmatrix} \begin{bmatrix} I_r & 0 \\ 0 & 0 \end{bmatrix} = \begin{bmatrix} \widehat{S}_{11} & 0 \\ 0 & 0 \end{bmatrix} \geq 0,$$

which implies that $\widehat{S}_{11} \geq 0$. By left-multiplying $I_2 \otimes N^T$ and right-multiplying $I_2 \otimes N$ on (2.75a), we have

$$\begin{aligned}
&(I_2 \otimes N^T)[\mathrm{Sym}\{\Theta \otimes (A^TXE)\} + \Omega \otimes (A^TSA)](I_2 \otimes N) \\
&= \mathrm{Sym}\{\Theta \otimes (N^TA^TM^T\Lambda MEN)\} + \Omega \otimes (N^TA^TM^T\widehat{S}MAN) \\
&= \mathrm{Sym}\left\{\Theta \otimes \left(\begin{bmatrix} A_1^T & 0 \\ 0 & I_{n-r} \end{bmatrix} \begin{bmatrix} \Lambda_{11} & \Lambda_{12} \\ \star & \Lambda_{22} \end{bmatrix} \begin{bmatrix} I_r & 0 \\ 0 & 0 \end{bmatrix}\right)\right\} \\
&\quad + \Omega \otimes \left(\begin{bmatrix} A_1^T & 0 \\ 0 & I_{n-r} \end{bmatrix} \begin{bmatrix} \widehat{S}_{11} & \widehat{S}_{12} \\ \star & \widehat{S}_{22}. \end{bmatrix} \begin{bmatrix} A_1 & 0 \\ 0 & I_{n-r} \end{bmatrix}\right) \\
&= \mathrm{Sym}\left\{\Theta \otimes \begin{bmatrix} A_1^T\Lambda_{11} & 0 \\ \Lambda_{12}^T & 0 \end{bmatrix}\right\} + \Omega \otimes \begin{bmatrix} A_1^T\widehat{S}_{11}A_1 & A_1^T\widehat{S}_{12} \\ \star & \widehat{S}_{22} \end{bmatrix} \\
&< 0,
\end{aligned}$$

which can be rewritten as

$$\mathrm{Sym}\left\{\begin{bmatrix} \Theta \otimes (A_1^T\Lambda_{11}) & 0 \\ \Theta \otimes \Lambda_{12}^T & 0 \end{bmatrix}\right\} + \begin{bmatrix} \Omega \otimes (A_1^T\widehat{S}_{11}A_1) & \Omega \otimes (A_1^T\widehat{S}_{12}) \\ \star & \Omega \otimes \widehat{S}_{22} \end{bmatrix} < 0. \tag{2.76}$$

Consider the (1,1) block of (2.76), that is,

$$\mathrm{Sym}\{\Theta \otimes (A_1^T\Lambda_{11})\} + \Omega \otimes (A_1^T\widehat{S}_{11}A_1) < 0. \tag{2.77}$$

Since $\widehat{S}_{11} \geq 0$, $\Omega > 0$, and then $\mathrm{Sym}\{\Theta \otimes (A_1^T\Lambda_{11})\} < 0$. By Lemma 2.7, the FOS $D^\alpha x_1(t) = A_1x_1(t)$ is asymptotically stable. By Theorem 2 in [67], the asymptotical stability of the FOS $D^\alpha x_1(t) = A_1x_1(t)$ indicates that $|\arg(\mathrm{spec}(A_1))| > \frac{\alpha}{2}\pi$. By Theorem 2.5, $|\arg(\mathrm{spec}(E, A))| > \frac{\alpha}{2}\pi$. Therefore FOSS (2.23) is asymptotically stable.

Necessity. Since FOSS (2.23) is admissible, we can find invertible matrices $\Lambda_{11} > 0$, M, and N such that

$$MAN = \begin{bmatrix} A_1 & 0 \\ 0 & I_{n-r} \end{bmatrix}, \quad MEN = \begin{bmatrix} I_r & 0 \\ 0 & 0 \end{bmatrix},$$

$$\mathrm{Sym}\{\Theta \otimes (A_1^T \Lambda_{11})\} < 0.$$

There exists a small enough constant $\varepsilon > 0$ such that

$$\mathrm{Sym}\{\Theta \otimes (A_1^T \Lambda_{11})\} + \Omega \otimes (A_1^T \varepsilon \Lambda_{11} A_1) < 0. \tag{2.78}$$

Note that $X = M^T \Lambda M = M^T \left[\begin{smallmatrix} \Lambda_{11} & 0 \\ 0 & I_{n-r} \end{smallmatrix}\right] M$ and $S = M^T \widehat{S} M = M^T \left[\begin{smallmatrix} \widehat{S}_{11} & 0 \\ 0 & \widehat{S}_2 \end{smallmatrix}\right] M$, where $\widehat{S}_{11} = \varepsilon \Lambda_{11}$, and $\widehat{S}_{22}$ is any negative definite matrix. Then inequality (2.78) can be rewritten as

$$\begin{bmatrix} \mathrm{Sym}\{\Theta \otimes (A_1^T \Lambda_{11})\} & 0 \\ 0 & 0 \end{bmatrix} + \begin{bmatrix} \Omega \otimes (A_1^T \varepsilon \Lambda_{11} A_1) & 0 \\ 0 & \Omega \otimes \widehat{S}_{22} \end{bmatrix} < 0, \tag{2.79}$$

which can be transformed into

$$\mathrm{Sym}\left\{\Theta \otimes \begin{bmatrix} A_1^T \Lambda_{11} & 0 \\ 0 & 0 \end{bmatrix}\right\} + \Omega \otimes \begin{bmatrix} A_1^T \varepsilon \Lambda_{11} A_1 & 0 \\ 0 & \widehat{S}_{22} \end{bmatrix} < 0. \tag{2.80}$$

Substituting A and Λ into inequality (2.80), we have

$$\mathrm{Sym}\{\Theta \otimes (N^T A^T M^T \Lambda MEN)\} + \Omega \otimes (N^T A^T M^T \widehat{S} MAN) < 0,$$

which is equivalent to

$$(I_2 \otimes N^T)(\mathrm{Sym}\{\Theta \otimes (A^T XE)\} + \Omega \otimes (A^T SA))(I_2 \otimes N) < 0.$$

Therefore we have

$$\mathrm{Sym}\{\Theta \otimes (A^T XE)\} + \Omega \otimes (A^T SA) < 0.$$

Since $\widehat{S}_{11} = \varepsilon Y_{11} > 0$, we have

$$\begin{bmatrix} \widehat{S}_{11} & 0 \\ 0 & 0 \end{bmatrix} \geq 0,$$

which can be transformed into

$$N^T E^T M^T \widehat{S} MEN \geq 0,$$

i. e., $N^T E^T SEN \geq 0$. Thus $E^T SE \geq 0$. The proof is completed. □

Theorem 2.10. *The FOSS (2.23) with $\alpha \in (1, 2)$ is admissible if and only if there exist matrices $X \in \mathbb{R}^{n\times n}$, $X = X^T > 0$, and $S \in \mathbb{R}^{(n-r)\times n}$ satisfying*

$$\mathrm{Sym}\{\Theta \otimes (A^T XE) + \Omega \otimes (A^T E_0 S)\} < 0 \tag{2.81}$$

with $E^T E_0 = 0$, *where* $\Theta = \begin{bmatrix} \sin\theta & -\cos\theta \\ \cos\theta & \sin\theta \end{bmatrix}$, $\theta = (2-\alpha)\pi/2$, *and* $\Omega > 0$ *is a matrix satisfying* $\Omega \in \mathbb{C}^{2\times 2}$.

Proof. Sufficiency. We prove the nonimpulsiveness by contradiction. Assume that system (2.23) is impulsive. By Lemma 2.4, we may choose an infinite eigenvector $v^2 \in \mathbb{R}^n$ of order two satisfying $Ev^1 = 0$, $Ev^2 = Av^1$. If $(I_2 \otimes v^1)^*$ and $I_2 \otimes v^1$ are left-multiplied and right-multiplied on inequality (2.81), then

$$\begin{aligned}
&(I_2 \otimes v^1)^* \mathrm{Sym}\{\Theta \otimes (A^T XE) + \Omega \otimes (A^T E_0 S)\}(I_2 \otimes v^1)\\
&\quad = \Omega \otimes \mathrm{Sym}\{(v^1)^* A^T E_0 S v^1\}\\
&\quad = \Omega \otimes \mathrm{Sym}\{(v^2)^* E^T E_0 S v^1\} < 0,
\end{aligned}$$

which is contradictory to the fact that $E^T E_0 = 0$. As a consequence, the assumption is not true, which implies that the FOSS (2.23) is with nonimpulsiveness. By Theorems 2.4 and 2.1, system (2.23) is regular.

Suppose that λ is a finite eigenvalue of the FOSS (2.23), v is the corresponding eigenvector, and thus $Av = \lambda Ev$. From (2.81), we have

$$\begin{aligned}
&(I_2 \otimes v)^* \mathrm{Sym}\{\Theta \otimes (A^T XE) + \Omega \otimes (A^T E_0 S)\}(I_2 \otimes v)\\
&\quad = \Theta \otimes (\lambda v^* E^T XEv) + \Theta^T \otimes (\tilde{\lambda} v^* E^T XEv)\\
&\quad = (I_2 \otimes (v^* E^T))\, \mathrm{Sym}\{\Theta \otimes (\lambda X)\}(I_2 \otimes (Ev)) < 0,
\end{aligned}$$

which indicates that

$$\mathrm{Sym}\{\Theta \otimes (\lambda X)\} < 0. \tag{2.82}$$

It follows from Lemma 2.7 and inequality (2.82) that the FOSS (2.23) is stable, and thus the admissibility is proven.

Necessity. Since the FOSS (2.23) is admissible, there exist nonsingular matrices $N = [N_1 \quad N_2]$ and $M = \begin{bmatrix} M_1 \\ M_2 \end{bmatrix}$ satisfying

$$MAN = \begin{bmatrix} A_1 & 0 \\ 0 & I_{n-r} \end{bmatrix}, \quad MEN = \begin{bmatrix} I_r & 0 \\ 0 & 0 \end{bmatrix}.$$

Then by Lemma 2.7, there exists $X_1 = X_1^T > 0$, $X_1 \in \mathbb{R}^{n\times n}$, such that

$$\mathrm{Sym}\{\Theta \otimes (A_1^T X_1)\} < 0.$$

There exists a small enough constant $\varepsilon > 0$ that satisfies

$$\begin{aligned}
&\mathrm{Sym}\{\Theta \otimes (A_1^T X_1)\} + (\Omega \otimes (\varepsilon N_1^T N_2))\\
&\quad \bullet (\Omega^{-1} \otimes (2\varepsilon N_2^T N_2)^{-1})(\Omega \otimes (\varepsilon N_2^T N_1)) < 0.
\end{aligned} \tag{2.83}$$

Inequality (2.83) can be transformed into

$$\begin{bmatrix} \mathrm{Sym}\{\Theta \otimes (A_1^T X_1)\} & -\Omega \otimes (\varepsilon N_1^T N_2) \\ -\Omega \otimes (\varepsilon N_2^T N_1) & -\Omega \otimes (2\varepsilon N_2^T N_2) \end{bmatrix} < 0,$$

which can be rewritten as

$$\begin{aligned} \mathrm{Sym}\Bigg\{ & \Theta \otimes \left(\begin{bmatrix} A_1^T & 0 \\ 0 & I_{n-r} \end{bmatrix} \begin{bmatrix} X_1 & 0 \\ 0 & I_{n-r} \end{bmatrix} \begin{bmatrix} I_r & 0 \\ 0 & 0 \end{bmatrix} \right) \\ & + \Omega \otimes \left(\begin{bmatrix} 0 \\ I_{n-r} \end{bmatrix} (-\varepsilon N_2^T) \begin{bmatrix} N_1 & N_2 \end{bmatrix} \right) \Bigg\} < 0. \end{aligned} \tag{2.84}$$

Substituting A and E into inequality (2.84), we have

$$\mathrm{Sym}\left\{ \Theta \otimes \left(N^T A^T M^T \begin{bmatrix} X_1 & 0 \\ 0 & I_{n-r} \end{bmatrix} MEN \right) + \Omega \otimes (N^T A^T M_2^T (-\varepsilon N_2^T) N) \right\} < 0.$$

Let $X = M^T \left[\begin{smallmatrix} X_1 & 0 \\ 0 & I_{n-r} \end{smallmatrix}\right] M$, $E_0 = M_2^T$, and $S = -\varepsilon N_2^T$. Then

$$(I_2 \otimes N^T)\, \mathrm{Sym}\{\Theta \otimes (A^T XE) + \Omega \otimes (A^T E_0 S)\}(I_2 \otimes N) < 0,$$

which can be transformed into

$$\mathrm{Sym}\{\Theta \otimes (A^T XE) + \Omega \otimes (A^T E_0 S)\} < 0.$$

This ends the proof. □

Theorem 2.11. *The FOSS* (2.23) *with $\alpha \in (1, 2)$ is admissible if and only if there exist matrices $X \in \mathbb{C}^{n\times n}$, $Y \in \mathbb{C}^{n\times n}$, and $Z \in \mathbb{C}^{n\times n}$ satisfying*

$$\mathrm{Sym}\{e^{j\theta} A(XE^T + E^\dagger Y)\} < 0, \tag{2.85a}$$

$$-EXE^T - \mathrm{Sym}\{E^\ddagger Z\} < 0, \tag{2.85b}$$

where $\theta = (1 - \alpha)\pi/2$, and $E^\dagger$ and $E^\ddagger$ are given in (2.27).

Proof. Sufficiency. Denote

$$\begin{aligned} MAN &= \begin{bmatrix} A_1 & A_2 \\ A_3 & A_4 \end{bmatrix}, \quad N^{-1} X N^{-T} = \begin{bmatrix} X_1 & X_2 \\ X_3 & X_4 \end{bmatrix}, \\ N^{-1} E^\dagger M^{-1} &= \begin{bmatrix} 0_r & 0_{r\times(n-r)} \\ 0_{(n-r)\times r} & I_{n-r} \end{bmatrix}, \quad MYM^T = \begin{bmatrix} Y_1 & Y_2 \\ Y_3 & Y_4 \end{bmatrix}. \end{aligned} \tag{2.86}$$

By premultiplying and postmultiplying (2.85a) by M and M^T, we have

$$\mathrm{Sym}\{e^{j\theta}MAN(N^{-1}XN^{-T}N^TE^TM^T + N^{-1}E^{\dagger}M^{-1}MYM^T)\} < 0. \tag{2.87}$$

Substituting (2.86) into (2.87) implies

$$\mathrm{Sym}\left\{e^{j\theta}\begin{bmatrix} A_1X_1 + A_2(X_3+Y_3) & A_2Y_4 \\ A_3X_1 + A_4(X_3+Y_3) & A_4Y_4 \end{bmatrix}\right\} < 0, \tag{2.88}$$

indicating that

$$\mathrm{Sym}\{e^{j\theta}A_4Y_4\} < 0. \tag{2.89}$$

From Lemma 2.12, it follows that A_4 is nonsingular. Then the FOSS is impulse-free by Lemma 2.9. Denote

$$\hat{M} = \begin{bmatrix} I_r & -A_2A_4^{-1} \\ 0_{(n-r)\times r} & A_4^{-1} \end{bmatrix}, \quad \hat{N} = \begin{bmatrix} I_r & 0_{r\times(n-r)} \\ -A_4^{-1}A_3 & I_{n-r} \end{bmatrix}, \tag{2.90}$$

and we have

$$\hat{M}^{-1} = \begin{bmatrix} I_r & A_2 \\ 0_{(n-r)\times r} & A_4 \end{bmatrix}, \quad \hat{N}^{-1} = \begin{bmatrix} I_r & 0_{r\times(n-r)} \\ A_4^{-1}A_3 & I_{n-r} \end{bmatrix}. \tag{2.91}$$

Combining (2.90) and (2.91) gives

$$\begin{aligned} \hat{M}MAN\hat{N} &= \begin{bmatrix} A_1 - A_2A_4^{-1}A_3 & 0_{r\times(n-r)} \\ 0_{(n-r)\times r} & I_{n-r} \end{bmatrix}, \\ \hat{M}MEN\hat{N} &= \begin{bmatrix} I_r & 0_{r\times(n-r)} \\ 0_{(n-r)\times r} & 0_{n-r} \end{bmatrix}, \\ \hat{N}^{-1}N^{-1}E^{\dagger}M^{-1}\hat{M}^{-1} &= \begin{bmatrix} 0_r & 0_{r\times(n-r)} \\ 0_{(n-r)\times r} & A_4 \end{bmatrix}, \\ \hat{M}MYM^T\hat{M}^T &= \begin{bmatrix} \Psi_1 & \Psi_2 \\ \Psi_3 & \Psi_4 \end{bmatrix}, \\ \hat{N}^{-1}N^{-1}XN^{-T}\hat{N}^{-T} &= \begin{bmatrix} X_1 & \Psi_5 \\ \Psi_6 & \Psi_7 \end{bmatrix}, \end{aligned} \tag{2.92}$$

where Ψ_i $(i = 1,2,\ldots,7)$ are matrices unrelated to the proof process. Denote $MZM^T = \left[\begin{smallmatrix} Z_1 & Z_2 \\ Z_3 & Z_4 \end{smallmatrix}\right]$. By premultiplying and postmultiplying (2.85b) by M and M^T, we have

$$MENN^{-1}XN^{-T}N^TE^TM^T + \mathrm{Sym}\{ME^{\ddagger}ZM^T\} > 0, \tag{2.93}$$

indicating that

$$\begin{bmatrix} X_1 & Z_3^* \\ Z_3 & Z_4 + Z_4^* \end{bmatrix} > 0, \tag{2.94}$$

i. e., $X_1 > 0$. By premultiplying and postmultiplying (2.87) by $\hat{M}$ and $\hat{M}^T$, we have

$$\mathrm{Sym}\left\{e^{j\theta}\begin{bmatrix}(A_1 - A_2A_4^{-1}A_3)X_1 & 0_{r\times(n-r)} \\ \Psi_6 + A_4\Psi_3 & A_4\Psi_4\end{bmatrix}\right\} < 0, \tag{2.95}$$

from which it follows that $\mathrm{Sym}\{e^{j\theta}(A_1 - A_2A_4^{-1}A_3)X_1\} < 0$, $X_1 > 0$. By Lemmas 2.7 and 2.9, the FOSS is admissible.

Necessity. Suppose that the FOSS is admissible. For the matrix decomposition in (2.26), applying Lemmas 2.7, 2.12, and 2.9 indicates that there exist $P_1 > 0$ and a nonsingular matrix P_2 such that

$$\mathrm{Sym}\{e^{j\theta}(A_1 - A_2A_4^{-1}A_3)P_1\} < 0, \tag{2.96}$$

$$\mathrm{Sym}\{e^{j\theta}A_4P_2\} < 0. \tag{2.97}$$

There exist a small enough constant $\kappa > 0$ and matrices P_3 and P_4 satisfying

$$\mathrm{Sym}\left\{e^{j\theta}\begin{bmatrix}(A_1 - A_2A_4^{-1}A_3)P_1 & 0_{r\times(n-r)} \\ 0_{(n-r)\times r} & A_4P_2\end{bmatrix}\right\} + \kappa\,\mathrm{Sym}\left\{e^{j\theta}\begin{bmatrix}0_r & 0_{r\times(n-r)} \\ P_3 + A_4P_4 & 0_{n-r}\end{bmatrix}\right\} < 0. \tag{2.98}$$

The admissibility of FOSS indicates that A_4 is nonsingular. Let

$$Y = M^{-1}\hat{M}^{-1}\begin{bmatrix}P_5 & P_6 \\ \kappa P_4 & P_2\end{bmatrix}\hat{M}^{-T}M^{-T}, \quad X = N\hat{N}\begin{bmatrix}P_1 & P_7 \\ \kappa P_3 & P_8\end{bmatrix}\hat{N}^TN^T, \tag{2.99}$$

where $\hat{M}$ and $\hat{N}$ are defined in (2.90), and P_5, P_6, P_7, P_8 are matrices of appropriate dimensions. Then

$$\begin{aligned}&\hat{M}M\,\mathrm{Sym}\{e^{j\theta}A(XE^T + E^\dagger Y)\}M^T\hat{M}^T \\ &= \mathrm{Sym}\left\{e^{j\theta}\begin{bmatrix}(A_1 - A_2A_4^{-1}A_3)P_1 & 0_{r\times(n-r)} \\ 0_{(n-r)\times r} & A_4P_2\end{bmatrix}\right\} \\ &\quad + \kappa\,\mathrm{Sym}\left\{e^{j\theta}\begin{bmatrix}0_r & 0_{r\times(n-r)} \\ P_3 + A_4P_4 & 0_{n-r}\end{bmatrix}\right\} < 0,\end{aligned} \tag{2.100}$$

indicating that inequality (2.85a) holds. Note that $P_1 > 0$. Then there exist a small enough constant $\kappa_2 > 0$ and matrices $P_9 > 0$ and P_{10} satisfying

$$-\begin{bmatrix}P_1 & 0_{r\times(n-r)} \\ 0_{(n-r)\times r} & P_9 + P_9^*\end{bmatrix} + \kappa_2\begin{bmatrix}0_r & P_{10}^* \\ P_{10} & 0_{n-r}\end{bmatrix} < 0. \tag{2.101}$$

Let

$$Z = M^{-1}\begin{bmatrix}P_{11} & P_{12} \\ -\kappa_2P_{10} & P_9\end{bmatrix}M^{-T}, \tag{2.102}$$

where P_{11} and P_{12} are matrices of appropriate dimensions. Then

$$-M(EXE^T + \mathrm{Sym}\{E^{\ddagger}Z\})M^T$$
$$= -\begin{bmatrix} P_1 & 0_{r\times(n-r)} \\ 0_{(n-r)\times r} & P_9 + P_9^* \end{bmatrix} + \kappa_2 \begin{bmatrix} 0_r & P_{10}^* \\ P_{10} & 0_{n-r} \end{bmatrix} < 0, \tag{2.103}$$

indicating that inequality (2.85b) holds. The proof is completed. □

Remark 2.13. Compared with the existing conclusions in [63], the novelty and advantages of the admissibility conditions of FOSSs with $\alpha \in (1, 2)$ in this chapter are summarized as follows.

1. There exist no equality constraints in Theorems 2.9, 2.10, and 2.11. By comparison condition (iii) of Theorem 2 in [63] is in terms of LMIs with equality constraints that would result in fragility and lead to numerical problems [101].
2. Theorem 2.10 is more general than the results in [63]. Note that in Theorem 2.10,

$$\Theta = \begin{bmatrix} \sin\theta & -\cos\theta \\ \cos\theta & \sin\theta \end{bmatrix}, \quad \theta = (2-\alpha)\pi/2,\ 1 < \alpha < 2,$$

   and then

$$\sin\theta > 0, \quad \sin\theta + \frac{\cos^2\theta}{\sin\theta} = \frac{1}{\sin\theta} > 0.$$

   By using the Schur complement, we have $\Theta > 0$. When Ω in Theorem 2.10 is chosen as Θ, condition (i) of Theorem 2 in [63] can be recovered. Therefore Theorem 2.10 is more general than the admissibility conditions in the form of strict LMIs in [63]. The usefulness of the notation Ω lies in that more potential choices can be provided when doing robust admissibility analysis or designing the system controllers based on Theorem 2.10.

Remark 2.14. The admissibility conditions, i. e., Theorems 2.9, 2.10, and 2.11, have obvious superiority compared with the conclusions in [96, 119]. In [96], the admissibility problems for FOSSs with $1 < \alpha < 2$ are solved by dividing α into k ($k \in \mathbb{Z}^+$, $k \geq 2$) equal parts and applying the conclusions of the case $0 < \alpha < 1$. Therefore the conclusions of the case $1 < \alpha < 2$ in [96] are not direct methods and more complex than Theorems 2.9, 2.10, and 2.11. In [119], the results contain an assumption on the location of the eigenvalues. Theorems 2.9, 2.10, and 2.11 are more general than the results in [119] as there is no assumption on the location of the eigenvalues.

Remark 2.15. Necessary and sufficient conditions of the regularity, nonimpulsiveness, stability, and admissibility of FOSSs with fractional order $\alpha \in (1, 2)$ are proposed in this chapter. Note that when $\alpha = 1$, the FOSSs degenerate to the IOSSs [101, 18] and the results in this chapter are also applicable, that is to say, the conclusions of this chapter include the conclusions of IOSSs.

The effectiveness of Theorems 2.9, 2.10, and 2.11 can be verified through numerical simulations, and we choose Theorem 2.11 as an example to illustrate it.

Example 2.2. The fractional-order electrical circuit in Figure 2.7 contains source voltages e_1, e_2, resistances R_1, R_2, R_3, and super-inductances L_1, L_2, L_3. As shown in [78], this fractional-order electrical circuit can be described by the FOSS $ED^{\alpha}x(t) = Ax(t) + Bu(t)$, where $A = \begin{bmatrix} -R_1 & 0 & -R_3 \\ 0 & -R_2 & -R_3 \\ 1 & 1 & -1 \end{bmatrix}$, $B = \begin{bmatrix} 1 & 0 \\ 0 & 1 \\ 0 & 0 \end{bmatrix}$, $E = \begin{bmatrix} L_1 & 0 & L_3 \\ 0 & L_2 & L_3 \\ 0 & 0 & 0 \end{bmatrix}$, $x(t) = \begin{bmatrix} i_1(t) & i_2(t) & i_3(t) \end{bmatrix}^T$, $u(t) = \begin{bmatrix} e_1 & e_2 \end{bmatrix}^T$. Consider the open-loop FOSS $ED^{\alpha}x(t) = Ax(t)$ with parameters

$$E = \begin{bmatrix} 5 & 0 & 2 \\ 0 & 3 & 2 \\ 0 & 0 & 0 \end{bmatrix}, \quad A = \begin{bmatrix} -3 & 0 & -4 \\ 0 & -1 & -4 \\ 1 & 1 & -1 \end{bmatrix}, \quad \alpha = 1.5. \tag{2.104}$$

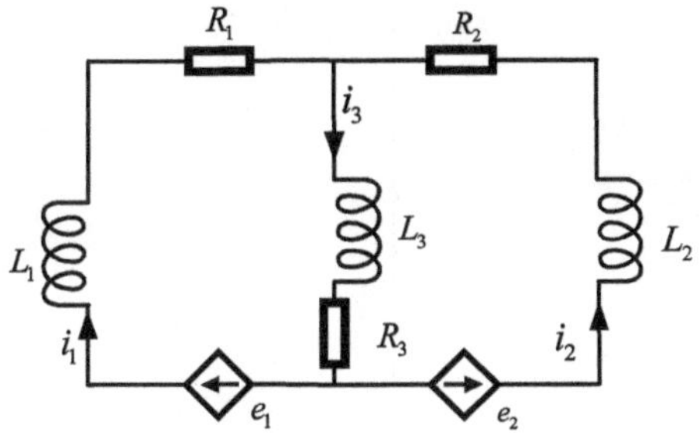

Figure 2.7: The fractional-order electrical circuit in Example 2.2.

The aim is to check whether the nominal FOSS $ED^{\alpha}x(t) = Ax(t)$ with parameters (2.104) is admissible.

Applying Theorem 2(iii) in [63] or Theorem 10 in [119], there exist no feasible solutions. Therefore it is impossible to determine whether the nominal FOSS $ED^{\alpha}x(t) = Ax(t)$ with parameters (2.104) is admissible or not by using Theorem 2(iii) in [63] or Theorem 10 in [119].

However, by Theorem 2.11 there exist the following feasible solutions:

$$X = \begin{bmatrix} 0.9833 & \star & \star \\ -3.0829 + 1.6580j & -0.7566 & \star \\ 0.7945 - 1.1159j & 3.9443 + 2.3852j & 3.9484 \end{bmatrix},$$

$$Y = \begin{bmatrix} 0 & \star & \star \\ 0 & 0 & \star \\ 1.5268 + 1.6711j & 2.9864 + 6.6373j & -3.3023 \end{bmatrix},$$

$$Z = \begin{bmatrix} 0 & \star & \star \\ 0 & 0 & \star \\ 0 & 0 & 24.0865 \end{bmatrix}.$$

By Theorem 2.11, the FOSS $ED^{\alpha}x(t) = Ax(t)$ with parameters (2.104) is admissible.

For simulation study, we have the following facts for the nominal FOSS $ED^{\alpha}x(t) = Ax(t)$ with parameters (2.104). (i) There is $c_0 \in \mathbb{C}$ such that $\det(c_0^{\alpha}E - A) \neq 0$; (ii) $\deg(\det(\lambda E - A)) = 2 = \operatorname{rank}(E)$, $\lambda \in \mathbb{C}$; (iii) The eigenvalues of the system are $\{-1.2527, -0.4893\}$, which are plotted in Figure 2.8 and are located in the stable area. The state evolution is plotted in Figure 2.9, in which all states approach zero. Therefore it is verified that the nominal FOSS $ED^{\alpha}x(t) = Ax(t)$ with parameters (2.104) is indeed admissible.

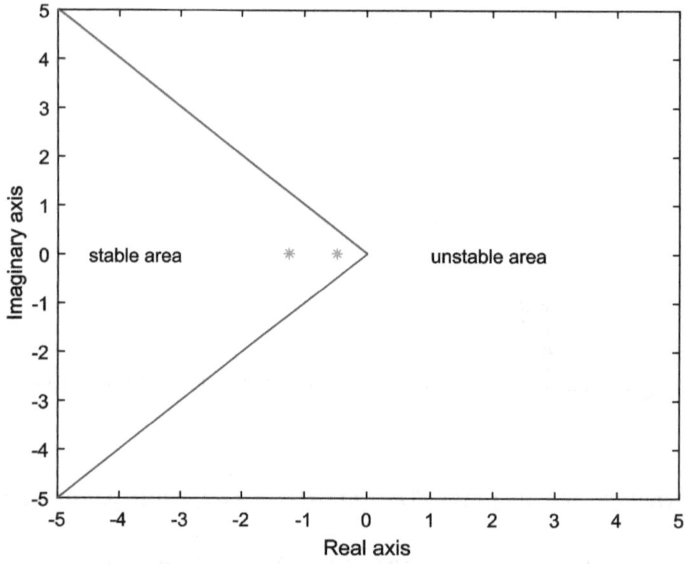

Figure 2.8: The eigenvalue location of the nominal FOSS $ED^{\alpha}x(t) = Ax(t)$ with parameters (2.104) in Example 2.2.

2.6 Chapter summary

The problems of the regularity, nonimpulsiveness, stability, and admissibility of the FOSSs with fractional order $\alpha \in (0, 2)$ have been investigated in this chapter. Firstly, the existence, uniqueness, and structure of the time domain solutions of FOSSs have been analyzed based on the Kronecker equivalent standard form. A necessary and sufficient condition for the regularity of FOSSs has been proposed on the basis of the above analysis. The regularity conditions proposed in this chapter can be applied to both linear and nonlinear FOSSs, which are more extensively applicable compared with the results in [69, 106]. Secondly, the time domain solutions of FOSSs have been derived, which are more complete and accurate compared with the existing results in [69, 106, 38, 116]. Then necessary and sufficient conditions for nonimpulsiveness and stability have been

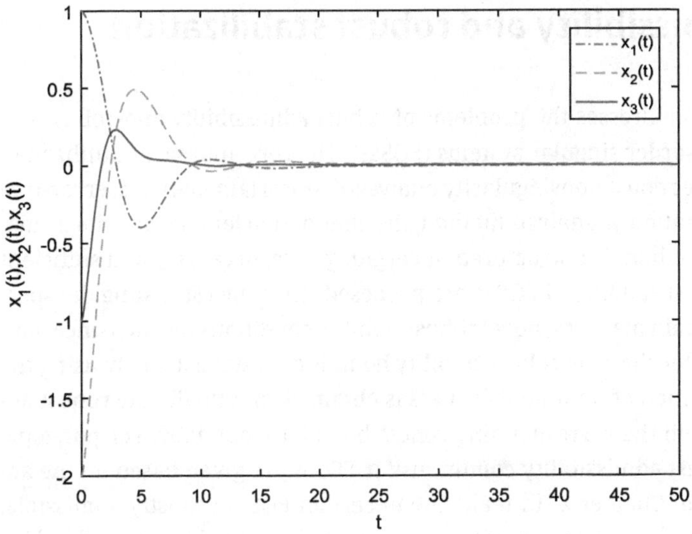

Figure 2.9: The state evolution of the nominal FOSS $ED^{\alpha}x(t) = Ax(t)$ with parameters (2.104) in Example 2.2.

obtained based on the time domain solutions. Thirdly, novel necessary and sufficient conditions for the admissibility of FOSSs have been derived, which are advantageous compared with the existing results.

3 Robust admissibility and robust stabilization

Abstract: This chapter addresses the problems of robust admissibility and robust stabilization of fractional-order singular systems (FOSSs). The core approaches utilized in this chapter include the robust nonsingularity analysis of uncertain matrices, error-free determinant transformation, μ-analysis method, the elimination lemma, and linear matrix inequality (LMI) method. For structured uncertainty case, necessary and sufficient conditions for the robust stability of FOSSs are proposed. Then, by estimating the spectral radius of the uncertain matrices, novel robust stability conditions for such uncertain FOSSs are obtained. After that, the robust stability bounds are established. By using the same method the condition of nonimpulsiveness is obtained, and finally the robust admissibility condition and the corresponding robust bound are obtained. For polytopic uncertainties, the robust admissibility conditions for FOSSs are given based on the admissibility conditions in Chapter 2. To make the uncertain FOSSs robustly admissible, the methods of designing the static output feedback controllers are obtained with wider application scope, which are direct, concise, and more relaxed compared with the existing results.

3.1 Introduction

Robust admissibility analysis and robust stabilization are important research problems for FOSSs, where the admissibility of FOSSs includes regularity, nonimpulsiveness, and stability. When actual FOSSs are uncertain due to factors such as component aging, environmental changes such as ambient temperature and pressure, and modeling errors, it is necessary to consider whether the uncertain FOSSs can still maintain regularity, nonimpulsiveness, and stability, that is, whether the uncertain FOSSs meet robust admissibility. If the robust admissibility of FOSSs is not met, then feedback controllers are to be designed.

In the field of robust control, scholars focus on studying typical types of uncertainties such as structured uncertainties and polytopic uncertainties. Structured uncertainties are more general than interval uncertainties and norm-bounded uncertainties [56]. Polytopic uncertainties exist in numerous practical systems, and most uncertain systems can be approximated by systems with polytopic uncertainties [13, 72, 104]. In the case where FOSSs are subjected to structured uncertainties or polytopic uncertainties, the robust admissibility analysis is an open problem. Furthermore, not all the internal states of FOSSs can be sensed, which prompts us to utilize output feedback controllers instead of state feedback controllers to stabilize the uncertain FOSSs.

In this chapter, we consider the problems of robust admissibility and robust stabilization for uncertain FOSSs. The structured uncertainty case and polytopic uncertainty case will be studied. The robust admissibility conditions of FOSSs with structured uncer-

https://doi.org/10.1515/9783111628646-003

tainties will be given. The robust admissibility conditions of FOSSs with polytopic uncertainties will be given by using the proposed decoupled methods and LMI method, which are easier to design closed-loop controllers. Then, for uncertain FOSSs with $\alpha \in (1,2)$ and $\alpha \in (0,1)$, the output feedback controllers are designed.

3.2 Robust admissibility

3.2.1 Structured uncertainty case

Consider the uncertain FOSS

$$ED^{\alpha}x(t) = Ax(t), \tag{3.1}$$

where $x(t) \in \mathbb{R}^n$ is the state vector, $\alpha \in (0,2)$ is the fractional order, $x(t)$ is sufficiently differentiable, D^{α} represents the Caputo fractional-order derivative, $E \in \mathbb{R}^{n\times n}$ is the constant real matrix with $\mathrm{rank}(E) = r \le n$, and $A \in \mathbb{R}^{n\times n}$ is a matrix with structured uncertainties and is described as

$$A = A_0 + \sum_{i=1}^{q} k_i A_i, \tag{3.2}$$

where $k_i \in \mathbb{R}$ $(i = 1,2,\ldots,q)$ are unknown time-invariant parameters in the range $(-\bar{k}_i, \bar{k}_i)$, i. e., $|k_i| < \bar{k}_i$, $\bar{k}_i > 0$. Denoting $\phi = \max\{\bar{k}_i, i = 1,2,\ldots,q\}$, we have $|k_i| < \bar{k}_i \le \phi$.

Assume that the nominal FOSS

$$ED^{\alpha}x(t) = A_0x(t), \tag{3.3}$$

is admissible. The problems to be addressed in this subsection are as follows.

(1) *The first problem is the robust admissibility criteria.* Given $\bar{k}_i$, one objective is to establish the robust admissibility conditions for the uncertain FOSS

$$ED^{\alpha}x(t) = Ax(t). \tag{3.4}$$

(2) *The second problem is the robust admissibility bound.* Based on the robust admissibility conditions, another objective is to seek the robust admissibility bound ϕ such that system (3.4) is robustly admissible for $|k_i| < \bar{k}_i \le \phi$.

Before deriving the main results, we need the following definitions and lemmas.

Definition 3.1 ([112]). For a complex matrix $\Omega \in \mathbb{C}^{n\times n}$, the structured singular value of Ω is

$$\mu_{\Delta}(\Omega) = \min\{\sigma_{\max}(\Delta) \mid \det(I_n + \Omega\Delta) = 0,\ \Delta \in \mathbf{\Delta}_{\infty}\}^{-1},$$

where Δ_∞ is the set of structured block-diagonal matrices of the form Δ. Specially, if no $\Delta \in \Delta_\infty$ makes $I + \Omega\Delta$ singular, then, by definition, $\mu_\Delta(\Omega) = 0$.

Lemma 3.1 ([112]). *Let δ be a real constant. Then, for $\sigma_{\max}(\Delta) \leq \delta$, $\det(I_n + \Omega\Delta) \neq 0$ if and only if $\mu_\Delta(\Omega) \cdot \delta < 1$.*

Let $L_E \in \mathbb{R}^{(n-r)\times n}$ and $R_E \in \mathbb{R}^{n\times(n-r)}$ be the maximum left and right annihilator matrices of a matrix E [50], respectively. Namely, L_E and R_E satisfy

$$\begin{aligned} L_E E &= 0, \quad \operatorname{rank}(L_E) + \operatorname{rank}(E) = n, \\ E R_E &= 0, \quad \operatorname{rank}(R_E) + \operatorname{rank}(E) = n. \end{aligned}$$

From [50] we can obtain the following lemma.

Lemma 3.2 ([50]). *FOSS $ED^\alpha x(t) = Ax(t)$ is impulse-free if and only if $\operatorname{rank}(L_E A R_E) = n-r$.*

Firstly, we give necessary and sufficient conditions for the robust stability of the uncertain FOSS (3.1).

Theorem 3.1. *Assume that the nominal FOSS* (3.3) *is asymptotically stable. For given $\bar{k}_i$, system* (3.1) *is robustly stable for $|k_i| < \bar{k}_i$ if and only if*

$$\det\left((jw)^\alpha E - \left(A_0 + \sum_{i=1}^{q} k_i A_i \right) \right) \neq 0 \quad \forall w \in (-\infty, +\infty) \tag{3.5}$$

for all $|k_i| < \bar{k}_i$, $i = 1, 2, \ldots, q$.

Proof. Necessity. If system (3.4) is robustly stable for $|k_i| < \bar{k}_i$, then all zeros of the characteristic equation $\det(s^\alpha E - A) = 0$ are in the open left half part of the complex plane, which implies that $\det(s^\alpha E - A) = 0$ has no zeros in the closed right half part of the complex plane. Thus, for all $w \in (-\infty, +\infty)$,

$$\det\left((jw)^\alpha E - \left(A_0 + \sum_{i=1}^{q} k_i A_i \right) \right) \neq 0$$

for all $|k_i| < \bar{k}_i$, $i = 1, 2, \ldots, q$.

Sufficiency. The asymptotical stability of the nominal FOSS (3.3) indicates that all zeros of the characteristic equation $\det(s^\alpha E - A_0) = 0$ are in the open left half part of the complex plane. The proof of sufficiency is by contradiction. If system (3.4) is not robustly stable, then we can choose a set of parameter ranges $\bar{k}_1, \bar{k}_2, \ldots, \bar{k}_q$ such that some zeros of the characteristic equation

$$\det\left((jw)^\alpha E - \left(A_0 + \sum_{i=1}^{q} k_i A_i \right) \right) = 0 \tag{3.6}$$

could be located in the closed right half part of the complex plane for $|k_i| < \bar{k}_i$, $i = 1, 2, \ldots, q$. Since the zeros of characteristic equation (3.6) continuously change with k_i, there must exist k_i such that some zeros lie on the imaginary axis. Namely,

$$\exists w \in (-\infty, +\infty), \quad \det\left((jw)^\alpha E - \left(A_0 + \sum_{i=1}^{q} k_i A_i \right) \right) = 0,$$

which contradicts with (3.5). Therefore system (3.4) is robustly stable for $|k_i| < \bar{k}_i$. This ends the proof. □

Note that the traversal interval of the frequency variable w in Theorem 3.1 is $(-\infty, +\infty)$. To convert the infinite frequency interval into the finite frequency interval, we develop the following spectral radius estimation method.

Lemma 3.3. *Let $\gamma = (\|A_0\| + \bar{k}_i \sum_{i=1}^{q} \|A_i\|)^{\frac{1}{\alpha}}$. For all $|k_i| < \bar{k}_i$, $i = 1, 2, \ldots, q$, if $|w| > \gamma(\rho(E))^{-\frac{1}{\alpha}}$, the following condition always holds:*

$$\det\left((jw)^\alpha E - \left(A_0 + \sum_{i=1}^{q} k_i A_i \right) \right) \neq 0. \tag{3.7}$$

Proof. We can estimate the spectral radius of $\Psi = A_0 + \sum_{i=1}^{q} k_i A_i$ as follows:

$$\rho(\Psi) \leq \left\| A_0 + \sum_{i=1}^{q} k_i A_i \right\| \leq \|A_0\| + \sum_{i=1}^{q} |k_i| \cdot \|A_i\|.$$

For all $|k_i| < \bar{k}_i$, $i = 1, 2, \ldots, q$, if $|w| > \gamma(\rho(E))^{-\frac{1}{\alpha}}$, then

$$\rho(\Psi) < \|A_0\| + \bar{k}_i \sum_{i=1}^{q} \|A_i\| = \gamma^\alpha < |w|^\alpha \rho(E),$$

indicating that inequality (3.7) holds. This ends the proof. □

By using Lemma 3.3, Theorem 3.1 can be transformed into the following theorem.

Theorem 3.2. *Assume that the nominal FOSS* (3.3) *is asymptotically stable. For given $\bar{k}_i$, system* (3.4) *is robustly stable for $|k_i| < \bar{k}_i$ if and only if*

$$\det\left((jw)^\alpha E - \left(A_0 + \sum_{i=1}^{q} k_i A_i \right) \right) \neq 0 \quad \forall w \in (-h, +h) \tag{3.8}$$

for all $|k_i| < \bar{k}_i$, $i = 1, 2, \ldots, q$, where $h = (\|A_0\| + \bar{k}_i \sum_{i=1}^{q} \|A_i\|)^{\frac{1}{\alpha}} (\rho(E))^{-\frac{1}{\alpha}}$.

To establish the robust stability bound ϕ, we first need a positive real constant ϕ_u that makes FOSS $ED^\alpha x(t) = (A_0 + \phi_u \sum_{i=1}^{q} A_i)x(t)$ or $ED^\alpha x(t) = (A_0 - \phi_u \sum_{i=1}^{q} A_i)x(t)$ unstable. The asymptotical stability of the system $ED^\alpha x(t) = (A_0 + \phi_u \sum_{i=1}^{q} A_i)x(t)$ or $ED^\alpha x(t) =$

$(A_0 - \phi_u \sum_{i=1}^{q} A_i)x(t)$ can be checked by using Lemma 2.5. Obviously, we have $\phi \le \phi_u$. Referring to Lemma 3.3 and Theorem 3.2, we can obtain the following lemma and theorem.

Lemma 3.4. *Let* $\beta = (\|A_0\| + \phi_u \sum_{i=1}^{q} \|A_i\|)^{\frac{1}{\alpha}}$. *For all* $|k_i| < \bar{k}_i$, $i = 1, 2, \ldots, q$, *if* $|w| > \beta(\rho(E))^{-\frac{1}{\alpha}}$, *the following condition always holds:*

$$\det\left((jw)^{\alpha} E - \left(A_0 + \sum_{i=1}^{q} k_i A_i \right) \right) \ne 0. \tag{3.9}$$

Theorem 3.3. *Assume that the nominal FOSS* (3.3) *is asymptotically stable. For given* $\bar{k}_i$, *system* (3.4) *is robustly stable for* $|k_i| < \bar{k}_i \le \phi$ *if and only if*

$$\det\left((jw)^{\alpha} E - \left(A_0 + \sum_{i=1}^{q} k_i A_i \right) \right) \ne 0 \quad \forall w \in (-h_u, +h_u) \tag{3.10}$$

for all $|k_i| < \bar{k}_i \le \phi$, $i = 1, 2, \ldots, q$, *where*

$$h_u = \left(\|A_0\| + \phi_u \sum_{i=1}^{q} \|A_i\| \right)^{\frac{1}{\alpha}} (\rho(E))^{-\frac{1}{\alpha}}.$$

The robust stability condition of the FOSSs with structured uncertainties is proposed in Theorem 3.3 by analyzing the nonsingularity of uncertain matrices, based on which the robust stability bound can be established by using the μ-analysis method.

Theorem 3.4. *Assume that the nominal FOSS* (3.3) *is asymptotically stable. System* (3.4) *is robustly stable for* $|k_i| < \bar{k}_i$ *if any of the two following conditions holds:*

(i)

$$\bar{k}_i < \bar{k}_{\max} = \mu_{\Delta_2(k)}^{-1}(J(w)) \quad \forall w \in (-h_u, +h_u); \tag{3.11}$$

(ii)

$$\bar{k}_i < \bar{k}_{\max} = \mu_{\Delta_2(k)}^{-1}(\hat{J}(w)) \quad \forall w \in (-h_u, +h_u), \tag{3.12}$$

where

$$J(w) = \begin{bmatrix} A_1 R_0(w)^{-1} & A_1 R_0(w)^{-1} & \cdots & A_1 R_0(w)^{-1} \\ A_2 R_0(w)^{-1} & A_2 R_0(w)^{-1} & \cdots & A_2 R_0(w)^{-1} \\ \vdots & \vdots & \vdots & \vdots \\ A_q R_0(w)^{-1} & A_q R_0(w)^{-1} & \cdots & A_q R_0(w)^{-1} \end{bmatrix},$$

$$\hat{J}(w) = \begin{bmatrix} R_0(w)^{-1} A_1 & R_0(w)^{-1} A_2 & \cdots & R_0(w)^{-1} A_q \\ R_0(w)^{-1} A_1 & R_0(w)^{-1} A_2 & \cdots & R_0(w)^{-1} A_q \\ \vdots & \vdots & \vdots & \vdots \\ R_0(w)^{-1} A_1 & R_0(w)^{-1} A_2 & \cdots & R_0(w)^{-1} A_q \end{bmatrix},$$

$$h_u = \left(\left(\|A_0\| + \phi_u \sum_{i=1}^{q} \|A_i\| \right) \rho^{-1}(E) \right)^{\frac{1}{\alpha}}, \quad R_0(w) = (jw)^\alpha E - A_0,$$

$$\Delta_2(k) = \operatorname{diag}\{k_1 I_n, k_2 I_n, \ldots, k_q I_n\}, \quad i = 1, \ldots, q.$$

Proof. By Theorem 3.3, system (3.4) is robustly stable for $|k_i| < \bar{k}_i$ if and only if for all $w \in (-h_u, h_u)$, we have

$$\det\left((jw)^\alpha E - \left(A_0 + \sum_{i=1}^{q} k_i A_i \right) \right) \neq 0, \tag{3.13}$$

i. e.,

$$\det\left(((jw)^\alpha E - A_0) - \sum_{i=1}^{q} k_i A_i \right) \neq 0. \tag{3.14}$$

Since the nominal FOS (3.3) is asymptotically stable, all zeros of the characteristic equation $\det(s^\alpha E - A_0) = 0$ are in the open left half part of the complex plane, which indicates that $\det((jw)^\alpha E - A_0) \neq 0$ for all $w \in (-h_u, +h_u)$. Denote $R_0(w) = (jw)^\alpha E - A_0$, which is nonsingular. Inequality (3.14) is equivalent to

$$\det\left(R_0 - [k_1 I_n \quad k_2 I_n \quad \cdots \quad k_q I_n] \begin{bmatrix} A_1 \\ A_2 \\ \vdots \\ A_q \end{bmatrix} \right) \neq 0, \tag{3.15}$$

i. e.,

$$\det\left(I_n - [k_1 I_n \quad k_2 I_n \quad \cdots \quad k_q I_n] \begin{bmatrix} A_1 \\ A_2 \\ \vdots \\ A_q \end{bmatrix} R_0^{-1} \right) \neq 0. \tag{3.16}$$

Inequality (3.16) can be rewritten as

$$\det\left(I_{nq} - \begin{bmatrix} A_1 \\ A_2 \\ \vdots \\ A_q \end{bmatrix} R_0^{-1} [k_1 I_n \quad k_2 I_n \quad \cdots \quad k_q I_n] \right) \neq 0, \tag{3.17}$$

i. e.,

$$\det\left(I_{nq}-\begin{bmatrix}A_1\\A_2\\\vdots\\A_q\end{bmatrix}R_0^{-1}\begin{bmatrix}I_n & I_n & \cdots & I_n\end{bmatrix}\begin{bmatrix}k_1I_n & 0 & \cdots & 0\\0 & k_2I_n & \cdots & 0\\\vdots & \vdots & \vdots & \vdots\\0 & 0 & \cdots & k_qI_n\end{bmatrix}\right)\neq 0. \tag{3.18}$$

Inequality (3.18) can be written as

$$\det(I_{nq}-J(w)\Delta_2(k))\neq 0. \tag{3.19}$$

From (3.11), we have $\sigma_{\max}(\Delta_2(k)) < \mu^{-1}_{\Delta_2(k)}(J(w))$. By Lemma 3.1, inequality (3.19) holds, which implies that system (3.4) is robustly stable for $|k_i| < \bar{k}_i$. By using the same method, if (3.12) holds, then system (3.4) is also robustly stable for $|k_i| < \bar{k}_i$. This completes the proof. □

Remark 3.1. The robust stability conditions for uncertain FOSSs are given in Theorems 3.2 and 3.4 by checking the finite frequency interval. By comparison the results in [92] are based on the traversal of the frequency value in the infinite frequency interval, which may not be numerically tractable.

By using a similar method as that of the proof in Theorem 3.4, according to Lemma 3.2, the following conditions for robust nonimpulsiveness of uncertain FOSS can be established.

Theorem 3.5. *Assume that the nominal FOSS* (3.3) *is impulse-free. For given* $\bar{k}_i$, *system* (3.4) *is robustly impulse-free for* $|k_i| < \bar{k}_i$ *if and only if any of the following two conditions holds:*

(i)

$$\mu_\Delta(H) < 1, \tag{3.20}$$

(ii)

$$\mu_\Delta(\hat{H}) < 1, \tag{3.21}$$

where

$$H=\begin{bmatrix}L_EA_1R_E(L_EA_0R_E)^{-1}\bar{k}_1 & L_EA_1R_E(L_EA_0R_E)^{-1}\bar{k}_2 & \cdots & L_EA_1R_E(L_EA_0R_E)^{-1}\bar{k}_q\\L_EA_2R_E(L_EA_0R_E)^{-1}\bar{k}_1 & L_EA_2R_E(L_EA_0R_E)^{-1}\bar{k}_2 & \cdots & L_EA_2R_E(L_EA_0R_E)^{-1}\bar{k}_q\\\vdots & \vdots & \vdots & \vdots\\L_EA_qR_E(L_EA_0R_E)^{-1}\bar{k}_1 & L_EA_qR_E(L_EA_0R_E)^{-1}\bar{k}_2 & \cdots & L_EA_qR_E(L_EA_0R_E)^{-1}\bar{k}_q\end{bmatrix},$$

$$\hat{H}=\begin{bmatrix}(L_EA_0R_E)^{-1}\bar{k}_1L_EA_1R_E & (L_EA_0R_E)^{-1}\bar{k}_2L_EA_2R_E & \cdots & (L_EA_0R_E)^{-1}\bar{k}_qL_EA_qR_E\\(L_EA_0R_E)^{-1}\bar{k}_1L_EA_1R_E & (L_EA_0R_E)^{-1}\bar{k}_2L_EA_2R_E & \cdots & (L_EA_0R_E)^{-1}\bar{k}_qL_EA_qR_E\\\vdots & \vdots & \vdots & \vdots\\(L_EA_0R_E)^{-1}\bar{k}_1L_EA_1R_E & (L_EA_0R_E)^{-1}\bar{k}_2L_EA_2R_E & \cdots & (L_EA_0R_E)^{-1}\bar{k}_qL_EA_qR_E\end{bmatrix},$$

$$\Delta=\operatorname{diag}\{k_1/\bar{k}_1I_{n-r}, k_2/\bar{k}_2I_{n-r},\ldots,k_q/\bar{k}_qI_{n-r}\},\quad i=1,\ldots,q.$$

By Theorem 3.5, the robust bounds of k_i for system (3.4) are robustly impulse-free and thus can be derived by the following theorem.

Theorem 3.6. *Assume that the nominal FOSS* (3.3) *is impulse-free. System* (3.4) *is robustly impulse-free for* $|k_i| < \bar{k}_i$ *if any of the following two conditions holds:*

$$\text{(i)}\quad \bar{k}_i < \bar{k}_{\max} = \mu^{-1}_{\Delta(k)}(Z), \tag{3.22}$$

$$\text{(ii)}\quad \bar{k}_i < \bar{k}_{\max} = \mu^{-1}_{\Delta(k)}(\hat{Z}), \tag{3.23}$$

where

$$Z = \begin{bmatrix} L_EA_1R_E(L_EA_0R_E)^{-1} & L_EA_1R_E(L_EA_0R_E)^{-1} & \cdots & L_EA_1R_E(L_EA_0R_E)^{-1} \\ L_EA_2R_E(L_EA_0R_E)^{-1} & L_EA_2R_E(L_EA_0R_E)^{-1} & \cdots & L_EA_2R_E(L_EA_0R_E)^{-1} \\ \vdots & \vdots & \vdots & \vdots \\ L_EA_qR_E(L_EA_0R_E)^{-1} & L_EA_qR_E(L_EA_0R_E)^{-1} & \cdots & L_EA_qR_E(L_EA_0R_E)^{-1} \end{bmatrix},$$

$$\hat{Z} = \begin{bmatrix} (L_EA_0R_E)^{-1}L_EA_1R_E & (L_EA_0R_E)^{-1}L_EA_2R_E & \cdots & (L_EA_0R_E)^{-1}L_EA_qR_E \\ (L_EA_0R_E)^{-1}L_EA_1R_E & (L_EA_0R_E)^{-1}L_EA_2R_E & \cdots & (L_EA_0R_E)^{-1}L_EA_qR_E \\ \vdots & \vdots & \vdots & \vdots \\ (L_EA_0R_E)^{-1}L_EA_1R_E & (L_EA_0R_E)^{-1}L_EA_2R_E & \cdots & (L_EA_0R_E)^{-1}L_EA_qR_E \end{bmatrix},$$

$$\Delta(k) = \operatorname{diag}\{k_1I_{n-r}, k_2I_{n-r}, \dots, k_qI_{n-r}\}, \quad i = 1, \dots, q.$$

By Theorems 3.3 and 3.5, we can obtain the robust admissibility conditions for system (3.4).

Theorem 3.7. *Assume that the nominal FOSS* (3.3) *is admissible. For given* $\bar{k}_i$, *system* (3.4) *is robustly admissible for* $|k_i| < \bar{k}_i$ *if and only if the following two conditions simultaneously hold:*

(i)

$$\mu_\Delta(H) < 1, \tag{3.24}$$

or

$$\mu_\Delta(\hat{H}) < 1; \tag{3.25}$$

(ii)

$$\det\left((jw)^\alpha E - \left(A_0 + \sum_{i=1}^{q} k_iA_i\right)\right) \neq 0 \quad \forall w \in (-h_u, +h_u), \tag{3.26}$$

where

$$H = \begin{bmatrix} L_EA_1R_E(L_EA_0R_E)^{-1}\bar{k}_1 & L_EA_1R_E(L_EA_0R_E)^{-1}\bar{k}_2 & \cdots & L_EA_1R_E(L_EA_0R_E)^{-1}\bar{k}_q \\ L_EA_2R_E(L_EA_0R_E)^{-1}\bar{k}_1 & L_EA_2R_E(L_EA_0R_E)^{-1}\bar{k}_2 & \cdots & L_EA_2R_E(L_EA_0R_E)^{-1}\bar{k}_q \\ \vdots & \vdots & \vdots & \vdots \\ L_EA_qR_E(L_EA_0R_E)^{-1}\bar{k}_1 & L_EA_qR_E(L_EA_0R_E)^{-1}\bar{k}_2 & \cdots & L_EA_qR_E(L_EA_0R_E)^{-1}\bar{k}_q \end{bmatrix},$$

$$\hat{H} = \begin{bmatrix} (L_EA_0R_E)^{-1}\bar{k}_1L_EA_1R_E & (L_EA_0R_E)^{-1}\bar{k}_2L_EA_2R_E & \cdots & (L_EA_0R_E)^{-1}\bar{k}_qL_EA_qR_E \\ (L_EA_0R_E)^{-1}\bar{k}_1L_EA_1R_E & (L_EA_0R_E)^{-1}\bar{k}_2L_EA_2R_E & \cdots & (L_EA_0R_E)^{-1}\bar{k}_qL_EA_qR_E \\ \vdots & \vdots & \vdots & \vdots \\ (L_EA_0R_E)^{-1}\bar{k}_1L_EA_1R_E & (L_EA_0R_E)^{-1}\bar{k}_2L_EA_2R_E & \cdots & (L_EA_0R_E)^{-1}\bar{k}_qL_EA_qR_E \end{bmatrix},$$

$$\Delta = \operatorname{diag}\{k_1/\bar{k}_1I_{n-r}, k_2/\bar{k}_2I_{n-r}, \dots, k_q/\bar{k}_qI_{n-r}\},$$

$$h_u = \left(\left(\|A_0\| + \phi_u \sum_{i=1}^{q} \|A_i\|\right)\rho^{-1}(E)\right)^{\frac{1}{\alpha}}, \quad i = 1, \dots, q.$$

Based on Theorems 3.4 and 3.6, the robust bounds of k_i for robustly admissible system (3.4) can be derived by the following theorem.

Theorem 3.8. *Assume that the nominal FOSS* (3.3) *is admissible. System* (3.4) *is robustly admissible for* $|k_i| < \bar{k}_i$ *if the two following conditions simultaneously hold:*

(i)

$$\bar{k}_i < \bar{k}_{\max} = \mu^{-1}_{\Delta(k)}(Z), \tag{3.27}$$

or

$$\bar{k}_i < \bar{k}_{\max} = \mu^{-1}_{\Delta(k)}(\hat{Z}). \tag{3.28}$$

(ii)

$$\bar{k}_i < \bar{k}_{\max} = \mu^{-1}_{\Delta_2(k)}(J(w)) \quad \forall w \in (-h_u, +h_u), \tag{3.29}$$

or

$$\bar{k}_i < \bar{k}_{\max} = \mu^{-1}_{\Delta_2(k)}(\hat{J}(w)) \quad \forall w \in (-h_u, +h_u), \tag{3.30}$$

where

$$Z = \begin{bmatrix} L_EA_1R_E(L_EA_0R_E)^{-1} & L_EA_1R_E(L_EA_0R_E)^{-1} & \cdots & L_EA_1R_E(L_EA_0R_E)^{-1} \\ L_EA_2R_E(L_EA_0R_E)^{-1} & L_EA_2R_E(L_EA_0R_E)^{-1} & \cdots & L_EA_2R_E(L_EA_0R_E)^{-1} \\ \vdots & \vdots & \vdots & \vdots \\ L_EA_qR_E(L_EA_0R_E)^{-1} & L_EA_qR_E(L_EA_0R_E)^{-1} & \cdots & L_EA_qR_E(L_EA_0R_E)^{-1} \end{bmatrix},$$

$$\hat{Z} = \begin{bmatrix} (L_EA_0R_E)^{-1}L_EA_1R_E & (L_EA_0R_E)^{-1}L_EA_2R_E & \cdots & (L_EA_0R_E)^{-1}L_EA_qR_E \\ (L_EA_0R_E)^{-1}L_EA_1R_E & (L_EA_0R_E)^{-1}L_EA_2R_E & \cdots & (L_EA_0R_E)^{-1}L_EA_qR_E \\ \vdots & \vdots & \vdots & \vdots \\ (L_EA_0R_E)^{-1}L_EA_1R_E & (L_EA_0R_E)^{-1}L_EA_2R_E & \cdots & (L_EA_0R_E)^{-1}L_EA_qR_E \end{bmatrix},$$

$$J(w) = \begin{bmatrix} A_1R_0(w)^{-1} & A_1R_0(w)^{-1} & \cdots & A_1R_0(w)^{-1} \\ A_2R_0(w)^{-1} & A_2R_0(w)^{-1} & \cdots & A_2R_0(w)^{-1} \\ \vdots & \vdots & \vdots & \vdots \\ A_qR_0(w)^{-1} & A_qR_0(w)^{-1} & \cdots & A_qR_0(w)^{-1} \end{bmatrix},$$

$$\hat{J}(w)=\begin{bmatrix} R_0(w)^{-1}A_1 & R_0(w)^{-1}A_2 & \cdots & R_0(w)^{-1}A_q \\ R_0(w)^{-1}A_1 & R_0(w)^{-1}A_2 & \cdots & R_0(w)^{-1}A_q \\ \vdots & \vdots & \vdots & \vdots \\ R_0(w)^{-1}A_1 & R_0(w)^{-1}A_2 & \cdots & R_0(w)^{-1}A_q \end{bmatrix},$$

$$\Delta(k)=\mathrm{diag}\{k_1I_{n-r},k_2I_{n-r},\dots,k_qI_{n-r}\},\quad h_u=\left(\left(\|A_0\|+\phi_u\sum_{i=1}^{q}\|A_i\|\right)\rho^{-1}(E)\right)^{\frac{1}{\alpha}},$$

$$R_0(w)=(jw)^{\alpha}E-A_0,\quad \Delta_2(k)=\mathrm{diag}\{k_1I_n,k_2I_n,\dots,k_qI_n\},\quad i=1,\dots,q.$$

Proof. Since the nominal FOSS (3.3) is admissible, it is regular by Lemma 2.3. By Definition 1 in [106], $\det((jw)^{\alpha}E-A_0)$ is not identically zero. Therefore $\det((jw)^{\alpha}E-(A_0+\sum_{i=1}^{q}k_iA_i))$ remains not identically zero, resulting in that system (3.4) is robustly regular.

Also, note that the nominal FOSS (3.3) is admissible, and thus it is impulse-free and stable by Lemma 2.3. According to Theorems 3.6 and 2.80, if condition (i) holds, then system (3.4) is robustly impulse-free and stable. The proof is completed. □

From Theorem 3.8, as long as w is determined, the robust admissibility bound r can be found such that the uncertain FOSSs are robustly admissible. Therefore the interval $(-h_u,+h_u)$ can be divided into $m\in N^{+}$ equivalents if m is large enough. Because for all $w\in(-h_u,+h_u)$, inequality (3.27) (or inequality (3.28)) and inequality (3.29) (or inequality (3.30)) need to be satisfied, all w in the interval must be traversed. Denote the robust bounds obtained using inequalities (3.27)–(3.30) by $\bar{k}_{1\max}$, $\bar{k}_{2\max}$, $\bar{k}_{3\max}$, $\bar{k}_{4\max}$, respectively. The robust admissibility bound r can be established by Algorithm 1.

Now we provide an example to illustrate the effectiveness of Algorithm 1.

Example 3.1. Consider the uncertain FOSS

$$ED^{\alpha}x(t)=(A_0+k_1A_1)x(t), \tag{3.31}$$

where $\alpha=1.5$, $|k_i|<\bar{k}_i\le\phi$ $(i=1)$, and

$$A_0=\begin{bmatrix}-1.5 & 0.5 & 1\\ -1 & -3 & 1\\ -0.5 & -1.5 & -2.5\end{bmatrix},\quad A_1=\begin{bmatrix}1&1&1\\1&1&1\\1&1&2\end{bmatrix},\quad E=\begin{bmatrix}1&0&0\\0&1&0\\0&0&0\end{bmatrix}.$$

The objective in this example is to establish the robust admissibility bound r such that uncertain FOSS (3.31) is robustly admissible for $|k_i|<\bar{k}_i\le\phi$ $(i=1)$. Analyze the nominal FOSS $ED^{\alpha}x(t)=A_0x(t)$: (i) $\det(s^{\alpha}E-A_0)$ is not identically zero, implying that the nominal FOSS is regular. (ii) Take the maximum left and right annihilator matrices of matrix E as $L_E=\begin{bmatrix}0&0&1\end{bmatrix}$ and $R_E=\begin{bmatrix}0\\0\\1\end{bmatrix}$. Then $\mathrm{rank}(L_EA_0R_E)=1$, implying that the nominal FOSS is impulse-free. (iii) The eigenvalues of the system are $\{-1.6388,-3.6612\}$, implying that the nominal FOSS is stable. Thus the nominal FOSS is admissible.

Algorithm 1 The algorithm for solving the robust admissibility bound of the uncertain FOSS (3.4).

Input: System parameters: α, A_0, $\{A_i\}$, $i = 1, 2, \ldots, q$, a positive real constant: ϕ_u, the number of interval segmentation: m
Output: The robust admissibility bound of uncertain FOSS (3.4): ϕ

function BOUND($\alpha, A_0, \{A_i\}, \phi_u, m$)
  $h_u \leftarrow ((\|A_0\| + \phi_u \sum_{i=1}^{q} \|A_i\|)\rho^{-1}(E))^{\frac{1}{\alpha}}, g(i) \leftarrow -h_u + \frac{2h_u}{m}(i-1), (i = 1, 2, \ldots, m+1)$
  $\phi_{1\max} \leftarrow \mu^{-1}_{\Delta(k)}(Z)$ (i. e., applying (3.27) in Theorem 3.8 to calculate and obtain $\phi_{3\max}$)
  $\phi_{2\max} \leftarrow \mu^{-1}_{\Delta(k)}(\hat{Z})$ (i. e., applying (3.28) in Theorem 3.8 to calculate and obtain $\phi_{3\max}$)
  $Bound1(i) \leftarrow \max(\phi_{1\max}, \phi_{2\max})$ (i. e., taking the larger one between $\phi_{1\max}$ and $\phi_{2\max}$ and storing it in an array $Bound1$)
  for $i = 1$ to $m + 1$ **do**
    $w \leftarrow g(i), \phi_{3\max} \leftarrow \mu^{-1}_{\Delta_2(k)}(J(w))$ (i. e., applying (3.29) in Theorem 3.8 to calculate and obtain $\phi_{3\max}$)
    $w \leftarrow g(i), \phi_{4\max} \leftarrow \mu^{-1}_{\Delta_2(k)}(\hat{J}(w))$ (i. e., applying (3.30) in Theorem 3.8 to calculate and obtain $\phi_{4\max}$)
    $Bound2(i) \leftarrow \max(\phi_{3\max}, \phi_{4\max})$ (i. e., taking the larger one between $\phi_{3\max}$ and $\phi_{4\max}$ and storing it in an array $Bound2$)
  end for
  $Bound3 \leftarrow \min(Bound2(i))$ $(i = 1, 2, \ldots, m+1)$ (i. e., taking the minimum element in the array $Bound2$ and assigning it to $Bound3$)
  $\phi \leftarrow \min(Bound1, Bound3)$ (i. e., taking the minimum one between $Bound1$ and $Bound3$)
  return ϕ
end function

Using Algorithm 1 and taking ϕ_u as 1, the eigenvalues of the system $ED^\alpha x(t) = (A_0 + \phi_u A_1)x(t)$ are $\{-3.8117, 1.3117\}$, which indicates that the system is unstable and the ϕ_u meets the requirement. Then taking $m = 10000$ in Algorithm 1, we can obtain the robust bound $\phi = 0.8203$. For $\bar{k}_1 = 0.8203$ and $k_1 \in (-0.8203, 0.8203)$, the eigenvalues and stability area of uncertain FOSS (3.31) are shown in Figure 3.1. From Figure 3.1 we can see that all eigenvalues are located in the stability area, some of which are very close to the stability boundary. Randomly select the uncertain parameter $k_1 = -0.5$. With it, the state trajectory of the uncertain FOSS is shown in Figure 3.2, in which the state trajectory tends to zero. Meanwhile, (i) $\det(s^\alpha E - (A_0 + k_1 A_1))$ is not identically zero, implying that the system is regular; (ii) $\operatorname{rank}(L_E(A_0 + k_1 A_1)R_E) = 1$, implying that the system is impulse-free. Thus, when the uncertain parameter $k_1 = -0.5$, system (3.31) is admissible.

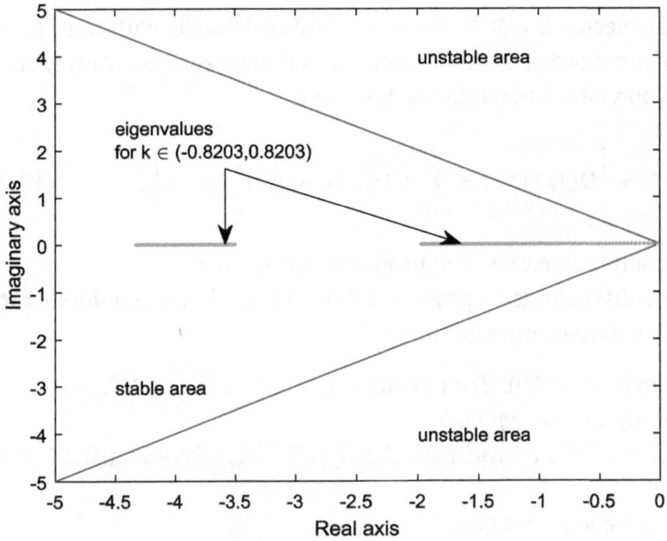

Figure 3.1: The eigenvalue distribution and the stability area of the uncertain FOSS (3.31) with $k_1 \in (-0.8203, 0.8203)$ in Example 3.1.

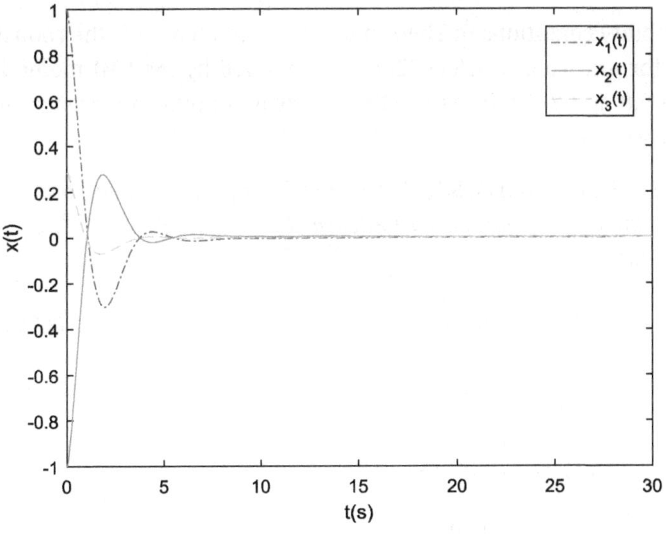

Figure 3.2: The state trajectory of the uncertain FOSS with the selected parameter $k_1 = -0.5$ in Example 3.1.

3.2.2 Polytopic uncertainty case

Consider the following FOSS with polytopic uncertainties:

$$ED^{\alpha}x(t) = A(\lambda)x(t), \tag{3.32}$$

where $x(t) \in \mathbb{R}^n$ is the state vector, $E \in \mathbb{R}^{n\times n}$ is the constant real matrix with $\operatorname{rank}(E) = r \leq n$, $\alpha \in (0,2)$ is the fractional-order, D^α is the Caputo fractional-order derivative, and $A(\lambda)$ is a real matrix with polytopic uncertainties described as

$$\Omega(\lambda) = A(\lambda) \in \mathcal{R}, \quad \mathcal{R} = \left\{\Omega(\lambda) \mid \Omega(\lambda) = \sum_{k=0}^{l} \lambda_k \Omega_k,\ \lambda_k > 0,\ \sum_{k=0}^{l} \lambda_k = 1\right\}, \tag{3.33}$$

where $\Omega_k = (A_k)$ is a constant real matrix of appropriate dimensions.

The objective of this subsection is to propose robust admissibility conditions for uncertain FOSS (3.32) with polytopic uncertainties.

Lemma 3.5 (The elimination lemma [87]). *For matrices $\Pi_1 \in \mathbb{C}^{n\times m}$, $\Pi_2 \in \mathbb{C}^{k\times n}$, and $\Pi_1 \in \mathbb{C}^{n\times n}$, the following two conditions are equivalent:*

(1) *There exists a matrix $X \in \mathbb{C}^{m\times k}$ such that $\operatorname{Sym}\{\Pi_1 X \Pi_2\} + \Pi_3 < 0$, where $\operatorname{Sym}\{\Pi_1 X \Pi_2\} = \Pi_1 X \Pi_2 + (\Pi_1 X \Pi_2)^*$.*

(2) *Both of the following two conditions hold:*
   (i) $N_{\Pi_1} \Pi_3 N_{\Pi_1}^* < 0$.
   (ii) $N_{\Pi_2}^* \Pi_3 N_{\Pi_2} < 0$, *where N_{Π_1} and N_{Π_2} are orthogonal complement matrices of Π_1 and Π_2, respectively.*

Based on the admissibility conditions in Theorems 2.6–2.11 in Chapter 2, the robust admissibility conditions for uncertain FOSS (3.32) can be derived by the LMI method. For simplicity, Theorems 2.8 and 2.11 will be used for derivation; other theorems for derivation can be similarly used.

Theorem 3.9. *When $1 < \alpha < 2$, uncertain FOSS (3.32), of which the polytopic uncertainties satisfy (3.33), is robustly admissible if there exist matrices $X_k \in \mathbb{C}^{n\times n}$, $Y_k \in \mathbb{C}^{n\times n}$, $Z_k \in \mathbb{C}^{n\times n}$, and $G \in \mathbb{C}^{n\times n}$ such that*

$$-EX_kE^T - \operatorname{Sym}\{E^\ddagger Z_k\} < 0, \tag{3.34a}$$

$$\operatorname{Sym}\left\{\begin{bmatrix} A_k G & A_k G \\ e^{j\theta}(X_k E^T + E^\dagger Y_k) - G & -G \end{bmatrix}\right\} < 0, \tag{3.34b}$$

where $k = 1, 2, \ldots, l$, $\theta = (1-\alpha)\pi/2$, and $E^\dagger$ and $E^\ddagger$ are given in (2.27).

Proof. From inequality (3.34b) and the fact that $\lambda_k > 0$, $\sum_{k=0}^{l} \lambda_k = 1$, we have

$$\sum_{k=0}^{l} \lambda_k \operatorname{Sym}\left\{\begin{bmatrix} A_k G & A_k G \\ e^{j\theta}(X_k E^T + E^\dagger Y_k) - G & -G \end{bmatrix}\right\} < 0, \tag{3.35}$$

which can be written as

$$\operatorname{Sym}\left\{\begin{bmatrix} A(\lambda) G & A(\lambda) G \\ e^{j\theta}(X(\lambda) E^T + E^\dagger Y(\lambda)) - G & -G \end{bmatrix}\right\} < 0. \tag{3.36}$$

Inequality (3.36) yields

$$\begin{bmatrix} 0_n & \star \\ e^{j\theta}(X(\lambda)E^T + E^\dagger Y(\lambda)) & 0_n \end{bmatrix} + \operatorname{Sym}\left\{\begin{bmatrix} A(\lambda) \\ -I_n \end{bmatrix} G \begin{bmatrix} I_n & I_n \end{bmatrix}\right\} < 0. \tag{3.37}$$

Denoting

$$\Pi_1 = \begin{bmatrix} A(\lambda) \\ -I_n \end{bmatrix}, \quad \Pi_2 = \begin{bmatrix} I_n & I_n \end{bmatrix}, \quad \Pi_3 = \begin{bmatrix} 0_n & \star \\ e^{j\theta}(X(\lambda)E^T + E^\dagger Y(\lambda)) & 0_n \end{bmatrix},$$

inequality (3.37) can be rewritten as $\operatorname{Sym}\{\Pi_1 G \Pi_2\} + \Pi_3 < 0$. The orthogonal complement matrices of Π_1 and Π_2 can be chosen as $N_{\Pi_1} = [I_n \quad A(\lambda)]$ and $N_{\Pi_2} = \left[\begin{smallmatrix} I_n \\ -I_n \end{smallmatrix}\right]$. By Lemma 3.5 we have

$$\begin{aligned} N_{\Pi_1}\Pi_3 N_{\Pi_1}^* &= \begin{bmatrix} I_n & A(\lambda) \end{bmatrix} \begin{bmatrix} 0_n & \star \\ e^{j\theta}(X(\lambda)E^T + E^\dagger Y(\lambda)) & 0_n \end{bmatrix} \begin{bmatrix} I_n & A(\lambda) \end{bmatrix}^T \\ &< 0. \end{aligned} \tag{3.38}$$

Inequality (3.38) is equivalent to

$$\operatorname{Sym}\{e^{j\theta} A(\lambda)(X(\lambda)E^T + E^\dagger Y(\lambda))\} < 0. \tag{3.39}$$

Similarly, inequality (3.34a) implies

$$-EX(\lambda)E^T - \operatorname{Sym}\{E^\ddagger Z(\lambda)\} < 0. \tag{3.40}$$

By inequalities (3.39) and (3.40) the uncertain FOSS (3.32) is robustly admissible by Theorem 2.11 in Chapter 2. □

Similarly, the robust admissibility conditions for FOSS with polytopic uncertainties and $0 < \alpha < 1$ can be studied.

Theorem 3.10. *When $0 < \alpha < 1$, uncertain FOSS (3.32), of which the polytopic uncertainties satisfy (3.33), is robustly admissible if there exist matrices X_k, Y_k, Z_k, and G such that*

$$-EX_k E^T - \operatorname{Sym}\{E^\ddagger Z_k\} < 0, \tag{3.41a}$$

$$\operatorname{Sym}\left\{\begin{bmatrix} A_k G & A_k G \\ (e^{j\theta}X_k + e^{-j\theta}\bar{X}_k)E^T + E^\dagger Y_k - G & -G \end{bmatrix}\right\} < 0, \tag{3.41b}$$

where $k = 1, 2, \ldots, l$, $\theta = (1-\alpha)\pi/2$, and $E^\dagger$ and $E^\ddagger$ are given in (2.27).

Proof. By Theorem 2.8 in Chapter 2 the proof can be completed by using similar methods as those in Theorem 3.9 and thus is omitted for brevity. □

It should be emphasized that although the robust admissibility conditions proposed in Theorems 3.9 and 3.10 are based on Theorems 2.8 and 2.11, the conservative results, such as $\mathrm{Sym}\{e^{j\theta}A(\lambda)(XE^T + E^\dagger Y)\} < 0$ or $\mathrm{Sym}\{A(\lambda)((e^{j\theta}X + e^{-j\theta}\bar{X})E^T + E^\dagger Y)\} < 0$, are not given. The reasons lie in that the inequality $\mathrm{Sym}\{e^{j\theta}A(\lambda)(XE^T + E^\dagger Y)\} < 0$ or $\mathrm{Sym}\{A(\lambda)((e^{j\theta}X + e^{-j\theta}\bar{X})E^T + E^\dagger Y)\} < 0$ strictly restricts the matrices X and Y. Namely, the same matrix variables X and Y must satisfy all l inequalities generated by l vertices of the polytopic uncertainties, which will lead to relatively strong conservatism. Instead, the elimination lemma (Lemma 3.5) is applied to decoupling A and $XE^T + E^\dagger Y$ or $(e^{j\theta}X + e^{-j\theta}\bar{X})E^T + E^\dagger Y$ in this subsection. The methods applied in Theorems 3.9 and 3.10 will allow us to find the corresponding X_k and Y_k for different vertices of the polytopic uncertainties, which indicates the weak conservatism of the proposed results.

Remark 3.2. Note that polytopic uncertainties exist in numerous practical systems and most uncertain systems can be approximated by systems with polytopic uncertainties [13, 72, 104]. The robust admissibility conditions for FOSSs with polytopic uncertainties and fractional orders $1 < \alpha < 2$ and $0 < \alpha < 1$ are proposed in Theorems 3.9 and 3.10, respectively. By comparison most existing works on the robust admissibility of FOSSs focus on norm-bounded uncertainties such as the results in [36, 116, 62], where the methods cannot be directly applied to study the robust admissibility of FOSSs with polytopic uncertainties. The results proposed in this subsection can be effectively applied to solve the robust control problems of FOSSs with polytopic uncertainties, for example, the robust stabilization in Section 3.3 in this chapter.

Now we provide a numerical example to demonstrate the practicality of the proposed results. Since the verification methods are similar for $1 < \alpha < 2$ and $0 < \alpha < 1$, a numerical example of FOSSs with fractional order $0 < \alpha < 1$ is not presented for brevity.

Example 3.2. Consider uncertain FOSS (3.32) with the following parameters:

$$E = \begin{bmatrix} 1 & 1 \\ 2 & 2 \end{bmatrix}, \quad A_1 = \begin{bmatrix} -4 & -2 \\ -2 & -3 \end{bmatrix}, \quad A_2 = \begin{bmatrix} -5 & -2 \\ -2 & -3 \end{bmatrix}, \quad A_3 = \begin{bmatrix} -4 & -2 \\ -2 & -3.3 \end{bmatrix},$$
$$A_4 = \begin{bmatrix} -5 & -2 \\ -2 & -3.3 \end{bmatrix}, \quad \alpha = 1.5, \quad u(t) = 0. \tag{3.42}$$

The aim is to check whether uncertain FOSS (3.32) with parameters (3.42) is robustly admissible.

By Theorem 3.9 there exist the following feasible solutions:

$$X_1 = \begin{bmatrix} -0.5678 & \star \\ 1.4611 - 0.5807j & 2.0289 \end{bmatrix}, \quad X_2 = \begin{bmatrix} -0.2480 & \star \\ 1.4967 - 0.7164j & 1.7447 \end{bmatrix},$$
$$X_3 = \begin{bmatrix} -0.8020 & \star \\ 1.5643 - 0.5240j & 2.3664 \end{bmatrix}, \quad X_4 = \begin{bmatrix} -0.4557 & \star \\ 1.6010 - 0.6454j & 2.0566 \end{bmatrix},$$

$$Y_1 = \begin{bmatrix} 3.4658 & \star \\ -4.7642 - 0.2904j & 1.5156 \end{bmatrix}, \quad Y_2 = \begin{bmatrix} 4.8223 & \star \\ -5.8186 - 0.3582j & 1.7037 \end{bmatrix},$$

$$Y_3 = \begin{bmatrix} 3.5974 & \star \\ -5.1817 - 0.2620j & 1.6915 \end{bmatrix}, \quad Y_4 = \begin{bmatrix} 4.8961 & \star \\ -6.1523 - 0.3227j & 1.8521 \end{bmatrix},$$

$$Z_1 = \begin{bmatrix} 4.3023 & \star \\ -7.4409 & 2.6449 \end{bmatrix}, \quad Z_2 = \begin{bmatrix} 4.1702 & \star \\ -7.5984 & 2.7566 \end{bmatrix},$$

$$Z_3 = \begin{bmatrix} 3.9190 & \star \\ -7.8978 & 2.9692 \end{bmatrix}, \quad Z_4 = \begin{bmatrix} 3.7830 & \star \\ -8.0599 & 3.0842 \end{bmatrix}.$$

By Theorem 3.9 uncertain FOSS (3.32) with parameters (3.42) is robustly admissible.

We randomly selected a set of uncertain parameters:

$$\lambda_1 = 0.1, \quad \lambda_2 = 0.1, \quad \lambda_3 = 0.4, \quad \lambda_4 = 0.4. \tag{3.43}$$

FOSS (3.32) with parameters (3.42) and (3.43) can be written as

$$\begin{bmatrix} 1 & 1 \\ 2 & 2 \end{bmatrix} D^\alpha x(t) = \begin{bmatrix} -4.5 & -2 \\ -2 & -3.24 \end{bmatrix} x(t). \tag{3.44}$$

From the view of simulation study, for system (3.44), we have

$$\exists c_0 \in \mathbb{C}, \quad \det\left(c_0^\alpha \begin{bmatrix} 1 & 1 \\ 2 & 2 \end{bmatrix} - \begin{bmatrix} -4.5 & -2 \\ -2 & -3.24 \end{bmatrix} \right) \neq 0,$$

and $\deg(\det(\lambda[\begin{smallmatrix} 1 & 1 \\ 2 & 2 \end{smallmatrix}] - [\begin{smallmatrix} -4.5 & -2 \\ -2 & -3.24 \end{smallmatrix}])) = 1 = \operatorname{rank}([\begin{smallmatrix} 1 & 1 \\ 2 & 2 \end{smallmatrix}])$, $\lambda \in \mathbb{C}$, implying that the system is regular and impulse-free. Furthermore, the eigenvalue of the system is $\{-1.6955\}$, which is plotted in Figure 3.3 and located in the stability area. The state evolution is plotted in Figure 3.4, in which all states approach zero. Therefore we verified that the uncertain FOSS (3.32) with parameters (3.42) and (3.43) is indeed admissible.

Remark 3.3. The parameters $A_1, A_2, A_3, A_4, \alpha$ and the research problems in Example 3.2 are the same as those in Example 1 in [59]. However, the fractional-order derivative coefficient matrix E in [59] must be nonsingular, indicating that the methods proposed in [59] are particular cases of the methods proposed in this subsection.

3.3 Robust stabilization

In Section 3.2, the μ-analysis method and LMI method are used to solved the robust admissibility problems of uncertain FOSSs. As stated in Remark 3.2, the results proposed by the LMI method in Subsection 3.2.2 are more suitable for designing controllers to solve the stabilization problems of uncertain FOSSs. Thus in this section, we will solve the robust stabilization problems of uncertain FOSSs based on the results in Subsection 3.2.2.

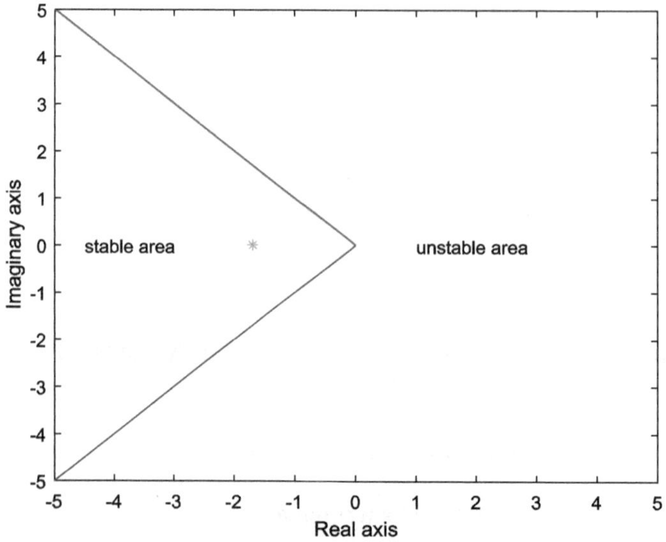

Figure 3.3: The eigenvalue locations of the uncertain FOSS (3.32) with parameters (3.42) and (3.43) in Example 3.2.

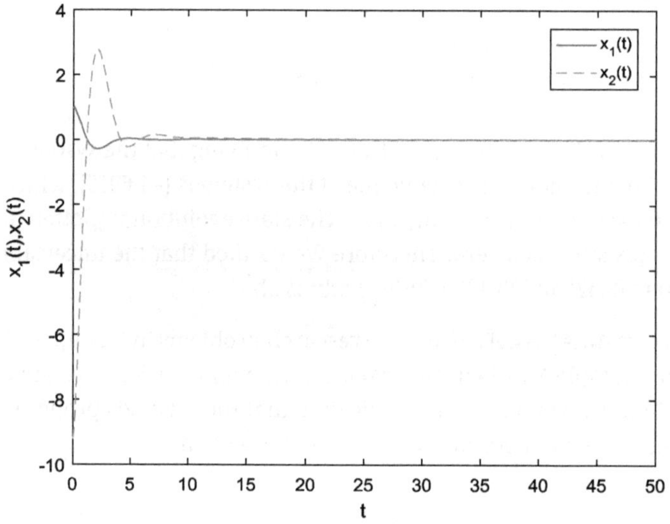

Figure 3.4: The state evolution of the uncertain FOSS (3.32) with parameters (3.42) and (3.43) in Example 3.2.

Consider the following FOSS with polytopic uncertainties:

$$ED^{\alpha}x(t) = A(\lambda)x(t) + B(\lambda)u(t), \tag{3.45}$$

where $x(t) \in \mathbb{R}^n$ and $u(t) \in \mathbb{R}^p$ are the state vector and the control input, $E \in \mathbb{R}^{n\times n}$ is the constant real matrix with $\mathrm{rank}(E) = r \le n$, $\alpha \in (0,2)$ is the fractional order, D^α is the Caputo fractional-order derivative, and $A(\lambda)$ and $B(\lambda)$ are real matrices with polytopic uncertainties described as

$$\Omega(\lambda) = (A(\lambda), B(\lambda)) \in \mathcal{R}, \quad \mathcal{R} = \left\{\Omega(\lambda) \mid \Omega(\lambda) = \sum_{k=0}^{l} \lambda_k \Omega_k,\ \lambda_k > 0,\ \sum_{k=0}^{l} \lambda_k = 1\right\}, \tag{3.46}$$

where $\Omega_k = (A_k, B_k)$ are constant real matrices of appropriate dimensions.

To make uncertain FOSS (3.45) satisfy robust admissibility, an output feedback controller can be designed:

$$u(t) = Ky(t), \tag{3.47}$$

where $y(t) = Cx(t)$ is the measured output, and $y(t) \in \mathbb{R}^m$ with known constant matrices $C \in \mathbb{R}^{m\times n}$. Without loss of generality, we assume that $m \le n$. Let $\mathrm{rank}(C) = q$. Then $q \le m \le n$. The SVD of C can be described as

$$C = U \begin{bmatrix} S & 0 & 0 \\ 0 & 0 & 0 \end{bmatrix} V^T, \tag{3.48}$$

where $S = \mathrm{diag}(\sigma_1, \sigma_2, \dots, \sigma_q)$ with σ_i $(i = 1, 2, \dots, q)$ the nonzero singular values of C. With controller (3.47), FOSS (3.45) can be transformed into

$$ED^\alpha x(t) = (A(\lambda) + B(\lambda)KC)x(t). \tag{3.49}$$

The issues on checking whether there exists the controller (3.47) and determining the gain matrix K can be solved by Theorems 3.11 and 3.12 as follows.

3.3.1 Case $\alpha \in (1, 2)$

Theorem 3.11. *For* $1 < \alpha < 2$, *uncertain FOSS* (3.49), *of which the polytopic uncertainties satisfy* (3.46), *is robustly admissible if there exist matrices* $X_k \in \mathbb{C}^{n\times n}$, $Y_k \in \mathbb{C}^{n\times n}$, $Z_k \in \mathbb{C}^{n\times n}$, $G_{22} \in \mathbb{C}^{(m-q)\times(m-q)}$, $G_{33} \in \mathbb{C}^{(n-m)\times(n-m)}$, *a real matrix* $T \in \mathbb{R}^{p\times m}$, *and a nonsingular real matrix* $G_{11} \in \mathbb{R}^{q\times q}$ *such that*

$$-EX_kE^T - \mathrm{Sym}\{E^{\ddagger} Z_k\} < 0, \tag{3.50a}$$

$$\mathrm{Sym}\left\{\begin{bmatrix} A_kG + B_kTC & A_kG + B_kTC \\ e^{j\theta}(X_kE^T + E^{\dagger}Y_k) - G & -G \end{bmatrix}\right\} < 0, \tag{3.50b}$$

where

$$G = V \begin{bmatrix} G_{11} & 0 & 0 \\ 0 & G_{22} & 0 \\ 0 & 0 & G_{33} \end{bmatrix} V^T,$$

$k = 1, 2, \ldots, l$, $\theta = (1-\alpha)\pi/2$, $E^\dagger$ and $E^\ddagger$ are given in (2.27), and U and V are determined in (3.48). Then controller (3.47) can be chosen as $u(t) = Ky(t) = T\hat{G}^{-1}y(t)$, where $\hat{G} = U\left[\begin{smallmatrix} SG_{11}S^{-1} & 0 \\ 0 & \Upsilon \end{smallmatrix}\right]U^{-1}$, and $\Upsilon \in \mathbb{R}^{(m-q)\times(m-q)}$ is a nonsingular matrix.

Proof. Since $K = T\hat{G}^{-1}$, we have $T = K\hat{G}$. Note that

$$\begin{aligned}\hat{G}C &= U\begin{bmatrix} SG_{11}S^{-1} & 0 \\ 0 & \Upsilon \end{bmatrix}U^{-1} \times U\begin{bmatrix} S & 0 & 0 \\ 0 & 0 & 0 \end{bmatrix}V^T \\ &= U\begin{bmatrix} SG_{11} & 0 & 0 \\ 0 & 0 & 0 \end{bmatrix}V^T \end{aligned}\tag{3.51}$$

and

$$\begin{aligned} CG &= U\begin{bmatrix} S & 0 & 0 \\ 0 & 0 & 0 \end{bmatrix}V^T \times V\begin{bmatrix} G_{11} & 0 & 0 \\ 0 & G_{22} & 0 \\ 0 & 0 & G_{33} \end{bmatrix}V^T \\ &= U\begin{bmatrix} SG_{11} & 0 & 0 \\ 0 & 0 & 0 \end{bmatrix}V^T. \end{aligned}\tag{3.52}$$

Then $\hat{G}C = CG$. Substituting $T = K\hat{G}$ into inequality (3.50b) yields

$$\operatorname{Sym}\left\{\begin{bmatrix} A_kG + B_kK\hat{G}C & A_kG + B_kK\hat{G}C \\ e^{j\theta}(X_kE^T + E^\dagger Y_k) - G & -G \end{bmatrix}\right\} < 0, \tag{3.53}$$

which, combined with $\hat{G}C = CG$, implies

$$\operatorname{Sym}\left\{\begin{bmatrix} A_kG + B_kKCG & A_kG + B_kKCG \\ e^{j\theta}(X_kE^T + E^\dagger Y_k) - G & -G \end{bmatrix}\right\} < 0. \tag{3.54}$$

Therefore

$$\operatorname{Sym}\left\{\begin{bmatrix} (A_k + B_kKC)G & (A_k + B_kKC)G \\ e^{j\theta}(X_kE^T + E^\dagger Y_k) - G & -G \end{bmatrix}\right\} < 0. \tag{3.55}$$

By inequalities (3.50a) and (3.55) uncertain FOSS (3.49) is robustly admissible by Theorem 3.9. This completes the proof. □

3.3.2 Case $\alpha \in (0, 1)$

By using similar methods as those in Theorem 3.11, the case $0 < \alpha < 1$ can be studied as follows.

Theorem 3.12. *For $0 < \alpha < 1$, uncertain FOSS (3.49), of which the polytopic uncertainties satisfy (3.46), is robustly admissible if there exist matrices $X_k \in \mathbb{C}^{n\times n}$, $Y_k \in \mathbb{C}^{n\times n}$, $Z_k \in$*

$\mathbb{C}^{n\times n}$, $G_{22} \in \mathbb{C}^{(n-m)\times(n-m)}$, *a real matrix* $T \in \mathbb{R}^{p\times m}$, *and a nonsingular real matrix* $G_{11} \in \mathbb{R}^{m\times m}$ *such that*

$$-EX_kE^T - \mathrm{Sym}\{E^{\ddagger}Z_k\} < 0, \tag{3.56a}$$

$$\mathrm{Sym}\left\{\begin{bmatrix} A_kG + B_kTC & A_kG + B_kTC \\ \Xi & -G \end{bmatrix}\right\} < 0, \tag{3.56b}$$

where

$$\Xi = (e^{j\theta}X_k + e^{-j\theta}\bar{X}_k)E^T + E^{\dagger}Y_k - G,$$

$$G = V\begin{bmatrix} G_{11} & 0 & 0 \\ 0 & G_{22} & 0 \\ 0 & 0 & G_{33} \end{bmatrix}V^T,$$

$k = 1, 2, \ldots, l$, $\theta = (1-\alpha)\pi/2$, $E^{\dagger}$ *and* $E^{\ddagger}$ *are given in* (2.27), *and* U *and* V *are determined in* (3.48). *Then controller* (3.47) *can be chosen as* $u(t) = Ky(t) = T\hat{G}^{-1}y(t)$, *where* $\hat{G} = U\left[\begin{smallmatrix} SG_{11}S^{-1} & 0 \\ 0 & \Upsilon \end{smallmatrix}\right]U^{-1}$, *and* $\Upsilon \in \mathbb{R}^{(m-q)\times(m-q)}$ *is a nonsingular matrix.*

Proof. Based on Theorem 3.10, the proof can be completed by using the similar methods as those in Theorem 3.11 and thus is omitted for brevity. □

Remark 3.4. Theorems 3.11 and 3.12 deal with the output feedback controller synthesis of FOSSs with polytopic uncertainties and propose the gain matrix, thus improving the existing results. The details are as follows:

1. The robust stabilization of FOSSs are solved in [68, 116] via state feedback control. Since it is usually impossible to sense all the internal states in actual systems, it is practical to stabilize uncertain FOSSs by output feedback technique as proposed in this chapter.
2. The input matrix B^T of FOSSs in [36] and the output matrix C of FOSSs in [63, 120] are assumed to be of full row rank. By comparison, there is no assumption for the input or output matrices to be with full row rank in this chapter, indicating that the methods in this chapter are more general than those in [36, 63, 120].
3. A two-step design method is applied for the output feedback control of FOSSs in [96, 97]. As pointed in [49], this method needs to design virtual state feedback gains in the first step followed by the second step, which is the indirect method and may lead to no feasible solutions if the parameters in the first step are improperly selected.
4. The output feedback controllers established in [62] include the BMI and LMI forms. The methods in the BMI form are difficult to be solved via the current MATLAB LMI Toolbox. The methods in the LMI form in [62] are rather complicated, and the improper selection of the parameter J may lead to no feasible solutions. By comparison, the methods proposed in this chapter are more concise. Furthermore, the matrix Υ in Theorems 3.11 and 3.12 can be any nonsingular real matrix, indicating that the controllers in this chapter are more relaxed.

Now we provide a numerical example to verify the practicality of the proposed controllers. Since the verification methods are similar for $1 < \alpha < 2$ and $0 < \alpha < 1$, a numerical example of FOSSs with fractional order $0 < \alpha < 1$ is not presented for brevity.

Example 3.3. Consider uncertain FOSS (3.45) with the following parameters:

$$\begin{aligned} E &= \begin{bmatrix} 2 & 0 & 0 \\ 0 & 4 & 0 \\ 0 & 0 & 0 \end{bmatrix}, \quad A_1 = \begin{bmatrix} 0.6 & 0.6 & 1.2 \\ -0.8 & -2.4 & 1.2 \\ -0.4 & -1.2 & -2.0 \end{bmatrix}, \quad B_1 = \begin{bmatrix} 1 & 0 \\ 1 & 1 \\ 1 & 0 \end{bmatrix}, \\ A_2 &= \begin{bmatrix} -1.5 & 0.5 & 1 \\ 0.4 & -2.5 & 0.9 \\ 0.2 & -1.3 & -2.0 \end{bmatrix}, \quad B_2 = \begin{bmatrix} 0.5 & 0 \\ 1.2 & -1 \\ 0.8 & 0 \end{bmatrix}, \quad \alpha = 1.2. \end{aligned} \tag{3.57}$$

Let the output matrix be $C = \left[\begin{smallmatrix} 1 & 0 & 1 \\ 0 & 0 & 0 \end{smallmatrix}\right]$. The aim is to check whether there exists an output feedback controller (3.47) such that uncertain FOSS (3.45) with parameters (3.57) is robustly admissible and to obtain the gain matrix K.

By applying Theorem 3.11 there exist the following feasible solutions:

$$X_1 = \begin{bmatrix} 7.1680 & \star & \star \\ -0.6036 - 0.6139j & 6.7504 & \star \\ -2.9351 - 0.8668j & -3.8577 - 0.4387j & 0 \end{bmatrix},$$

$$Y_1 = \begin{bmatrix} 0 & \star & \star \\ 0 & 0 & \star \\ -1.4675 - 0.4334j & -0.9644 - 0.1097j & 43.9705 \end{bmatrix},$$

$$X_2 = \begin{bmatrix} 14.3521 & \star & \star \\ 2.1787 - 0.2438j & 5.7275 & \star \\ -5.0419 + 0.1343j & 2.8491 + 0.1495j & 0 \end{bmatrix},$$

$$Y_2 = \begin{bmatrix} 0 & \star & \star \\ 0 & 0 & \star \\ -2.5210 + 0.0672j & 0.7123 + 0.0374j & 42.0757 \end{bmatrix},$$

$$Z_1 = \begin{bmatrix} 0 & \star & \star \\ 0 & 0 & \star \\ 0 & 0 & 23.9372 \end{bmatrix}, \quad Z_2 = \begin{bmatrix} 0 & \star & \star \\ 0 & 0 & \star \\ 0 & 0 & 23.9372 \end{bmatrix},$$

$$M = \begin{bmatrix} -8.6173 & \star \\ 10.6876 & 0 \end{bmatrix}, \quad G_{11} = 9.4404, \quad G_{22} = 9.5407, \quad G_{33} = 15.8110.$$

By Theorem 3.11 there exists an output feedback controller (3.47) such that uncertain FOSS (3.45) with parameters (3.57) is robustly admissible. Choosing $\Upsilon = 1$, the gain matrix is $K = \left[\begin{smallmatrix} -81.3506 & 10.6876 \\ 100.8949 & 0 \end{smallmatrix}\right]$.

To show the effectiveness of the designed controller, the admissibility of the open-loop uncertain FOSS (3.45) and the closed-loop uncertain FOSS (3.49) with parameters (3.57) is verified by simulation. For example, choose $\lambda_1 = 0.9$ and $\lambda_2 = 0.1$. When $u(t) = 0$, the admissibility of the open-loop FOSS (3.45) with parameters (3.57) is not satisfied because the eigenvalues of the open-loop system are $\{-0.7948, 0.1100\}$, some of which are located in the unstability area. When $u(t) = \begin{bmatrix} -81.3506 & 10.6876 \\ 100.8949 & 0 \end{bmatrix} y(t)$, we have that there is $c_0 \in \mathbb{C}$ such that $\det(c_0^\alpha E - (A(\lambda) + B(\lambda)KC)) \neq 0$ and $\deg(\det(\lambda E - (A(\lambda) + B(\lambda)KC))) = 2 = \operatorname{rank}(E), \lambda \in \mathbb{C}$. Furthermore, the eigenvalues of the closed-loop system are $\{-0.8832 - 0.5663j, -0.8832 + 0.5663j\}$, both of which are located in the stability area plotted in Figure 3.5. The state evolutions of the open-loop and closed-loop systems with parameters (3.57) and with $\lambda_1 = 0.9$ and $\lambda_2 = 0.1$ are plotted in Figure 3.6, where we can see that the state evolutions of the open-loop system are divergent and the state evolutions of the closed-loop system are convergent. Therefore the designed controller is shown to be effective.

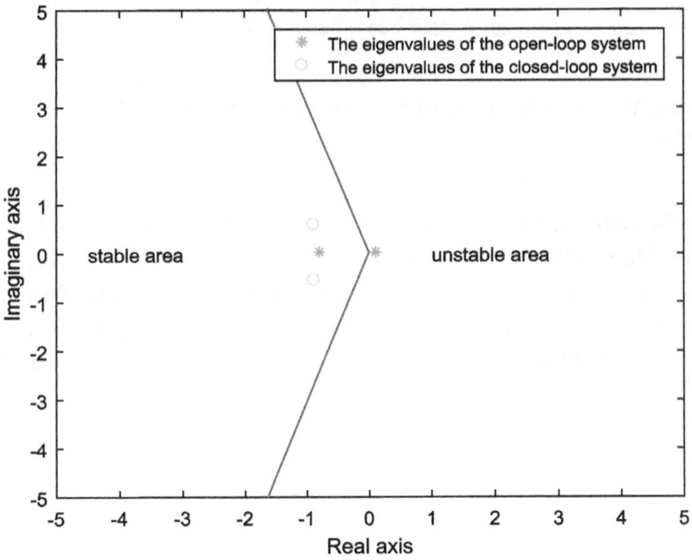

Figure 3.5: The eigenvalue locations of the open-loop system and the closed-loop system with the parameters in (3.57) and with $\lambda_1 = 0.9, \lambda_2 = 0.1$ in Example 3.3.

3.4 Chapter summary

In this chapter, we considered the problems of robust admissibility and robust stabilization for uncertain FOSSs. The structured and polytopic uncertainty cases have been studied. The robust admissibility conditions of FOSSs with structured uncertainties have been established by using robust nonsingularity analysis of uncertain matrices, error-free determinant transformation, and μ-analysis methods. The robust admissibility conditions of FOSSs with polytopic uncertainties have been given by using the proposed de-

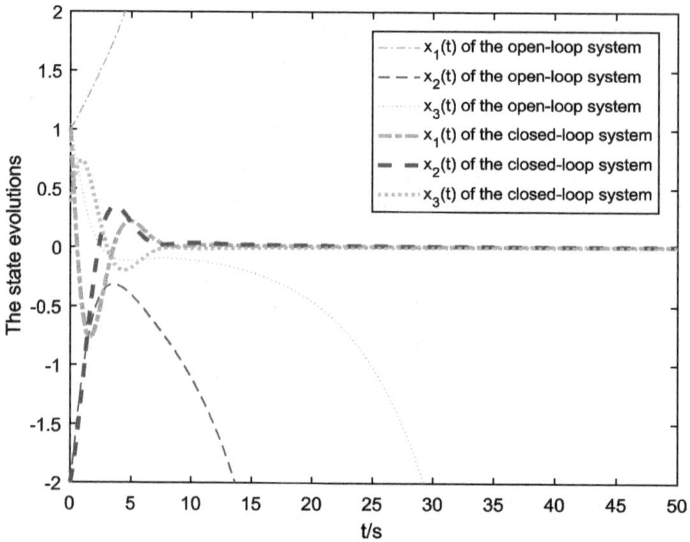

Figure 3.6: The state evolution of the open-loop and closed-loop systems with parameters (3.57) and with $\lambda_1 = 0.9$ and $\lambda_2 = 0.1$ in Example 3.3.

coupled methods and LMI method, which are easier to design closed-loop controllers. Since it is almost impossible to sense all system states, for uncertain FOSSs with $\alpha \in (1,2)$ and $\alpha \in (0,1)$, the output feedback controllers have been designed. The design methods are direct, concise, and in terms of LMIs, which have no assumption on the input or output matrices being of full row rank, improving the existing methods in [36, 120, 63, 96, 62, 97].

4 Robust H_∞ control

Abstract: In this chapter, we study the robust H_∞ control problems of uncertain fractional-order singular systems (FOSSs). Based on the generalized Kalman–Yakubovich–Popov (GKYP) lemma and the linear matrix inequality (LMI) method, the bounded real lemmas of nominal FOSSs in the sense of the H_∞ norm are given, which can ensure both the system admissibility and H_∞ performance. By using the above results, for FOSSs with norm-bounded uncertainties, the state feedback H_∞ controllers are designed so that the closed-loop systems satisfy both the robust admissibility and H_∞ performance.

4.1 Introduction

H_∞ control is one of the hot topics in the field of control theory. It can be summarized as minimizing the H_∞ norm of the system transfer function matrix or making it less than a predetermined positive number. Since Zames proposed the H_∞ control problem in 1981, scholars in the field of control have conducted extensive in-depth research, leading to many significant achievements. Meanwhile, with the increasing attention to FOSSs by more and more scholars, research on H_∞ control of FOSSs has become a recent focus in the field of control.

In [20] and [46], the bounded real lemmas in the sense of the L_∞ and H_∞ norms of fractional-order systems (FOSs) are proposed, respectively. Based on this lemma, an H_∞ state feedback controller in the form of LMI is designed. However, for the systems in [20, 46], the uncertain parameters are not considered. To address the robust stabilization problems of uncertain FOSs, an observer-based robust H_∞ controller using a parameterized approach is proposed in [4]. Similarly, in [32], new LMI conditions are provided to analyze whether FOSs satisfy the stability and H_∞ performance. To tackle the H_∞ control problem of FOSs in finite-frequency ranges, a method for designing output feedback controllers based on the Kalman–Yakubovich–Popov (KYP) lemma is presented in [93], ensuring both the stability and H_∞ performance for closed-loop FOSs in finite-frequency ranges.

The results mentioned above all address the H_∞ norm calculation and control problems of nonsingular FOSs. To analyze the H_∞ problems of FOSSs, in [3], an algorithm for calculating the H_∞ norm of FOSSs is proposed based on the method of level sets of the maximum singular value of the transfer function. This algorithm can effectively compute the H_∞ norm of the system transfer function but cannot be directly used for robustness analysis and controller design. In [45], a bounded real lemma in the sense of the L_∞ norm for FOSSs is presented based on the GKYP lemma, and a corresponding state feedback controller is designed. However, its results can only ensure the L_∞ performance of FOSSs without guaranteeing the admissibility. Motivated by the above

https://doi.org/10.1515/9783111628646-004

discussions, in this chapter, we investigate the robust H_∞ control problems for uncertain FOSSs. Firstly, the bounded real lemmas of nominal FOSSs in the sense of the H_∞ norm are given, which can ensure both the system admissibility and H_∞ performance. Secondly, the state feedback H_∞ controllers are designed so that the closed-loop FOSSs with norm-bounded uncertainties satisfy both the robust admissibility and H_∞ performance.

4.2 Preliminaries

Consider the FOSS

$$\begin{aligned} ED^\alpha x(t) &= Ax(t) + B_1 u(t) + Bw(t), \\ z(t) &= Cx(t) + B_2 u(t), \end{aligned} \tag{4.1}$$

where $x(t) \in \mathbb{R}^n$, $z(t) \in \mathbb{R}^q$, $w(t) \in \mathbb{R}^p$, and $u(t) \in \mathbb{R}^m$ are the state vector, controlled output, disturbance input, and control input, respectively, $E \in \mathbb{R}^{n\times n}$ is a constant matrix with $\operatorname{rank}(E) = r \leq n$, D^α represents the Caputo fractional derivative with order $\alpha \in (0,2)$, and A, B, C, B_1, B_2 are uncertainty matrices satisfying

$$\begin{aligned} &A = A_0 + \Delta A, \quad B = B_0 + \Delta B, \quad B_1 = B_{10} + \Delta B_1, \quad C = C_0 + \Delta C, \\ &B_2 = B_{20} + \Delta B_2, \quad \Delta\Gamma = D_\Gamma F_\Gamma G_\Gamma, \quad \Gamma \in \{A, B, C, B_1, B_2\}, \end{aligned}$$

where $A_0, B_0, C_0, B_{10}, B_{20}, D_\Gamma, G_\Gamma$ are known real matrices, and F_Γ is the matrix with the norm-bounded uncertainties satisfying $F_\Gamma^T F_\Gamma \leq I$.

Consider the state feedback controller

$$u(t) = Kx(t). \tag{4.2}$$

Utilizing the state feedback controller (4.2) to control system (4.1), we have

$$\begin{aligned} ED^\alpha x(t) &= (A + B_1 K)x(t) + Bw(t), \\ z(t) &= (C + B_2 K)x(t), \end{aligned} \tag{4.3}$$

with the transfer function

$$G_K(s) = (C + B_2 K)(s^\alpha E - (A + B_1 K))^{-1} B. \tag{4.4}$$

When both the uncertain parameters and control input $u(t)$ in system (4.1) are zero, the system is transformed into the nominal FOSS

$$\begin{aligned} ED^\alpha x(t) &= A_0 x(t) + B_0 w(t), \\ z(t) &= C_0 x(t), \end{aligned} \tag{4.5}$$

with the transfer function

$$G_N(s) = C_0(s^\alpha E - A_0)^{-1} B_0. \tag{4.6}$$

Next, we give some necessary definitions and lemmas. The definition of the H_∞ norm of the transfer function of FOSSs is as follows.

Definition 4.1 ([46]). For FOSS (4.5), the H_∞ norm of the transfer function G_N is defined as

$$\|G_N\|_\infty = \sup_{\mathrm{Re}(s)\geq 0} \sigma_{\max}(G_N(s)).$$

This definition is equivalent to the following one.

Definition 4.2 ([20]). For the stable FOSS (4.5), the H_∞ norm of the transfer function G_N is defined as

$$\|G_N\|_\infty = \sup_{w\geq 0} \sigma_{\max}(G_N(jw)).$$

The definition of the L_∞ norm of the transfer function of FOSSs is as follows.

Definition 4.3 ([46]). For FOSS (4.5), the L_∞ norm of the transfer function G_N is defined as

$$\|G_N\|_{L_\infty} = \sup_{w\geq 0} \sigma_{\max}(G_N(jw)).$$

Lemma 4.1 ([20]). *Consider the FOSs with transfer function $G(s) = C(s^\alpha I - A)B + D$, $0 < \alpha < 1$. Then $\|G\|_\infty < \gamma$ if there exists $P \in \mathbb{H}_n > 0$ satisfying*

$$\begin{bmatrix} \mathrm{Sym}\{AX\} & X^* C^T & B \\ CX & -\gamma I & D \\ B^T & D^T & -\gamma I \end{bmatrix} < 0,$$

where $X = e^{j\theta}P + e^{-j\theta}\bar{P}$, $\theta = \frac{\pi}{2}(1-\alpha)$.

Considering that $\|G\|_\infty = \|G^T\|_\infty$, we can obtain the following lemma.

Lemma 4.2. *Consider the FOSs with transfer function $G(s) = C(s^\alpha I - A)B + D$, $0 < \alpha < 1$. Then $\|G\|_\infty < \gamma$ if there exists $P \in \mathbb{H}_n > 0$ satisfying*

$$\begin{bmatrix} \mathrm{Sym}\{A^T X\} & X^* B & C^T \\ B^T X & -\gamma I & D^T \\ C & D & -\gamma I \end{bmatrix} < 0,$$

where $X = e^{j\theta}P + e^{-j\theta}\bar{P}$, $\theta = \frac{\pi}{2}(1-\alpha)$.

Lemma 4.3 ([20, 46]). *Consider the FOSs with transfer function* $G(s) = C(s^\alpha I - A)^{-1}B + D$, $1 \le \alpha < 2$. *Then* $\|G\|_\infty < \gamma$ *if and only if there exists* $P \in \mathbb{H}_n > 0$ *satisfying*

$$\begin{bmatrix} \mathrm{Sym}\{AX\} & X^*C^T & B \\ CX & -\gamma I & D \\ B^T & D^T & -\gamma I \end{bmatrix} < 0,$$

where $X = e^{j\theta}P$, $\theta = \frac{\pi}{2}(1-\alpha)$.

Considering that $\|G\|_\infty = \|G^T\|_\infty$, based on Lemma 4.3, we can obtain the following lemma.

Lemma 4.4. *Consider the FOSSs with transfer function* $G(s) = C(s^\alpha I - A)^{-1}B + D$, $1 \le \alpha < 2$. *Then* $\|G\|_\infty < \gamma$ *if and only if there exists* $P \in \mathbb{H}_n > 0$ *satisfying*

$$\begin{bmatrix} \mathrm{Sym}\{A^TX\} & X^*B & C^T \\ B^TX & -\gamma I & D^T \\ C & D & -\gamma I \end{bmatrix} < 0,$$

where $X = e^{j\theta}P$, $\theta = \frac{\pi}{2}(1-\alpha)$.

Lemma 4.5 ([55]). *Let* M *be a Hermitian matrix of appropriate dimension, and let* $\mathrm{Re}(M)$ *and* $\mathrm{Im}(M)$ *be its real and imaginary parts. Then* $M < 0$ *if and only if*

$$\begin{bmatrix} \mathrm{Re}(M) & \mathrm{Im}(M) \\ -\mathrm{Im}(M) & \mathrm{Re}(M) \end{bmatrix} < 0.$$

Lemma 4.6 ([55]). *Let* X *and* Y *be matrices of appropriate dimensions, and let* $\varepsilon > 0$. *Then*

$$\mathrm{Sym}\{X^*Y\} \le \varepsilon X^*X + \varepsilon^{-1}Y^*Y.$$

Lemma 4.7 (Schur complement lemma [110]). *For a Hermitian matrix*

$$M = \begin{bmatrix} X & Y \\ Y^* & Z \end{bmatrix},$$

where X, Y, Z *are matrices of appropriate dimensions, the following three conditions are equivalent:*

(1) $M < 0$;

(2) $X < 0, Z - Y^*X^{-1}Y < 0$;

(3) $Z < 0, X - YZ^{-1}Y^* < 0$.

Lemma 4.8 ([45]). *Let* $E, A \in \mathbb{R}^{n\times n}$, $B \in \mathbb{R}^{n\times m}$, $\Phi \in \mathbb{H}_{n+m}$, Ψ, *and* $\Omega \in \mathbb{H}_2$. *Define*

$$\Lambda(\Psi,\Omega) = \left\{\lambda \in \mathbb{C} \mid \begin{bmatrix} \lambda \\ 1 \end{bmatrix}^* \Omega \begin{bmatrix} \lambda \\ 1 \end{bmatrix} \ge 0,\ \begin{bmatrix} \lambda \\ 1 \end{bmatrix}^* \Psi \begin{bmatrix} \lambda \\ 1 \end{bmatrix} = 0\right\}. \tag{4.7}$$

Define $\bar{\Lambda}$ *as* Λ *if* Λ *is bounded and as* $\Lambda \cup \{\infty\}$ *otherwise. Let* $\Pi(\lambda) = \left[\begin{smallmatrix} (\lambda E - A)^{-1}B \\ I_m \end{smallmatrix}\right]$. *Consider the following conditions:*

(1) *For all* $\lambda \in \bar{\Lambda}(\Psi, \Omega)$,

$$\Pi(\lambda)^* \Phi \Pi(\lambda) < 0. \tag{4.8}$$

(2) *There exist* $P_1, P_2 \in \mathbb{H}_n$, $P_2 > 0$, *satisfying*

$$\begin{bmatrix} A & B \\ E & \mathbf{0} \end{bmatrix}^* (\Psi \otimes P_1 + \Omega \otimes P_2) \begin{bmatrix} A & B \\ E & \mathbf{0} \end{bmatrix} + \Phi < 0. \tag{4.9}$$

If Λ *represents curves in the complex plane, then conditions* (1) *and* (2) *are equivalent.*

4.3 Bounded real lemma

For the system orders $\alpha \in (0, 1)$ and $\alpha \in (1, 2)$ of FOSSs, we propose the corresponding bounded real lemmas for nominal FOSS (4.5), in this section.

4.3.1 Case $\alpha \in (0, 1)$

Theorem 4.1. *Let* $\gamma > 0$. *The FOSS* (4.5) *with fractional order* $\alpha \in (0, 1)$ *satisfies the admissibility and its transfer function satisfies* $\|G_N\|_\infty < \gamma$ *if any of the following two conditions holds:*

(i) *There exists a matrix* $P \in \mathbb{C}^{n \times n}$ *satisfying*

$$E^T P = P^* E \geq 0, \tag{4.10}$$

$$\begin{bmatrix} \mathrm{Sym}\{A_0^T X\} & X^* B_0 & C_0^T \\ B_0^T X & -\gamma I & 0 \\ C_0 & 0 & -\gamma I \end{bmatrix} < 0, \tag{4.11}$$

where $X = e^{j\theta} P + e^{-j\theta} \bar{P}$, $\theta = \frac{\pi}{2}(1 - \alpha)$.

(ii) *There exists a matrix* $Q \in \mathbb{C}^{n \times n}$ *satisfying*

$$EQ = Q^* E^T \geq 0, \tag{4.12}$$

$$\begin{bmatrix} \mathrm{Sym}\{A_0 Y\} & Y^* C_0^T & B_0 \\ C_0 Y & -\gamma I & 0 \\ B_0^T & 0 & -\gamma I \end{bmatrix} < 0, \tag{4.13}$$

where $Y = e^{j\theta} Q + e^{-j\theta} \bar{Q}$, $\theta = \frac{\pi}{2}(1 - \alpha)$.

Proof. Considering that $\|G_N\|_\infty = \|G_N^T\|_\infty$, we only prove (i). The proof consists of two parts. The first part is the proof of admissibility, and the second part is the proof that $\|G_N\|_\infty < \gamma$.

Step 1. By inequality (4.11), there exists a matrix $P \in \mathbb{C}^{n\times n}$ satisfying

$$\mathrm{Sym}\{A_0^T(e^{j\theta}P + e^{-j\theta}\bar{P})\} < 0. \tag{4.14}$$

From inequalities (4.10) and (4.14), it follows that system (4.5) is admissible by Theorem 2.6 in Chapter 2.

Step 2. According to Lemma 4.7, inequality (4.11) is equivalent to

$$A_0^T X + X^* A_0 + \gamma^{-1} C_0^T C_0 + \gamma^{-1} X^* B_0 B_0^T X < 0. \tag{4.15}$$

Let $\Xi = \gamma A_0^T X + \gamma X^* A_0 + X^* B_0 B_0^T X$. Then inequality (4.15) can be written as

$$-\Xi - C_0^T C_0 > 0. \tag{4.16}$$

Pre- and postmultiplying $B_0^T(\lambda^* E - A_0)^{-T}$ and $(\lambda E - A_0)^{-1}B_0$ by $\lambda = e^{j\frac{\pi}{2}\alpha}w^\alpha$ for $w \geq 0$, we transform inequality (4.16) into

$$-B_0^T(\lambda^* E - A_0)^{-T} C_0^T C_0 (\lambda E - A_0)^{-1} B_0 - B_0^T(\lambda^* E - A_0)^{-T}\Xi(\lambda E - A_0)^{-1}B_0 \geq 0. \tag{4.17}$$

Denote $\Delta = \gamma^2 I - B_0^T(\lambda^* E - A_0)^{-T} C_0^T C_0 (\lambda E - A_0)^{-1} B_0$, where $\gamma^2 I > 0$. From (4.17), we have

$$\Delta - B_0^T(\lambda^* E - A_0)^{-T}\Xi(\lambda E - A_0)^{-1}B_0 > 0. \tag{4.18}$$

Inequality (4.18) is equivalent to

$$\begin{bmatrix} B_0^T(\lambda^* E - A_0)^{-T} & I \end{bmatrix} \begin{bmatrix} -\Xi & 0 \\ 0 & \Delta \end{bmatrix} \begin{bmatrix} (\lambda E - A_0)^{-1}B_0 \\ I \end{bmatrix} > 0. \tag{4.19}$$

Let $\Pi(\lambda) = \left[\begin{smallmatrix} (\lambda E - A_0)^{-1}B_0 \\ I \end{smallmatrix}\right]$, $\Phi = \left[\begin{smallmatrix} \Xi & 0 \\ 0 & -\Delta \end{smallmatrix}\right]$. Then, inequality (4.19) can be rewritten as

$$\Pi(\lambda)^* \Phi \Pi(\lambda) < 0, \tag{4.20}$$

where $\lambda = e^{j\frac{\pi}{2}\alpha}w^\alpha$, $w \geq 0$. The λ in inequality (4.20) can also be described by $\Lambda(\Psi, \Omega)$ in (4.7), and $\Psi = \left[\begin{smallmatrix} 0 & e^{j\theta} \\ e^{-j\theta} & 0 \end{smallmatrix}\right]$ and $\Omega = \left[\begin{smallmatrix} 0 & e^{-j\theta} \\ e^{j\theta} & 0 \end{smallmatrix}\right]$ with $\alpha \in (0, 1)$. By Lemma 4.8, inequality (4.20) is equivalent to

$$\begin{bmatrix} A_0 & B_0 \\ E & 0 \end{bmatrix}^* (\Psi \otimes P + \Omega \otimes Q) \begin{bmatrix} A_0 & B_0 \\ E & 0 \end{bmatrix} + \Phi < 0, \tag{4.21}$$

where P, $Q \in \mathbb{H}_n$, $Q > 0$. Substituting Φ, Ψ and Ω into inequality (4.21), we get

$$\begin{bmatrix} \mathrm{Sym}\{E^T(e^{-j\theta}P + e^{j\theta}Q)A_0\} & E^T(e^{-j\theta}P + e^{j\theta}Q)B_0 \\ B_0^T(e^{j\theta}P + e^{-j\theta}Q)E & 0 \end{bmatrix} + \begin{bmatrix} \Gamma & 0 \\ 0 & -\Delta \end{bmatrix} < 0.$$

Therefore $\Delta > 0$, which leads to $\|G\|_\infty < \gamma$. The theorem is proved. □

Remark 4.1. Theorem 4.1 is the bounded real lemma for FOSSs with $\alpha \in (0,1)$. We can see that Theorem 4.1 contains equality constraints that can cause problems in numerical computations. This is because computers introduce rounding errors during the calculations, and equality constraints are fragile, potentially not being fully satisfied [116]. Therefore, from the numerical computation perspective, it is necessary to provide a bounded real lemma without equality constraints.

For $\alpha \in (0,1)$, we have the following bounded real lemma for FOSSs without equality constraints.

Theorem 4.2. *Let $\gamma > 0$. The FOSS (4.5) with fractional order $\alpha \in (0,1)$ satisfies the admissibility, and its transfer function satisfies $\|G_N\|_\infty < \gamma$ if any of the following two conditions holds:*

(i) *There exist matrices $P = P^* > 0$, $P \in \mathbb{C}^{n\times n}$, and $S \in \mathbb{R}^{(n-r)\times n}$ satisfying*

$$\begin{bmatrix} \mathrm{Sym}\{A_0^T X\} & X^* B_0 & C_0^T \\ B_0^T X & -\gamma I & 0 \\ C_0 & 0 & -\gamma I \end{bmatrix} < 0, \tag{4.22}$$

where $X = (e^{j\theta}P + e^{-j\theta}\bar{P})E + E_0 S$, $\theta = \frac{\pi}{2}(1-\alpha)$, with E_0 any matrix of full column rank satisfying $E^T E_0 = 0$.

(ii) *There exist matrices $Q = Q^* > 0$, $Q \in \mathbb{C}^{n\times n}$, and $S \in \mathbb{R}^{(n-r)\times n}$ satisfying*

$$\begin{bmatrix} \mathrm{Sym}\{A_0 Y\} & Y^* C_0^T & B_0 \\ C_0 Y & -\gamma I & 0 \\ B_0^T & 0 & -\gamma I \end{bmatrix} < 0, \tag{4.23}$$

where $Y = (e^{j\theta}Q + e^{-j\theta}\bar{Q})E^T + E_0 S$, $\theta = \frac{\pi}{2}(1-\alpha)$, with E_0 any matrix of full column rank satisfying $EE_0 = 0$.

Proof. Considering that $\|G_N\|_\infty = \|G_N^T\|_\infty$, we only prove (i). The proof consists of two parts. The first part is the proof of admissibility, and the second part is the proof of $\|G_N\|_\infty < \gamma$.

Step 1. By inequality (4.22), there exist matrices $P = P^* > 0$, $P \in \mathbb{C}^{n\times n}$, and $S \in \mathbb{R}^{(n-r)\times n}$ satisfying

$$\mathrm{Sym}\{A_0^T((e^{j\theta}P + e^{-j\theta}\bar{P})E + E_0 S)\} < 0. \tag{4.24}$$

From inequality (4.24), it follows that system (4.5) is admissible by Theorem 2 in [64].

Step 2. The proof of $\|G_N\|_\infty < \gamma$ is similar to that of Theorem 4.1, so we omit it. □

4.3.2 Case $\alpha \in (1, 2)$

Theorem 4.3. *Let $\gamma > 0$. The FOSS (4.5) with fractional order $\alpha \in (1, 2)$ satisfies the admissibility, and its transfer function satisfies $\|G_N\|_\infty < \gamma$ if and only if any of the following two conditions holds:*

(i) *There exists a matrix $P \in \mathbb{C}^{n\times n}$ satisfying*

$$E^T P = P^* E \geq 0, \tag{4.25}$$

$$\begin{bmatrix} \mathrm{Sym}\{e^{j\theta} A_0^T P\} & e^{-j\theta} P^* B_0 & C_0^T \\ e^{j\theta} B_0^T P & -\gamma I & 0 \\ C_0 & 0 & -\gamma I \end{bmatrix} < 0. \tag{4.26}$$

(ii) *There exists a matrix $Q \in \mathbb{C}^{n\times n}$ satisfying*

$$EQ = Q^* E^T \geq 0, \tag{4.27}$$

$$\begin{bmatrix} \mathrm{Sym}\{e^{j\theta} A_0 Q\} & e^{-j\theta} Q^* C_0^T & B_0 \\ e^{j\theta} C_0 Q & -\gamma I & 0 \\ B_0^T & 0 & -\gamma I \end{bmatrix} < 0, \tag{4.28}$$

where $\theta = \frac{\pi}{2}(1 - \alpha)$ in conditions (i)–(ii).

Proof. Considering that $\|G_N\|_\infty = \|G_N^T\|_\infty$, we only prove (i).

Sufficiency. The proof consists of two steps. In Step 1, we prove the admissibility, and in Step 2, we prove that $\|G\|_\infty < \gamma$.

Step 1. From inequality (4.26), it follows that there exists P such that $\mathrm{Sym}\{e^{j\theta} A_0^T P\} < 0$, which can be written as

$$\cos\theta(A_0^T P + P^* A_0) + j\sin\theta(A_0^T P - P^* A_0) < 0. \tag{4.29}$$

Converting the complex LMI (4.29) into the real LMI, we have

$$\begin{bmatrix} \cos\theta(A_0^T P + P^T A_0) & \sin\theta(A_0^T P - P^* A_0) \\ -\sin\theta(A_0^T P - P^* A_0) & \cos\theta(A_0^T P + P^T A_0) \end{bmatrix} < 0,$$

which is equivalent to

$$\mathrm{Sym}\{\Theta \otimes (A_0^T P)\} < 0, \quad \Theta = \begin{bmatrix} \cos\theta & \sin\theta \\ -\sin\theta & \cos\theta \end{bmatrix}. \tag{4.30}$$

With (4.25) and (4.30), the FOSS (4.5) is admissible by Theorem 2 in [63].

Step 2. By Lemma 4.7, inequality (4.26) is equivalent to

$$e^{j\theta} A_0^T P + e^{-j\theta} P^* A_0 + \gamma^{-1} C_0^T C_0 + \gamma^{-1} P^* B_0 B_0^T P < 0, \tag{4.31}$$

which can be rewritten as

$$-\Xi - C_0^T C_0 > 0, \tag{4.32}$$

where $\Xi = e^{j\theta}\gamma A_0^T P + e^{-j\theta}\gamma P^* A_0 + P^* B_0 B_0^T P$. Pre- and postmultiplying $B_0^T(\lambda^* E - A_0)^{-T}$ and $(\lambda E - A_0)^{-1}B_0$ by $\lambda = e^{j\frac{\pi}{2}\alpha} w^\alpha$ for $w \in \mathbb{R}$, respectively, we have

$$-B_0^T(\lambda^* E - A_0)^{-T} C_0^T C_0 (\lambda E - A_0)^{-1} B_0 - B_0^T(\lambda^* E - A_0)^{-T}\Xi(\lambda E - A_0)^{-1} B_0 \geq 0.$$

Note that $\gamma^2 I > 0$. Denote $\Delta = \gamma^2 I - B_0^T(\lambda^* E - A_0)^{-T} C_0^T C_0 (\lambda E - A_0)^{-1} B_0$. Then

$$\Delta - B_0^T(\lambda^* E - A_0)^{-T}\Xi(\lambda E - A_0)^{-1} B_0 > 0,$$

which can be transformed into

$$\begin{bmatrix} B_0^T(\lambda^* E - A_0)^{-T} & I \end{bmatrix} \begin{bmatrix} -\Xi & 0 \\ 0 & \Delta \end{bmatrix} \begin{bmatrix} (\lambda E - A_0)^{-1} B_0 \\ I \end{bmatrix} > 0. \tag{4.33}$$

Let $\Pi(\lambda) = \left[\begin{smallmatrix} (\lambda E - A_0)^{-1} B_0 \\ I \end{smallmatrix}\right]$. Inequality (4.33) can be rewritten as

$$\Pi(\lambda)^* \Phi \Pi(\lambda) < 0, \tag{4.34}$$

where $\Phi = \left[\begin{smallmatrix} \Xi & 0 \\ 0 & -\Delta \end{smallmatrix}\right]$. The λ in equality (4.34) can be described by $\Lambda(\Psi, \Omega)$ in Lemma 4.8, and we can take $\Psi = \left[\begin{smallmatrix} 0 & e^{j\theta} \\ e^{-j\theta} & 0 \end{smallmatrix}\right]$ and $\Omega = 0$. Then inequality (4.34) is equivalent to

$$\begin{bmatrix} A_0 & B_0 \\ E & 0 \end{bmatrix}^* (\Psi \otimes P + \Omega \otimes Q) \begin{bmatrix} A_0 & B_0 \\ E & 0 \end{bmatrix} + \Phi < 0. \tag{4.35}$$

Substituting Φ, Ψ and Ω into inequality (4.35), we get

$$\begin{bmatrix} e^{j\theta} A_0^T PE + e^{-j\theta} E^T P A_0 & e^{-j\theta} E^T P B_0 \\ e^{j\theta} B_0^T PE & 0 \end{bmatrix} + \begin{bmatrix} \Xi & 0 \\ 0 & -\Delta \end{bmatrix} < 0.$$

Therefore $\Delta > 0$, and then $\|G\|_\infty < \gamma$.

Necessity. Suppose that system (4.5) satisfies the admissibility and its transfer function satisfies $\|G\|_\infty < \gamma$. Then there exist nonsingular matrices M and N satisfying

$$E = M \begin{bmatrix} I & 0 \\ 0 & 0 \end{bmatrix} N, \quad A_0 = M \begin{bmatrix} A_1 & 0 \\ 0 & I \end{bmatrix} N,$$

where $A_1 \in \mathbb{R}^{r\times r}$ is stable. Denote

$$C_0 N^{-1} = \begin{bmatrix} C_1 & C_2 \end{bmatrix}, \quad M^{-1} B_0 = \begin{bmatrix} B_1 \\ B_2 \end{bmatrix}.$$

Then the transfer function of system (4.5) is

$$\begin{aligned} G_N(s) &= C_0(s^\alpha E - A_0)^{-1}B_0 \\ &= C_1(s^\alpha I - A_1)^{-1}B_1 - C_2B_2. \end{aligned}$$

By Lemma 4.4, there exists $P_1 \in \mathbb{H}_n > 0$ such that

$$\begin{bmatrix} e^{j\theta}A_1^TP_1 + e^{-j\theta}P_1^*A_1 & e^{-j\theta}P_1^*B_1 & C_1^T \\ e^{j\theta}B_1^TP_1 & -\gamma I & -B_2^TC_2^T \\ C_1 & -C_2B_2 & -\gamma I \end{bmatrix} < 0. \tag{4.36}$$

Let $W = e^{j\theta}A_1^TP_1 + e^{-j\theta}P_1^*A_1$. Then inequality (4.36) can be transformed into

$$\begin{bmatrix} W + \gamma^{-1}C_1^TC_1 & e^{-j\theta}P_1^*B_1 - \gamma^{-1}C_1^TC_2B_2 \\ e^{j\theta}B_1^TP_1 - \gamma^{-1}B_2^TC_2^TC_1 & -\gamma I + \gamma^{-1}B_2^TC_2^TC_2B_2 \end{bmatrix} < 0.$$

There exists a small enough $\varepsilon > 0$ satisfying

$$\begin{bmatrix} W + \gamma^{-1}C_1^TC_1 & e^{-j\theta}P_1^*B_1 - \gamma^{-1}C_1^TC_2B_2 \\ e^{j\theta}B_1^TP_1 - \gamma^{-1}B_2^TC_2^TC_1 & -\gamma I + \gamma^{-1}B_2^T\Omega B_2 \end{bmatrix} < 0,$$

where $\Omega = C_2^TC_2 + \varepsilon I > 0$. Let

$$P = M^{-T}\begin{bmatrix} P_1 & 0 \\ -\gamma^{-1}e^{-j\theta}C_2^TC_1 & -\gamma^{-1}e^{-j\theta}\Omega \end{bmatrix}N.$$

Substituting P into (4.25) and (4.26), we see that both (4.25) and (4.26) hold. The proof is completed. □

Theorem 4.3 is the bounded real lemma for FOSSs with $\alpha \in [1,2)$. We can see that Theorem 4.3 contains equality constraints. As mentioned in Remark 4.1, for numerical computations, it is necessary to provide a bounded real lemma without equality constraints.

Theorem 4.4. *Let $\gamma > 0$. The FOSS (4.5) with fractional order $\alpha \in (1,2)$ satisfies the admissibility, and its transfer function satisfies $\|G_N\|_\infty < \gamma$ if and only if any of the following two conditions holds:*

(i) *There exist matrices $P = P^* > 0$, $P \in \mathbb{C}^{n\times n}$, and $S \in \mathbb{R}^{(n-r)\times n}$ satisfying*

$$\begin{bmatrix} \mathrm{Sym}\{A_0^TX\} & X^*B_0 & C_0^T \\ B_0^TX & -\gamma I & 0 \\ C_0 & 0 & -\gamma I \end{bmatrix} < 0, \tag{4.37}$$

where $X = e^{j\theta}(PE+E_0S)$, $\theta = \frac{\pi}{2}(1-\alpha)$, with E_0 any matrix of full column rank satisfying $E^TE_0 = 0$.

(ii) *There exist matrices $Q = Q^* > 0$, $Q \in \mathbb{C}^{n\times n}$, and $S \in \mathbb{R}^{(n-r)\times n}$ satisfying*

$$\begin{bmatrix} \mathrm{Sym}\{A_0Y\} & Y^*C_0^T & B_0 \\ C_0Y & -\gamma I & 0 \\ B_0^T & 0 & -\gamma I \end{bmatrix} < 0, \tag{4.38}$$

where $Y = e^{j\theta}(QE^T + E_0S)$, $\theta = \frac{\pi}{2}(1-\alpha)$, with E_0 any matrix of full column rank satisfying $EE_0 = 0$.

Proof. Considering that $\|G_N\|_\infty = \|G_N^T\|_\infty$, we only prove (i).

Sufficiency. Note that $X = e^{j\theta}(PE + E_0S)$, $E^TE_0 = 0$, and $P > 0$ in condition (i). Then

$$\begin{aligned} E^T(PE + E_0S) &= E^TPE \geq 0, \\ (PE + E_0S)^*E &= E^TP^*E \geq 0. \end{aligned} \tag{4.39}$$

From (4.39) and $P = P^*$, it follows that

$$E^T(PE + E_0S) = (PE + E_0S)^*E \geq 0. \tag{4.40}$$

According to (4.40) and (4.37), the sufficiency is proved by Theorem 4.3.

Necessity. Suppose that system (4.5) satisfies the admissibility and its transfer function satisfies $\|G\|_\infty < \gamma$. Then there exist nonsingular matrices M and N satisfying

$$E = M\begin{bmatrix} I & 0 \\ 0 & 0 \end{bmatrix}N, \quad A_0 = M\begin{bmatrix} A_1 & 0 \\ 0 & I \end{bmatrix}N,$$

where $A_1 \in \mathbb{R}^{r\times r}$ is stable. Denote

$$C_0N^{-1} = \begin{bmatrix} C_1 & C_2 \end{bmatrix}, \quad M^{-1}B_0 = \begin{bmatrix} B_1 \\ B_2 \end{bmatrix}.$$

Then the transfer function of system (4.5) is

$$\begin{aligned} G_N(s) &= C_0(s^\alpha E - A_0)^{-1}B_0 \\ &= C_1(s^\alpha I - A_1)^{-1}B_1 - C_2B_2. \end{aligned}$$

By Lemma 4.4, there exists a matrix $P_1 \in \mathbb{H}_n > 0$ such that

$$\begin{bmatrix} e^{j\theta}A_1^TP_1 + e^{-j\theta}P_1^*A_1 & e^{-j\theta}P_1^*B_1 & C_1^T \\ e^{j\theta}B_1^TP_1 & -\gamma I & -B_2^TC_2^T \\ C_1 & -C_2B_2 & -\gamma I \end{bmatrix} < 0. \tag{4.41}$$

Denote $W = e^{j\theta}A_1^TP_1 + e^{-j\theta}P_1^*A_1$. Then equality (4.41) can transform into

$$\begin{bmatrix} W + \gamma^{-1}C_1^T C_1 & e^{-j\theta}P_1^* B_1 - \gamma^{-1}C_1^T C_2 B_2 \\ e^{j\theta}B_1^T P_1 - \gamma^{-1}B_2^T C_2^T C_1 & -\gamma I + \gamma^{-1}B_2^T C_2^T C_2 B_2 \end{bmatrix} < 0.$$

There exists a small enough $\varepsilon > 0$ satisfying

$$\begin{bmatrix} W + \gamma^{-1}C_1^T C_1 & e^{-j\theta}P_1^* B_1 - \gamma^{-1}C_1^T C_2 B_2 \\ e^{j\theta}B_1^T P_1 - \gamma^{-1}B_2^T C_2^T C_1 & -\gamma I + \gamma^{-1}B_2^T \Omega B_2 \end{bmatrix} < 0,$$

where $\Omega = C_2^T C_2 + \varepsilon I > 0$. The E_0 satisfying $E^T E_0 = 0$ can be taken as

$$E_0 = M^{-T} \begin{bmatrix} 0 \\ I_{n-r} \end{bmatrix} \Gamma, \tag{4.42}$$

where Γ is any nonsingular real matrix. Let

$$\begin{aligned} P &= M^{-T} \begin{bmatrix} P_1 & 0 \\ 0 & I_{n-r} \end{bmatrix} M^{-1} > 0, \\ S &= \Gamma^{-1} \begin{bmatrix} -\gamma^{-1}e^{-j\theta}C_2^T C_1 & -\gamma^{-1}e^{-j\theta}\Omega \end{bmatrix} N. \end{aligned} \tag{4.43}$$

Substituting E_0, P, and S into (4.37), we see that (4.37) holds. The proof is completed. □

Remark 4.2. When E in system (4.5) is taken as I, Theorems 4.1 and 4.2 are equivalent to the Lemmas 4.1 and 4.2, respectively. Theorems 4.3 and 4.4 are equivalent to Lemmas 4.3 and 4.4. Therefore the conclusions in this section are more general than those in [20, 46].

Remark 4.3. A bounded real lemma for FOSSs is provided in [45] in the sense of the L_∞ norm, but it does not guarantee the admissibility. In comparison, the bounded real lemmas presented in this section are in the sense of the H_∞ norm, which not only describe the H_∞ performance but also ensure the admissibility. The comparisons will be illustrated in Example 4.3.

Now we provide two numerical examples to verify the validity of the results in this section. Example 4.1 is used to verify the validity of Theorems 4.1–4.2, and Example 4.2 is used to verify the validity of Theorems 4.3–4.4.

Example 4.1. Consider the fractional-order singular electrical circuit [41] with inductances L_1 and L_2, resistances R_1 and R_2, and source voltages e_1 and e_2, where the voltage $u_L(t)$ on the coil with the inductance L is the αth-order derivative of its magnetic flux. Then the circuit can be described by

$$\begin{aligned} u_{L_1}(t) &= L_1 D^\alpha i_{L_1}(t), \\ u_{L_2}(t) &= L_2 D^\alpha i_{L_2}(t), \end{aligned}$$

where $i_{L_1}(t)$ and $i_{L_2}(t)$ are the currents in the coils. Let the measured output be $z(t) = C_0 x(t)$. Then the circuit in Figure 4.1 can be described as

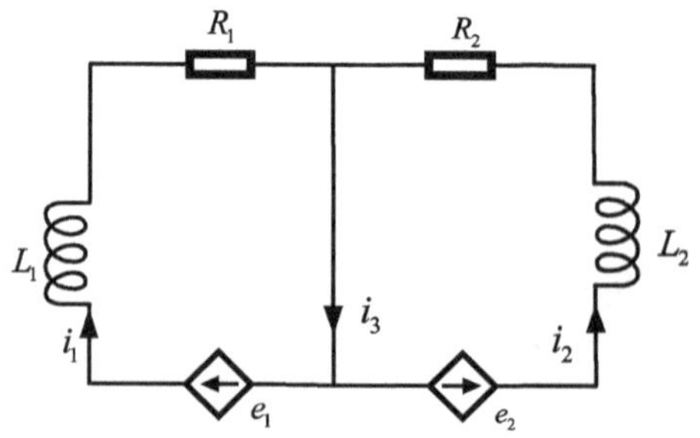

Figure 4.1: The fractional-order singular circuit model in Example 4.1 [41].

$$\begin{aligned} ED^{\alpha}x(t) &= A_0x(t) + B_0w(t), \\ z(t) &= C_0x(t), \end{aligned} \tag{4.44}$$

where

$$x(t) = \begin{bmatrix} i_1(t) \\ i_2(t) \\ i_3(t) \end{bmatrix}, \quad w(t) = \begin{bmatrix} e_1 \\ e_2 \end{bmatrix}, \quad E = \begin{bmatrix} L_1 & 0 & 0 \\ 0 & L_2 & 0 \\ 0 & 0 & 0 \end{bmatrix},$$

$$B_0 = \begin{bmatrix} 1 & 0 \\ 0 & 1 \\ 0 & 0 \end{bmatrix}, \quad A_0 = \begin{bmatrix} -R_1 & 0 & 0 \\ 0 & -R_2 & 0 \\ 1 & 1 & -1 \end{bmatrix}.$$

It can be shown that system (4.44) is a nominal FOSS studied in this chapter. Take $L_1 = 10H$, $L_2 = 5H$, $R_1 = 1\Omega$, $R_2 = 2\Omega$, $\alpha = 0.8$, $\gamma = 2$, $C_0 = \left[\begin{smallmatrix} 0 & 1 & 0 \\ 1 & 0 & 0 \end{smallmatrix}\right]$, and the following system parameters:

$$E = \begin{bmatrix} 10 & 0 & 0 \\ 0 & 5 & 0 \\ 0 & 0 & 0 \end{bmatrix}, \quad A_0 = \begin{bmatrix} -1 & 0 & 0 \\ 0 & -2 & 0 \\ 1 & 1 & -1 \end{bmatrix}, \quad B_0 = \begin{bmatrix} 1 & 0 \\ 0 & 1 \\ 0 & 0 \end{bmatrix}, \quad C_0 = \begin{bmatrix} 0 & 1 & 0 \\ 1 & 0 & 0 \end{bmatrix}. \tag{4.45}$$

The purpose is to investigate whether the system satisfies the admissibility and H_∞ performance.

From conditions (i)–(ii) in Theorem 4.1, it follows that the feasible solutions P and Q are

$$P = \begin{bmatrix} 1.6829 & 0.0425 - 0.0968\mathbf{i} & 0 \\ 0.085 + 0.1937\mathbf{i} & 3.3001 & 0 \\ 1 + 9261.9137\mathbf{i} & 1 + 9251.9170\mathbf{i} & 1 - 9285.1723\mathbf{i} \end{bmatrix},$$

$$Q = \begin{bmatrix} 1.5864 & 0 & 0 \\ 0 & 3.1402 & 0 \\ 1.5864 & 3.1402 & 10731.7058 \end{bmatrix}.$$

Thus system (4.45) is admissible, and its transfer function satisfies $\|G_N\|_\infty < \gamma = 2$.

When choosing $E_0 = \begin{bmatrix} 0 & 0 & 1 \end{bmatrix}^T$ in Theorem 4.2, the feasible solutions in condition (i) are

$$P = \begin{bmatrix} 0.1242 & -0.0166 + 0.0437\mathbf{i} & 0 \\ -0.0166 - 0.0437\mathbf{i} & 0.5044 & 0 \\ 0 & 0 & 39153.0379 \end{bmatrix},$$

$$S = \begin{bmatrix} -6999.3739 & -7008.3890 & 6966.5283 \end{bmatrix},$$

and the feasible solutions in condition (ii) are

$$Q = \begin{bmatrix} 0.1204 & -0.0013 & 0 \\ -0.0013 & 0.6522 & 0 \\ 0 & 0 & 51777.0980 \end{bmatrix},$$

$$S = \begin{bmatrix} -6.0934 & -252.0728 & 29406.5505 \end{bmatrix}.$$

Hence system (4.5) with parameter (4.45) is admissible, and its transfer function satisfies $\|G_N\|_\infty < \gamma = 2$.

By numerical simulations we get:

(1) The eigenvalues of system (4.5) are $\{-0.4, -0.1\}$, which are all in the stable areas, as we can see from Figure 4.2. Meanwhile, (i) $\det(s^\alpha E - A_0)$ is not 0, i. e., the system is regular; (ii) Take the left and right maximal zeroed submatrices $L_E = \begin{bmatrix} 0 & 0 & 1 \end{bmatrix}$ and $R_E = \begin{bmatrix} 0 \\ 0 \\ 1 \end{bmatrix}$ of the matrix E. Then $\operatorname{rank}(L_E A_0 R_E) = 1$, i. e., the system is impulse-free. Therefore the FOSS is admissible.

(2) The maximum singular value curve of the system transfer function is shown in Figure 4.3, from which we can see that the peak of the maximum singular value curve is $0.9980 < 2$.

From the numerical simulations (1) and (2), we see that the admissibility and inequality $\|G\|_\infty < \gamma = 2$ for the transfer function are both verified.

Remark 4.4. In Example 4.1 the circuit model is a typical modeling case of the FOSS. From the perspective of circuit physics models, the currents i_1 and i_2 are dynamically constrained due to the presence of superinductors L_1 and L_2, which generate fractional-order differential equation constraints. According to Kirchhoff's current law, the current i_3 is the sum of currents i_1 and i_2, constituting a static equality constraint. From the perspective of the system theory, traditional FOSs only have dynamic fractional-order differential equation constraints. However, for FOSSs, since the fractional-order derivative coefficient matrix E is singular, they include both dynamic fractional-order differential equation constraints and static algebraic equation constraints. Therefore traditional FOSs cannot model this circuit, necessitating the use of FOSSs for modeling.

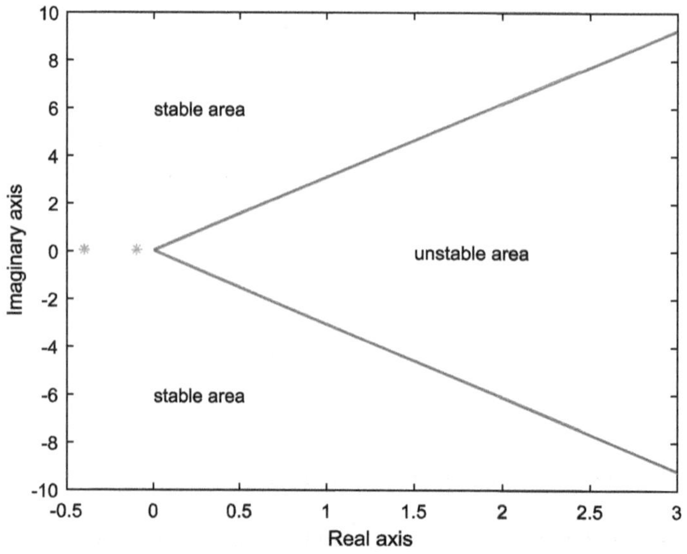

Figure 4.2: The location of the eigenvalues of FOSS (4.5) in Example 4.1.

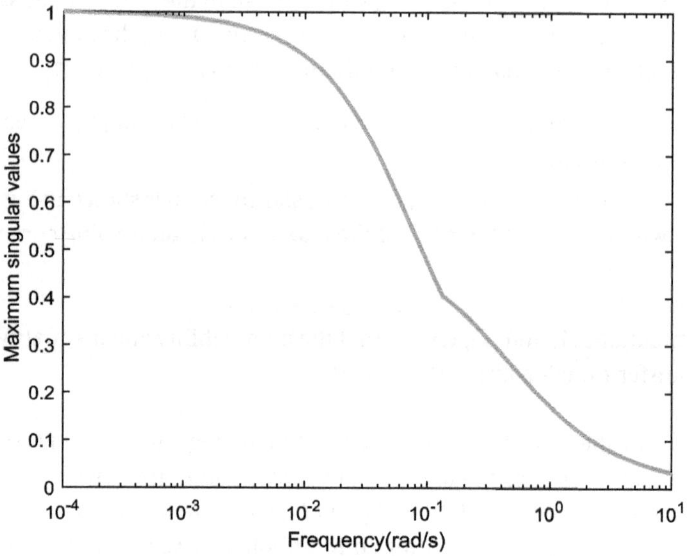

Figure 4.3: Maximum singular values of the FOSS (4.5) in Example 4.1.

Example 4.2. Consider the same circuit model and parameters as in Example 4.1. Here take $\alpha = 1.5$ and $\gamma = 2$. The problem is to determine whether it satisfies the admissibility and H_∞ performance.

By using SCILAB to solve conditions (i) and (ii) in Theorem 4.3, we get the following feasible solutions P and Q, respectively:

$$P = \begin{bmatrix} 1.4156 & -0.0146 - 0.0025\mathbf{i} & 0 \\ -0.0292 + 0.0049\mathbf{i} & 3.6046 & 0 \\ 3.9738 - 6896.1839\mathbf{i} & 2.4586 - 6893.8678\mathbf{i} & 1 + 6923.6122\mathbf{i} \end{bmatrix},$$

$$Q = \begin{bmatrix} 1.5144 & -0.0268 - 0.0084\mathbf{i} & 0 \\ -0.0536 + 0.0168\mathbf{i} & 3.4671 & 0 \\ -22.9008 + 0.7534\mathbf{i} & -25.0299 - 4.9885\mathbf{i} & 22880.6187 \end{bmatrix}.$$

Hence system (4.5) is admissible, and its transfer function satisfies $\|G_N\|_\infty < \gamma = 2$.

When choosing $E_0 = [0 \quad 0 \quad 1]^T$ in Theorem 4.4, the feasible solutions in condition (i) are

$$P = \begin{bmatrix} 0.1872 & 0.0149 + 0.0033\mathbf{i} & -0.1628 - 0.4899\mathbf{i} \\ 0.0149 - 0.0033\mathbf{i} & 0.7121 & 0.2151 - 1.0116\mathbf{i} \\ -0.1628 + 0.4899\mathbf{i} & 0.2151 + 1.0116\mathbf{i} & 41230.4653 \end{bmatrix},$$

$$S = [-9221.1245 \quad -9231.7565 \quad 9246.2187],$$

and the feasible solutions in condition (ii) are

$$Q = \begin{bmatrix} 0.1475 & 0 & 0.0011\mathbf{i} \\ 0 & 0.9226 & 0 \\ -0.0011\mathbf{i} & 0 & 57097.9788 \end{bmatrix},$$

$$S = [-5.295 \quad -16.4496 \quad 41262.8191].$$

Hence system (4.5) is admissible, and its transfer function satisfies $\|G_N\|_\infty < \gamma = 2$.

By numerical simulations we have the following:

(1) The eigenvalues of system (4.5) are $\{-0.4, -0.1\}$, which are all in the stable areas, as we see from Figure 4.4. Meanwhile, (i) $\det(s^\alpha E - A_0)$ is not constant 0, i. e., the system is regular; (ii) Take the left and right maximal zeroed submatrices $L_E = [0 \quad 0 \quad 1]$ and $R_E = \begin{bmatrix} 0 \\ 0 \\ 1 \end{bmatrix}$ of the matrix E. Then $\operatorname{rank}(L_E A_0 R_E) = 1$, i. e., the system is impulse-free. Therefore the FOSS is admissible.

(2) The maximum singular value curve of the system transfer function is shown in Figure 4.5, from which we can see that the peak of the maximum singular value curve is $1.4142 < 2$.

By numerical simulations (1) and (2), the admissibility and H_∞ performance are both verified.

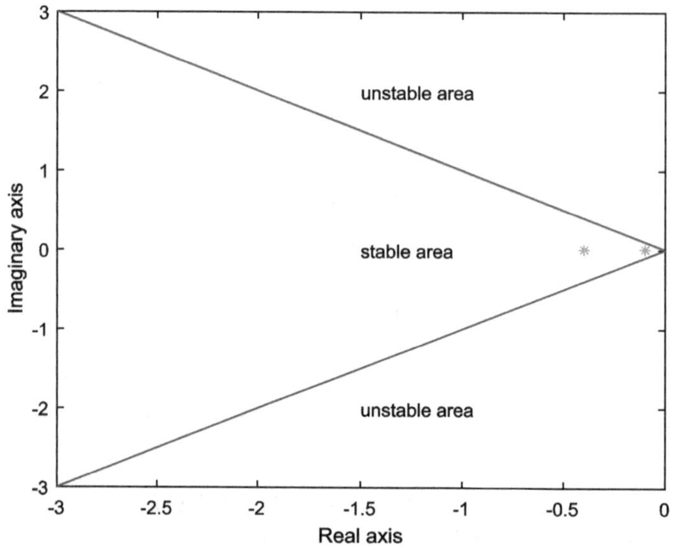

Figure 4.4: The location of the eigenvalues of FOSS (4.5) in Example 4.2.

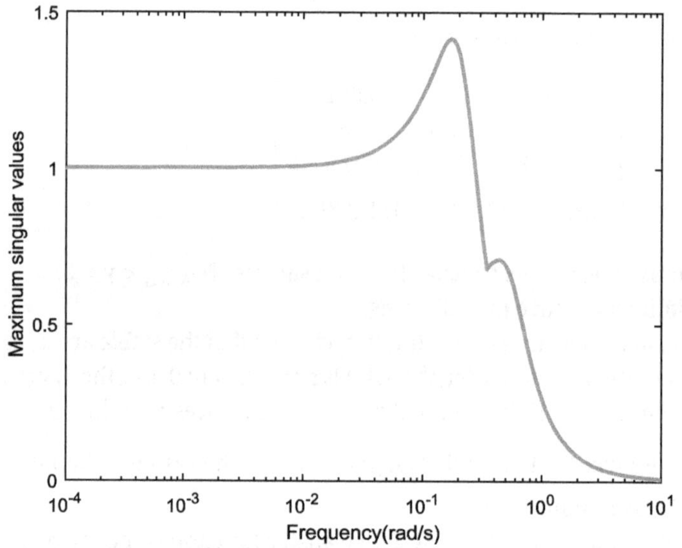

Figure 4.5: Maximum singular values of the FOSS (4.5) in Example 4.2.

4.4 Robust state feedback H_∞ control

For the cases where the system orders of FOSSs are $\alpha \in (0,1)$ and $\alpha \in (1,2)$, this section provides the design methods for the state feedback controller of the closed-loop system (4.3).

4.4.1 Case $\alpha \in (0, 1)$

Theorem 4.5. *Let $\gamma > 0$. The closed-loop system (4.3) satisfies the admissibility, and its transfer function satisfies $\|G_K\|_\infty < \gamma$ if there exist a matrix $Q \in \mathbb{C}^{n\times n}$, a matrix $J \in \mathbb{R}^{m\times n}$, and real scalar constants $\varepsilon_1 > 0$ and $\varepsilon_2 > 0$ satisfying:*

$$EQ = Q^*E^T \geq 0, \tag{4.46}$$

$$\begin{bmatrix} W_0 + \varepsilon_1 \Upsilon_1 \Upsilon_1^T + \varepsilon_2 \Upsilon_3 \Upsilon_3^T & \Upsilon_2^T & \Upsilon_4^T \\ \Upsilon_2 & -\varepsilon_1 I & 0 \\ \Upsilon_4 & 0 & -\varepsilon_2 I \end{bmatrix} < 0, \tag{4.47}$$

where

$$\Upsilon_1 = \begin{bmatrix} D_A & 0 & 0 \\ 0 & D_C & 0 \\ 0 & 0 & G_B^T \end{bmatrix}, \quad \Upsilon_2 = \begin{bmatrix} G_A Y & 0 & 0 \\ G_C Y & 0 & 0 \\ D_B^T & 0 & 0 \end{bmatrix},$$

$$\Upsilon_3 = \begin{bmatrix} D_{B_1} & 0 & 0 \\ 0 & D_{B_2} & 0 \\ 0 & 0 & 0 \end{bmatrix}, \quad \Upsilon_4 = \begin{bmatrix} G_{B_1} J & 0 & 0 \\ G_{B_2} J & 0 & 0 \\ 0 & 0 & 0 \end{bmatrix},$$

$$W_0 = \begin{bmatrix} \operatorname{Sym}\{A_0 Y + B_{10} J\} & (C_0 Y + B_{20} J)^T & B_0 \\ C_0 Y + B_{20} J & -\gamma I & 0 \\ B_0^T & 0 & -\gamma I \end{bmatrix},$$

and $\theta = \frac{\pi}{2}(1-\alpha)$, $Y = e^{j\theta}Q + e^{-j\theta}\bar{Q}$. (Here Y can be considered nonsingular; otherwise, choose a nonsingular real matrix $\hat{Y}$ satisfying $E\hat{Y} = \hat{Y}^T E^T \geq 0$. Then choose a real constant $\xi \in (0, 1)$ such that (i) ξ is not an eigenvalue of the matrix $-Y\hat{Y}^{-1}$; (ii) inequality (4.47) still holds when Y in inequality (4.47) is replaced by $Y + \xi\hat{Y}$. Then the $Y + \xi\hat{Y}$ that meets the conditions is nonsingular.) In (4.2), the controller can be taken $u(t) = JY^{-1}x(t)$.

Proof. Note that $Y = e^{j\theta}Q + e^{-j\theta}\bar{Q}$. We can seen that Y is a real matrix. For $\Gamma \in \{A, B_1, B, C, B_2\}$, $F_\Gamma^T F_\Gamma \leq I$. Denote

$$M_1 = \begin{bmatrix} F_A & 0 & 0 \\ 0 & F_C & 0 \\ 0 & 0 & F_B^T \end{bmatrix}, \quad M_2 = \begin{bmatrix} F_{B_1} & 0 & 0 \\ 0 & F_{B_2}^T & 0 \\ 0 & 0 & 0 \end{bmatrix}, \tag{4.48}$$

and then we have

$$M_1^T M_1 \leq I, \quad M_2^T M_2 < I. \tag{4.49}$$

By Lemma 4.6, for any real scalar constants $\varepsilon_1 > 0$ and $\varepsilon_2 > 0$, we have

$$\operatorname{Sym}\{\Upsilon_1 M_1 \Upsilon_2\} \leq \varepsilon_1 \Upsilon_1 \Upsilon_1^T + \varepsilon_1^{-1} \Upsilon_2^T M_1^T M_1 \Upsilon_2 \leq \varepsilon_1 \Upsilon_1 \Upsilon_1^T + \varepsilon_1^{-1} \Upsilon_2^T \Upsilon_2, \tag{4.50}$$

$$\mathrm{Sym}\{\Upsilon_3 M_2 \Upsilon_4\} \le \varepsilon_2 \Upsilon_3 \Upsilon_3^T + \varepsilon_2^{-1} \Upsilon_4^T M_2^T M_1 \Upsilon_4 < \varepsilon_2 \Upsilon_3 \Upsilon_3^T + \varepsilon_2^{-1} \Upsilon_4^T \Upsilon_4. \tag{4.51}$$

Inequality (4.47) is equivalent to

$$W_0 + \varepsilon_1 \Upsilon_1 \Upsilon_1^T + \varepsilon_1^{-1} \Upsilon_2^T \Upsilon_2 + \varepsilon_{s2} \Upsilon_3 \Upsilon_3^T + \varepsilon_{s2}^{-1} \Upsilon_4^T \Upsilon_4 < 0. \tag{4.52}$$

According to (4.50), (4.51), and (4.52), we have

$$W_0 + \mathrm{Sym}\{\Upsilon_1 M_1 \Upsilon_2\} + \mathrm{Sym}\{\Upsilon_3 M_2 \Upsilon_4\} < 0. \tag{4.53}$$

Substituting W_0, Υ_1, M_1, Υ_2, Υ_3, M_2, and Υ_4 into inequality (4.53), we have

$$\begin{bmatrix} \mathrm{Sym}\{AY + B_1 J\} & (CY + B_2 J)^T & B \\ CY + B_2 J & -\gamma I & 0 \\ B^T & 0 & -\gamma I \end{bmatrix} < 0. \tag{4.54}$$

The controller gain is $K = JY^{-1}$. Then $J = KY$. Substituting J into inequality (4.54), we have

$$\begin{bmatrix} \mathrm{Sym}\{(A + B_1 K)Y\} & ((C + B_2 K)Y)^* & B \\ (C + B_2 K)Y & -\gamma I & 0 \\ B^T & 0 & -\gamma I \end{bmatrix} < 0.$$

Comparing the transfer function G_N in (4.6) with the transfer function G_K in (4.4) and applying condition (ii) in Theorem 4.1, we end the proof of the theorem. □

Remark 4.5. In Theorem 4.5 a method of designing the state feedback H_∞ controller is proposed, and the reason why Y in the controller gain can be considered as a nonsingular matrix is also given. Here a method for solving $\hat{Y}$ and ξ in Theorem 4.5 is presented. When the solved Y is a singular matrix, a nonsingular real matrix $\hat{Y}$ satisfying $E\hat{Y} = \hat{Y}^T E^T \ge 0$ can be obtained by the SCILAB software. Note that the matrix $-Y\hat{Y}^{-1}$ has a finite number of eigenvalues. For a real number $\xi \in (0, 1)$, a feasible domain that does not include the eigenvalues of the matrix $-Y\hat{Y}^{-1}$ can always be found, assuming it as $0 < \lambda_1 < \xi < \lambda_2 < 1$. Then we can take $\hat{Y}$, Y, J, ε_1, and ε_2 into the following constraints:

$$\begin{gathered} \xi < \lambda_2, \\ \xi > \lambda_1, \\ \begin{bmatrix} W_0 + \varepsilon_1 \Upsilon_1 \Upsilon_1^T + \varepsilon_2 \Upsilon_3 \Upsilon_3^T & \Upsilon_2^T & \Upsilon_4^T \\ \Upsilon_2 & -\varepsilon_1 I & 0 \\ \Upsilon_4 & 0 & -\varepsilon_2 I \end{bmatrix} < 0, \end{gathered} \tag{4.55}$$

where

$$\Upsilon_1 = \begin{bmatrix} D_A & 0 & 0 \\ 0 & D_C & 0 \\ 0 & 0 & G_B^T \end{bmatrix}, \quad \Upsilon_2 = \begin{bmatrix} G_A(Y+\xi\hat{Y}) & 0 & 0 \\ G_C(Y+\xi\hat{Y}) & 0 & 0 \\ D_B^T & 0 & 0 \end{bmatrix},$$

$$\Upsilon_3 = \begin{bmatrix} D_{B_1} & 0 & 0 \\ 0 & D_{B_2} & 0 \\ 0 & 0 & 0 \end{bmatrix}, \quad \Upsilon_4 = \begin{bmatrix} G_{B_1}J & 0 & 0 \\ G_{B_2}J & 0 & 0 \\ 0 & 0 & 0 \end{bmatrix},$$

$$W_0 = \begin{bmatrix} \mathrm{Sym}\{A_0(Y+\xi\hat{Y}) + B_{10}J\} & (C_0(Y+\xi\hat{Y}) + B_{20}J)^T & B_0 \\ C_0(Y+\xi\hat{Y}) + B_{20}J & -\gamma I & 0 \\ B_0^T & 0 & -\gamma I \end{bmatrix}.$$

We can see that inequality (4.55) is the LMIs with respect to ξ, and a feasible solution can be found using the SCILAB software. Then Y can be replaced by $Y + \xi\hat{Y}$, which, as we can see, is nonsingular.

Theorem 3.5 contains an equality constraint. As stated in Remark 4.1, from the point of view of numerical computation, we need results without equality constraints, which we give by the following:

Theorem 4.6. *Let $\gamma > 0$. The closed-loop system* (4.3) *satisfies the admissibility, and its transfer function satisfies $\|G_K\|_\infty < \gamma$ if there exist matrices $Q = Q^* > 0$, $Q \in \mathbb{C}^{n\times n}$, $J \in \mathbb{R}^{m\times n}$, and $S \in \mathbb{R}^{(n-r)\times n}$ and real scalar constants $\varepsilon_1 > 0$ and $\varepsilon_2 > 0$ satisfying*

$$\begin{bmatrix} W_0 + \varepsilon_1 \Upsilon_1\Upsilon_1^T + \varepsilon_2\Upsilon_3\Upsilon_3^T & \Upsilon_2^T & \Upsilon_4^T \\ \Upsilon_2 & -\varepsilon_1 I & 0 \\ \Upsilon_4 & 0 & -\varepsilon_2 I \end{bmatrix} < 0, \tag{4.56}$$

where

$$\Upsilon_1 = \begin{bmatrix} D_A & 0 & 0 \\ 0 & D_C & 0 \\ 0 & 0 & G_B^T \end{bmatrix}, \quad \Upsilon_2 = \begin{bmatrix} G_A Y & 0 & 0 \\ G_C Y & 0 & 0 \\ D_B^T & 0 & 0 \end{bmatrix},$$

$$\Upsilon_3 = \begin{bmatrix} D_{B_1} & 0 & 0 \\ 0 & D_{B_2} & 0 \\ 0 & 0 & 0 \end{bmatrix}, \quad \Upsilon_4 = \begin{bmatrix} G_{B_1}J & 0 & 0 \\ G_{B_2}J & 0 & 0 \\ 0 & 0 & 0 \end{bmatrix},$$

$$W_0 = \begin{bmatrix} \mathrm{Sym}\{A_0 Y + B_{10}J\} & (C_0 Y + B_{20}J)^T & B_0 \\ C_0 Y + B_{20}J & -\gamma I & 0 \\ B_0^T & 0 & -\gamma I \end{bmatrix},$$

and $\theta = \frac{\pi}{2}(1-\alpha)$, $Y = (e^{j\theta}Q + e^{-j\theta}\bar{Q})E^T + E_0 S$, with E_0 any matrix of full column rank satisfying $EE_0 = 0$. (Here Y can be considered nonsingular. Otherwise, choose a nonsingular real matrix $\hat{Y}$ such that it satisfies $E\hat{Y} = \hat{Y}^T E^T \geq 0$. Then choose a real constant $\xi \in (0,1)$ such that (i) *ξ is not an eigenvalue of the matrix $-Y\hat{Y}^{-1}$;* (ii) *Inequality* (4.56) *still holds when Y in inequality* (4.56) *is replaced by $Y + \xi\hat{Y}$. Then $Y + \xi\hat{Y}$ that meets the conditions is nonsingular.) The controller in* (4.2) *can be obtained as $u(t) = JY^{-1}x(t)$.*

Proof. The proof of this theorem is similar to that of Theorem 4.5 and can be completed by using Theorem 4.2, so we omit it. □

4.4.2 Case $\alpha \in (1, 2)$

Theorem 4.7. *Let $\gamma > 0$. The closed-loop system* (4.3) *is admissible, and its transfer function satisfies $\|G_K\|_\infty < \gamma$ if there exist a nonsingular matrix $Q \in \mathbb{R}^{n\times n}$, a matrix $J \in \mathbb{R}^{m\times n}$, and real scalar constants $\varepsilon_1 > 0$ and $\varepsilon_2 > 0$ satisfying*

$$EQ = Q^* E^T \geq 0, \tag{4.57}$$

$$\begin{bmatrix} W_0 + \varepsilon_1 \Upsilon_1 \Upsilon_1^T + \varepsilon_2 \Upsilon_3 \Upsilon_3^T & \Upsilon_2^T & \Upsilon_4^T \\ \Upsilon_2 & -\varepsilon_1 I & 0 \\ \Upsilon_4 & 0 & -\varepsilon_2 I \end{bmatrix} < 0, \tag{4.58}$$

where

$$\Upsilon_1 = \begin{bmatrix} D_A & 0 & 0 \\ 0 & D_C & 0 \\ 0 & 0 & G_B^T \end{bmatrix}, \quad \Upsilon_2 = \begin{bmatrix} G_A Y & 0 & 0 \\ G_C Y & 0 & 0 \\ D_B^T & 0 & 0 \end{bmatrix},$$

$$\Upsilon_3 = \begin{bmatrix} D_{B_1} & 0 & 0 \\ 0 & D_{B_2} & 0 \\ 0 & 0 & 0 \end{bmatrix}, \quad \Upsilon_4 = \begin{bmatrix} G_{B_1} J & 0 & 0 \\ G_{B_2} J & 0 & 0 \\ 0 & 0 & 0 \end{bmatrix},$$

$$W_0 = \begin{bmatrix} \mathrm{Sym}\{A_0 Y + B_{10} \hat{J}\} & (C_0 Y + B_{20} \hat{J})^T & B_0 \\ C_0 Y + B_{20} \hat{J} & -\gamma I & 0 \\ B_0^T & 0 & -\gamma I \end{bmatrix},$$

and $\theta = \frac{\pi}{2}(1 - \alpha)$, $Y = e^{j\theta} Q$, $\hat{J} = e^{j\theta} J$. Then the controller in (4.2) *can be designed as $u(t) = JQ^{-1}x(t)$.*

Proof. The proof of this theorem is similar to that of Theorem 4.5 and can be completed by using Theorem 4.3. Here we omit it. □

In Theorem 4.7 the result has the equality constraints. We provide a result without equality constraints.

Theorem 4.8. *Let $\gamma > 0$. The closed-loop system* (4.3) *is admissible, and its transfer function satisfies $\|G_K\|_\infty < \gamma$ if there exist a nonsingular matrix $Q = Q^T > 0$, $Q \in \mathbb{R}^{n\times n}$, matrices $J \in \mathbb{R}^{m\times n}$ and $S \in \mathbb{R}^{(n-r)\times n}$, and real scalar constants $\varepsilon_1 > 0$ and $\varepsilon_2 > 0$ satisfying*

$$\begin{bmatrix} W_0 + \varepsilon_1 \Upsilon_1 \Upsilon_1^T + \varepsilon_2 \Upsilon_3 \Upsilon_3^T & \Upsilon_2^T & \Upsilon_4^T \\ \Upsilon_2 & -\varepsilon_1 I & 0 \\ \Upsilon_4 & 0 & -\varepsilon_2 I \end{bmatrix} < 0, \tag{4.59}$$

where

$$\Upsilon_1 = \begin{bmatrix} D_A & 0 & 0 \\ 0 & D_C & 0 \\ 0 & 0 & G_B^T \end{bmatrix}, \quad \Upsilon_2 = \begin{bmatrix} G_A Y & 0 & 0 \\ G_C Y & 0 & 0 \\ D_B^T & 0 & 0 \end{bmatrix},$$

$$\Upsilon_3 = \begin{bmatrix} D_{B_1} & 0 & 0 \\ 0 & D_{B_2} & 0 \\ 0 & 0 & 0 \end{bmatrix}, \quad \Upsilon_4 = \begin{bmatrix} G_{B_1}\hat{J} & 0 & 0 \\ G_{B_2}\hat{J} & 0 & 0 \\ 0 & 0 & 0 \end{bmatrix},$$

$$W_0 = \begin{bmatrix} \mathrm{Sym}\{A_0 Y + B_{10}\hat{J}\} & (C_0 Y + B_{20}\hat{J})^T & B_0 \\ C_0 Y + B_{20}\hat{J} & -\gamma I & 0 \\ B_0^T & 0 & -\gamma I \end{bmatrix},$$

and $\theta = \frac{\pi}{2}(1-\alpha)$, $Y = e^{j\theta}L$, $L = QE^T + E_0 S$, $\hat{J} = e^{j\theta}J$, *with* E_0 *an arbitrary column full-rank matrix satisfying* $EE_0 = 0$. *(Here* L *can be considered nonsingular. Otherwise, choose a nonsingular real matrix* $\hat{L}$ *satisfying* $E\hat{L} = \hat{L}^T E^T$. *Then choose a real constant* $\xi \in (0,1)$ *such that* (i) ξ *is not an eigenvalue of the matrix* $-L\hat{L}^{-1}$; (ii) *Inequality* (4.59) *still holds when* L *in inequality* (4.56) *is replaced by* $L + \xi\hat{L}$. *Then* $L + \xi\hat{L}$ *that meets the conditions is nonsingular.) The controller in* (4.2) *can be obtained as* $u(t) = JL^{-1}x(t)$.

Proof. The proof can be completed by Theorem 4.4 and is similar to that of Theorem 4.5. Here it is omitted. □

Now we use numerical examples to verify the validity of the results of this section. Example 4.3 compares the results with those in the literature [45] to highlight the superiority of the results. Example 4.4 is used to verify the validity of Theorems 4.5–4.6. The verification of Theorems 4.7–4.8 is similar to that of Example 4.4 and is omitted.

Example 4.3. Consider the FOSS (4.1) with the following parameters:

$$E = \begin{bmatrix} 2 & 0 & 0 \\ 0 & 5 & 0 \\ 0 & 0 & 0 \end{bmatrix}, \quad A_0 = \begin{bmatrix} 10 & 4 & -1 \\ 4 & 10 & -2 \\ 0 & 1 & -1 \end{bmatrix}, \quad B_{10} = B_0 = \begin{bmatrix} 1 \\ 1 \\ 0 \end{bmatrix}, \tag{4.60}$$

$C_0 = [0\,1\,0]$, $\alpha = 0.8$, $\gamma = 1$, $B_{20} = 0$, D_Γ, $G_\Gamma = 0$, and $\Gamma \in \{A, B, C, B_1, B_2\}$.

The purposes are to judge whether the closed-loop system satisfies (i) the L_∞ performance and (ii) the admissibility and H_∞ performance. By using Theorem 6 in [45], we obtain the following feasible solutions:

$$P = \begin{bmatrix} -2283.6093 & 1882.3589 & 0 \\ -0.1718 & -8.0384 & 0 \\ -0.1716 & -8.0386 & 1 \end{bmatrix}, \quad Q = \begin{bmatrix} 0 & -16070.6470 & 0 \end{bmatrix}.$$

The gain of the state feedback controller is

$$K = QP^{-1} = \begin{bmatrix} -0.1478 & 1964.6242 & 0 \end{bmatrix}. \tag{4.61}$$

From the point of view of numerical simulations, the eigenvalues of system (4.1) are 2.9902 and 396.4607, which are both within the unstable area as plotted in Figure 4.6. The maximum singular value curve of the transfer function is shown in Figure 4.7, where the peak value is less than $\gamma = 1$. From the numerical simulation results, we can see that the method in [45] fails to ensure the admissibility of FOSSs, and it only can ensure its L_∞ performance index.

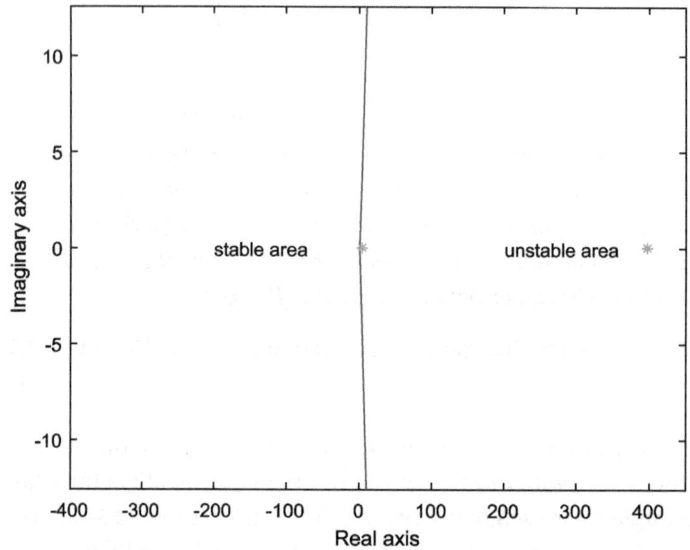

Figure 4.6: The location of the eigenvalues of the FOSS (4.1) with the effect of controller (4.61) in Example 4.3.

By applying Theorem 4.5, feasible solutions can be acquired:

$$Q = \begin{bmatrix} 176.8137 & 183.948 - 3.2653i & 0 \\ 92 + 1.3061i & 77.6422 & 0 \\ 0 & 6056.3139 & 6056.8246 \end{bmatrix},$$

$$J = \begin{bmatrix} -25107.4021 & -2253.5400 & 22898.5504 \end{bmatrix}.$$

The gain of the state feedback controller is

$$K = \begin{bmatrix} -296.9938 & 537.3921 & 1.9876 \end{bmatrix}. \tag{4.62}$$

From the point of view of numerical simulations, the eigenvalues of system (4.1) are −27.3751 and −6.6459, which are both within the stable area, as plotted in Figure 4.8. The

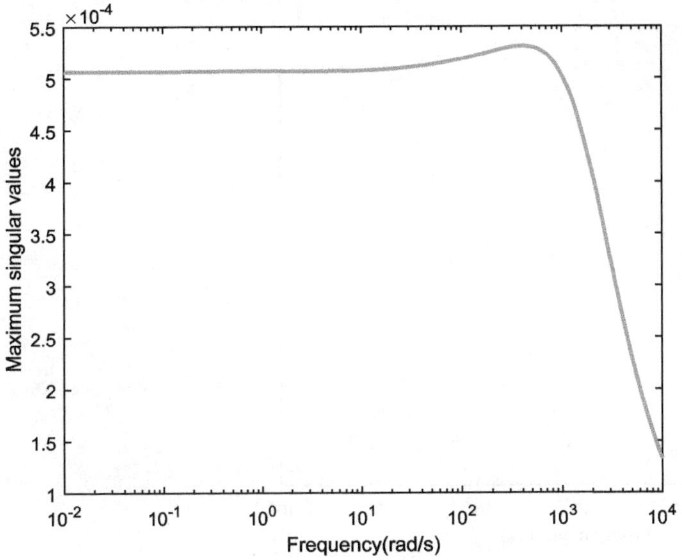

Figure 4.7: Maximum singular values of the FOSS (4.1) with the effect of controller (4.61) in Example 4.3.

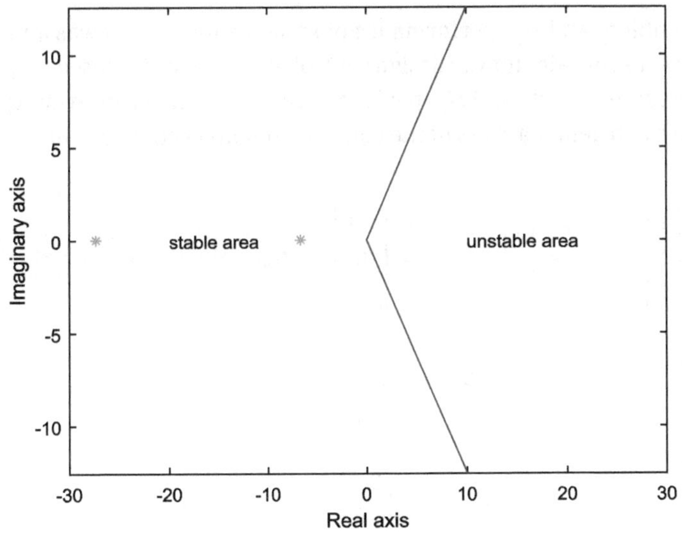

Figure 4.8: The location of the eigenvalues of the FOSS (4.1) with the effect of controller (4.62) in Example 4.3.

maximum singular value curve of the transfer function is shown in Figure 4.9, where the peak value is less than $\gamma = 1$. Based on the numerical simulation results, we can show that the controller design method in Theorem 4.5 can simultaneously ensure that the closed-loop system satisfies both the admissibility and H_∞ performance.

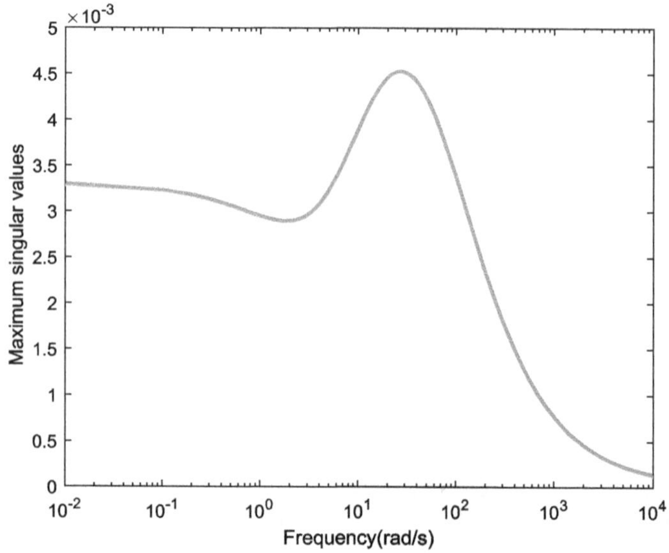

Figure 4.9: Maximum singular values of the FOSS (4.1) with the effect of controller (4.62) in Example 4.3.

Example 4.4. The admissibility and H_∞ performance of system with $\alpha = 0.8$ was analyzed in Example 4.1. Here we consider how to design controllers to make the closed-loop system (4.3) satisfy robust admissibility and H_∞ performance when there is uncertainty in the model. Choose the coefficient matrices of the control input $u(t)$ and uncertain parameter matrices as

$$\begin{gathered} B_{10} = \begin{bmatrix} 1 \\ 1 \\ 1 \end{bmatrix}, \quad B_{20} = \begin{bmatrix} 1 \\ 1 \end{bmatrix}, \quad D_{B1} = \begin{bmatrix} 0.1 \\ 0 \\ 0.1 \end{bmatrix}, \quad G_{B1} = 0.1, \\ D_{B2} = \begin{bmatrix} 0.1 \\ 0.1 \end{bmatrix}, \quad G_{B2} = 0.2, \quad D_A = \begin{bmatrix} 1.1 \\ 0.1 \\ 0 \end{bmatrix}, \quad G_A = \begin{bmatrix} 1 \\ 0.1 \\ 0 \end{bmatrix}^T, \\ D_B = \begin{bmatrix} 0.1 \\ 0.1 \\ 0.1 \end{bmatrix}, \quad G_B = \begin{bmatrix} 0.1 \\ 0.1 \end{bmatrix}^T, \quad D_C = \begin{bmatrix} 0.1 \\ 0.1 \end{bmatrix}, \quad G_C = \begin{bmatrix} 0.1 \\ 0.1 \\ 0.1 \end{bmatrix}^T. \end{gathered} \tag{4.63}$$

Firstly, we analyze the admissibility and H_∞ performance of the open-loop system. We can verify that the uncertain system (4.1) cannot satisfy both the admissibility and H_∞ performance when the control input $u(t) = 0$. For example, choose $F_A = F_B = F_C = 1$ and

$$A = \begin{bmatrix} 0.1 & 0.11 & 0 \\ 0.1 & -1.99 & 0 \\ 1 & 1 & -1 \end{bmatrix}, \quad B = \begin{bmatrix} 1.01 & 0.01 \\ 0.01 & 1.01 \\ 0.01 & 0.01 \end{bmatrix}, \quad C = \begin{bmatrix} 0.01 & 1.01 & 0.01 \\ 1.01 & 0.01 & 0.01 \end{bmatrix}. \tag{4.64}$$

Then the eigenvalues of the FOSS are −0.3985 and 0.0105, both in the unstable area, as plotted in Figure 4.10. The maximum singular value curve of the transfer function is shown in Figure 4.11, where the peak value is 10.3237 > 2. Hence its admissibility and H_∞ of FOSSs cannot be ensured here.

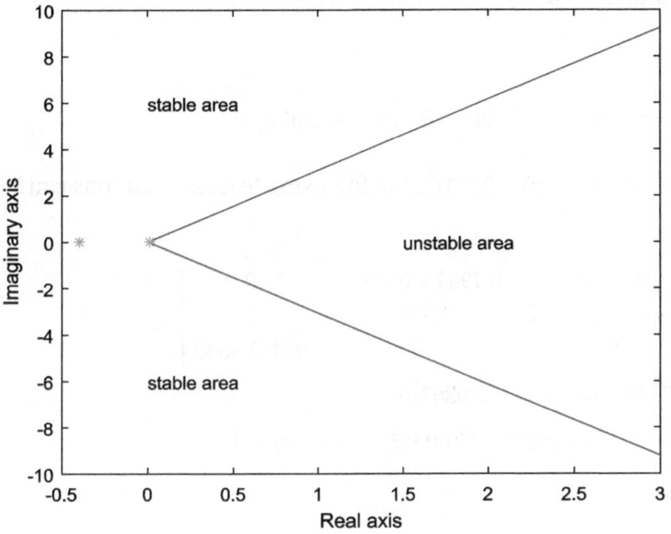

Figure 4.10: The location of the eigenvalues of the open-loop FOSS (4.1) in Example 4.4.

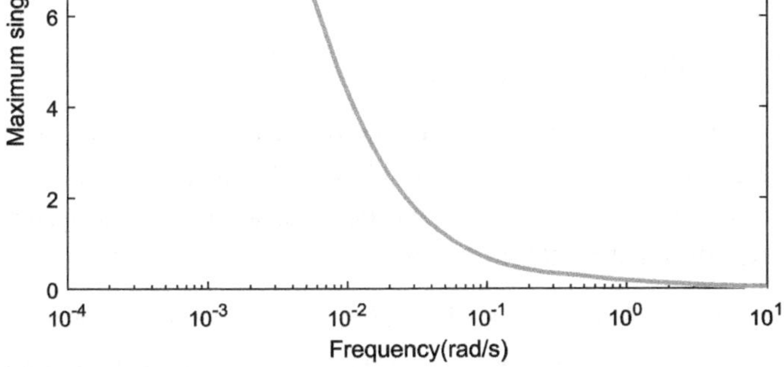

Figure 4.11: Maximum singular values of the open-loop FOSS (4.1) in Example 4.4.

By applying Theorem 4.5, feasible solutions can be acquired:

$$Q = \begin{bmatrix} 0.8742 & 0.9315 - 0.0181\mathbf{i} & 0 \\ 1.8631 + 0.0362\mathbf{i} & 7.0533 & 0 \\ -2.7486 & -8.0362 & 52.4299 \end{bmatrix},$$

$$J = \begin{bmatrix} -3.4052 & -8.1786 & -0.9335 \end{bmatrix}, \quad \varepsilon_1 = \varepsilon_2 = 1.$$

Hence the H_∞ control is

$$u(t) = Kx(t) = \begin{bmatrix} -1.0629 & -0.479 & -0.0094 \end{bmatrix} x(t).$$

By Theorem 4.6, choosing $E_0 = \begin{bmatrix} 0 & 0 & 1 \end{bmatrix}^T$, the following feasible solutions can be acquired:

$$Q = \begin{bmatrix} 0.0964 & 0.1955 + 0.0504\mathbf{i} & 0 \\ 0.1955 - 0.0504\mathbf{i} & 1.3834 & 0 \\ 0 & 0 & 98473.9542 \end{bmatrix},$$

$$J = \begin{bmatrix} -3.7731 & -8.087 & -0.9367 \end{bmatrix},$$

$$S = \begin{bmatrix} -6.6367 & -16.6698 & 101.0332 \end{bmatrix}, \quad \varepsilon_1 = \varepsilon_2 = 1.$$

Then the H_∞ control is

$$u(t) = Kx(t) = \begin{bmatrix} -0.9987 & -0.4971 & -0.0093 \end{bmatrix} x(t).$$

To verify the effectiveness of the designed state feedback H_∞ controller, we choose a set of parameters in the uncertain system and illustrate with the result obtained by Theorem 4.5. For example, we take $F_A = F_B = F_C = F_{B1} = F_{B2} = 1$ and

$$\begin{aligned} A &= \begin{bmatrix} 0.1 & 0.11 & 0 \\ 0.1 & -1.99 & 0 \\ 1 & 1 & -1 \end{bmatrix}, \quad B = \begin{bmatrix} 1.01 & 0.01 \\ 0.01 & 1.01 \\ 0.01 & 0.01 \end{bmatrix}, \\ C &= \begin{bmatrix} 0.01 & 1.01 \\ 1.01 & 0.01 \\ 0.01 & 0.01 \end{bmatrix}^T, \quad B_1 = \begin{bmatrix} 1.01 \\ 1 \\ 1.01 \end{bmatrix}, \quad B_2 = \begin{bmatrix} 1.02 \\ 1.02 \end{bmatrix}. \end{aligned} \tag{4.65}$$

Under the state-feedback control in Theorem 4.5, the eigenvalues of system (4.1) with parameters (4.65) are −0.5123 and −0.0797. As shown in Figure 4.12, the eigenvalues are all within the stable areas. Meanwhile, (i) $\det(s^\alpha E - A)$ is not constant 0, i. e., the system is regular; (ii) Take the left and right maximal zeroed submatrices $L_E = \begin{bmatrix} 0 & 0 & 1 \end{bmatrix}$, $R_E = \begin{bmatrix} 0 \\ 0 \\ 1 \end{bmatrix}$ of matrix E; then $\operatorname{rank}(L_E A R_E) = 1$, i. e., the system is impulse-free. Therefore the FOSS is admissible. In addition, the maximum singular value curve of the system transfer function is shown in Figure 4.13, from which we can see that the peak of the maximum singular value curve is 1.6249 < 2. Hence, under the designed controller, the closed-loop system satisfies both the admissibility and H_∞ performance.

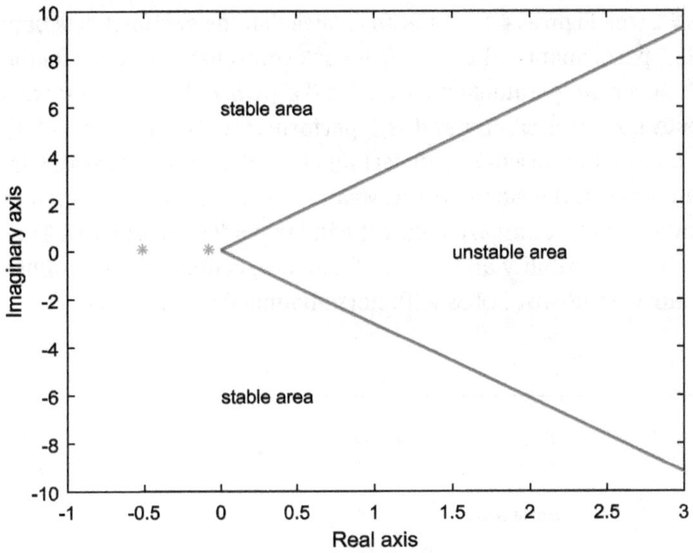

Figure 4.12: The location of the eigenvalues of the closed-loop FOSS (4.3) with parameters (4.65) in Example 4.4.

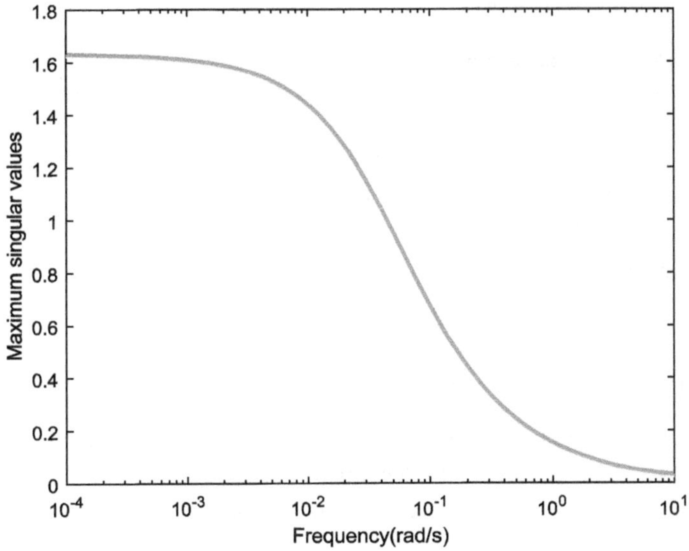

Figure 4.13: Maximum singular values of the closed-loop FOSS (4.3) with parameters (4.65) in Example 4.4.

Remark 4.6. In Example 4.4, we presented a numerical case of the FOSS with norm-bounded uncertainties and demonstrated that the system does not meet the admissibility or H_∞ performance with uncertain parameters $F_\Gamma = 1$, $\Gamma \in \{A, B, C\}$. However, when $F_\Gamma \in (-1, 1)$, some eigenvalues of the system are in the unstable areas, and some curves exceed the threshold $\gamma = 2$ (see Figures 4.14–4.15). To ensure that the system meets both the admissibility and H_∞ performance, the state feedback controllers are provided in Theorems 4.5–4.6. With uncertain parameters $F_\Gamma = 1$ for $\Gamma \in \{A, B, C, B_1, B_2\}$, the system successfully satisfies both the admissibility and H_∞ performance. When $F_\Gamma \in (-1, 1)$, the eigenvalue distributions of the closed-loop uncertain FOSS, shown in Figure 4.16, indicate that all eigenvalues are in the stable areas. Meanwhile, the maximum singular value curves demonstrate that all values are below the index $\gamma = 2$ in Figure 4.17. Thus the system meets both the admissibility and H_∞ performance, confirming the robustness of our designed controller against FOSSs with norm-bounded uncertainties.

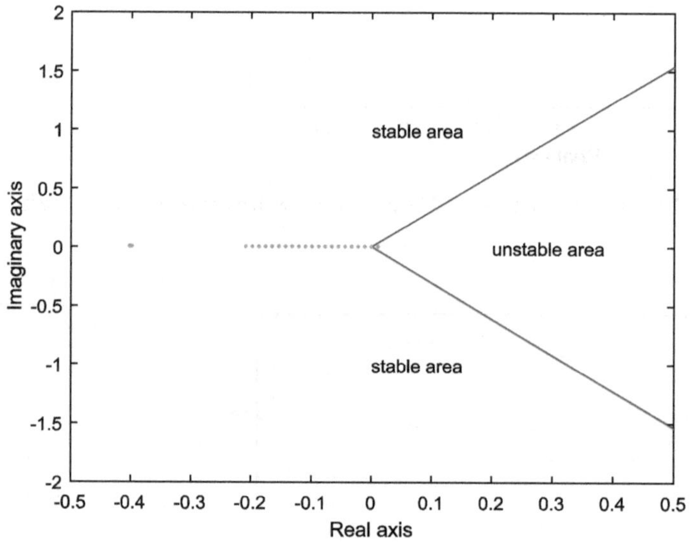

Figure 4.14: The distribution of the eigenvalues of the open-loop uncertain FOSS (4.1) in Example 4.4.

4.5 Chapter summary

This chapter addresses the robust H_∞ control problem for FOSSs with norm-bounded uncertainties. The results give the following main contributions:

(1) For nominal FOSSs with $\alpha \in (0, 1)$ and $\alpha \in [1, 2)$, bounded real lemmas in the sense of the H_∞ norm are provided, with the results for $\alpha \in [1, 2)$ being both sufficient and necessary. In contrast to the bounded real lemmas in the sense of the L_∞ norm in [45], which do not guarantee admissibility, the bounded real lemmas presented

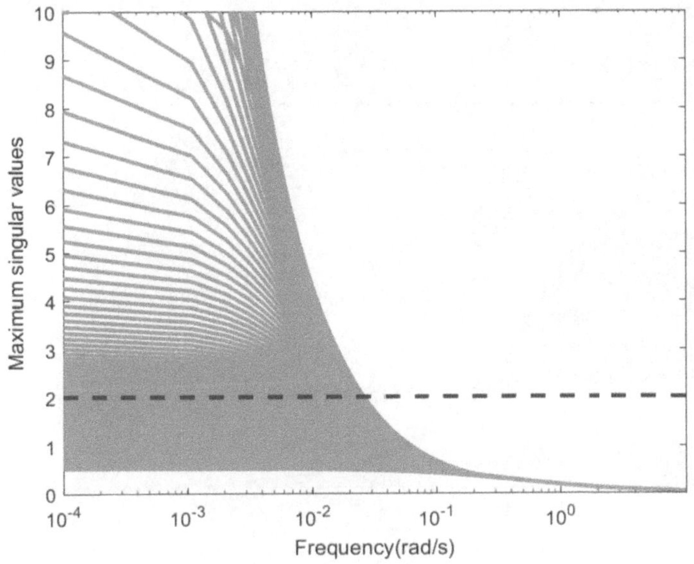

Figure 4.15: The distribution of maximum singular values of the open-loop uncertain FOSS (4.1) in Example 4.4.

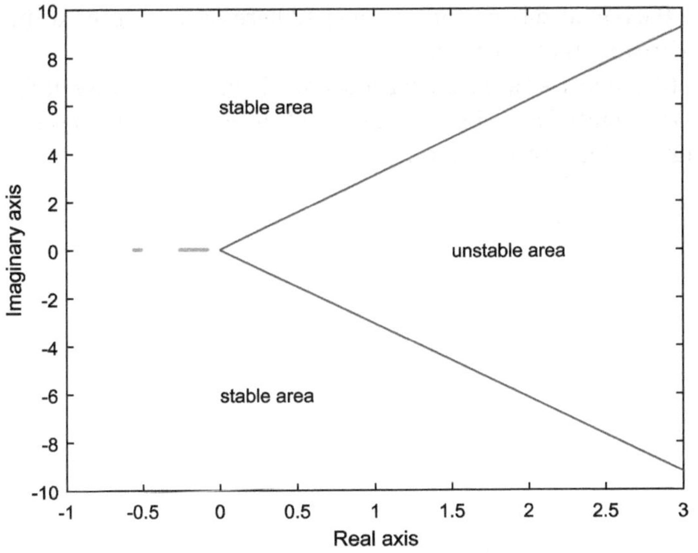

Figure 4.16: The distribution of the eigenvalues of the closed-loop uncertain FOSS (4.1) in Example 4.4.

in this paper are formulated in the sense of the H_∞ norm. This not only captures the H_∞ performance but also ensures the admissibility. Furthermore, when the coefficient matrix E is taken as the unit matrix, the results align with those in [20, 46],

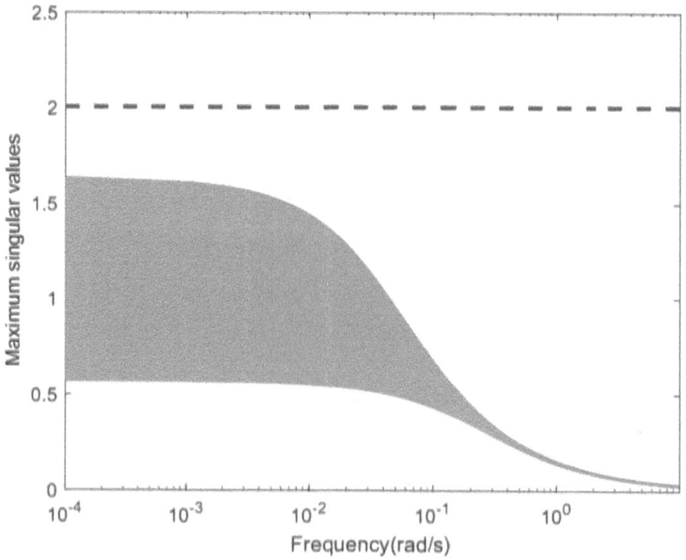

Figure 4.17: The distribution of maximum singular values of the closed-loop uncertain FOSS (4.1) in Example 4.4.

demonstrating that the bounded real lemmas proposed here are more general and applicable across a broader range of cases.

(2) For FOSSs with $\alpha \in (0,1)$ and $\alpha \in [1,2)$ with norm-bounded uncertainties, we design the state feedback H_∞ controllers, which enable the closed-loop system to satisfy both the admissibility and H_∞ performance.

5 Robust positive real control

Abstract: We study robust positive real control problems of uncertain fractional-order singular systems (FOSSs). Based on the Kalman–Yakubovich–Popov (KYP) lemma and the linear matrix inequality (LMI) method, we give conditions for the extended strictly positive realness of nominal fractional-order systems (FOSs). By using these results, we propose positive real lemmas of nominal FOSSs, which can ensure both the admissibility and extended strictly positive realness, and for FOSSs with norm-bounded uncertainties, we design the state feedback positive real controllers so that the closed-loop systems satisfy both the robust admissibility and extended strictly positive realness.

5.1 Introduction

Positive real control is an important issue in the field of robust control, which plays a meaningful role in system stability [90], hyper-stability [80], passivity [25], controller design [102], and so on. In [91] the definitions of positive realness, strictly positive realness and extended strictly positive realness are explicitly put forward in the context of state-space systems. Then some research results on the positive realness of integer-order systems (IOSs) have been proposed successively [99, 111, 82, 23, 24]. For the positive realness of FOSs, the work [123] gives the necessary and sufficient conditions for the positive realness of FOSs based on the generalized Kalman–Yakubovich–Popov (GKYP) lemma and designs the corresponding state feedback positive real controllers. To the authors' best knowledge, there are no research results on the extended strictly positive realness of FOSSs because of some technological challenges. On one hand, when deriving the conditions for the extended strictly positive realness, not only the strictly positive realness needs to be considered, but also the asymptotical stability needs to be taken into account. Conditions for the extended strictly positive realness are more complex than those for positive realness, which cannot ensure the asymptotical stability. For example, the results obtained in [123] can only ensure the positive realness of FOSs, but the asymptotical stability cannot be ensured. On the other hand, for normal FOSs, only the stability demands are to be considered. However, for FOSSs, since the system state derivative coefficient matrix may be singular, the regularity, nonimpulsiveness, and admissibility also need to be taken into account.

In this chapter, we study the robust positive real control problems of FOSSs with norm-bounded uncertainties. Positive real lemmas of nominal FOSSs are presented. Then a state feedback positive real controller is designed for FOSSs with norm-bounded uncertainties, ensuring that the closed-loop systems simultaneously satisfy the robust admissibility and extended strictly positive realness. Numerical examples are provided in each section to illustrate the effectiveness of the proposed results.

https://doi.org/10.1515/9783111628646-005

5.2 Preliminaries

Consider the following FOSS with norm-bounded uncertainties:

$$
\begin{aligned}
ED^{\alpha}x(t) &= Ax(t) + B_1u(t) + Bw(t),\\
z(t) &= Cx(t) + B_2u(t) + Dw(t),
\end{aligned}
\tag{5.1}
$$

where $x(t) \in \mathbb{R}^n$, $z(t) \in \mathbb{R}^q$, $w(t) \in \mathbb{R}^p$, and $u(t) \in \mathbb{R}^m$ are the state vector, control output, disturbance input, and control input, $\alpha \in (0,2)$ is the commensurate order, $E \in \mathbb{R}^{n\times n}$ is the known real matrix with $\operatorname{rank}(E) = r \le n$, D^{α} represents the Caputo fractional-order derivative, and A, B, C, D, B_1, B_2 are uncertain matrices satisfying

$$
\begin{aligned}
A &= A_0 + \Delta A, \quad B = B_0 + \Delta B, \quad B_1 = B_{10} + \Delta B_1,\\
C &= C_0 + \Delta C, \quad B_2 = B_{20} + \Delta B_2, \quad D = D_0 + \Delta D,\\
\Delta\Gamma &= D_{\Gamma}F_{\Gamma}G_{\Gamma}, \quad \Gamma \in \{A, B, C, D, B_1, B_2\},
\end{aligned}
$$

where $A_0, B_0, C_0, D_0, B_{10}, B_{20}, D_{\Gamma}, G_{\Gamma}$ are the known real matrices, and F_{Γ} is a matrix with norm-bounded uncertainties satisfying $F_{\Gamma}^T F_{\Gamma} \le I$. Consider the state feedback controller

$$
u(t) = Kx(t). \tag{5.2}
$$

Using the state feedback controller (5.2) to control system (5.1), we have

$$
\begin{aligned}
ED^{\alpha}x(t) &= (A + B_1K)x(t) + Bw(t),\\
z(t) &= (C + B_2K)x(t) + Dw(t).
\end{aligned}
\tag{5.3}
$$

The transfer function is

$$
G_K(s) = (C + B_2K)(s^{\alpha}E - (A + B_1K))^{-1}B + D. \tag{5.4}
$$

When the uncertain parameters and control input $u(t)$ in system (5.1) are both 0, the system can be described by the following nominal FOSS:

$$
\begin{aligned}
ED^{\alpha}x(t) &= A_0x(t) + B_0w(t),\\
z(t) &= C_0x(t) + D_0w(t).
\end{aligned}
\tag{5.5}
$$

The transfer function is

$$
G_N(s) = C_0(s^{\alpha}E - A_0)^{-1}B_0 + D_0. \tag{5.6}
$$

Definition 5.1 ([123]). (1) System (5.5) is positive real if its transfer function (5.6) is analytic in $\operatorname{Re}(s) > 0$ and satisfies $\operatorname{Sym}\{G(s)\} > 0$ for $\operatorname{Re}(s) > 0$.
(2) System (5.5) is strictly positive real if its transfer function (5.6) is analytic in $\operatorname{Re}(s) \ge 0$ and satisfies $\operatorname{Sym}\{G(jw)\} > 0$ for $w \ge 0$.
(3) System (5.5) is extended strictly positive real if it is strictly positive real and $\operatorname{Sym}\{G(j\infty)\} > 0$.

Remark 5.1. The above definitions provide the meanings of positive real, strictly positive real, and extended strictly positive real systems. We can see that positive realness only ensures that the system is analytic in the open right half-plane, meaning that the poles of the system may lie on the imaginary axis, which significantly affects the asymptotic stability of the system. In contrast, extended strictly positive realness requires the system to be analytic in the closed right half-plane, meaning that all poles of the system are located in the open left half-plane, which ensures that the system is asymptotically stable. Therefore this chapter focuses on the extended strictly positive realness of FOSSs.

Lemma 5.1 ([26]). *Let matrices $\Delta = \Delta^* \in \mathbb{C}^{n\times n}$, $\Gamma \in \mathbb{C}^{n\times m}$, and $\Pi \in \mathbb{C}^{k\times n}$. The following statements are equivalent.*

(i) *The following conditions are satisfied:*

$$\Gamma^{\perp}\Delta\Gamma^{\perp *} < 0, \quad \Pi^{*\perp}\Delta\Pi^{*\perp *} < 0.$$

(ii) *There exists a matrix X satisfying*

$$\Gamma X\Pi + (\Gamma X\Pi)^* + \Delta < 0.$$

Lemma 5.2 ([55]). *The FOS $D^{\alpha}x(t) = Ax(t)$, $0 < \alpha < 1$, is stable if and only if there exist matrices $P = P^* > 0$ and $Q = Q^* > 0$ satisfying*

$$\mathrm{Sym}\{e^{-j\theta}A^TP + e^{j\theta}A^TQ\} < 0,$$

where $\theta = \frac{\pi}{2}(1-\alpha)$.

Lemma 5.3. *The FOS $D^{\alpha}x(t) = Ax(t)$, $1 \le \alpha < 2$, is stable if and only if there exists a matrix $P = P^* > 0$ satisfying*

$$\mathrm{Sym}\{e^{-j\theta}A^TP\} < 0,$$

where $\theta = \frac{\pi}{2}(1-\alpha)$.

Proof. By Theorem 7 in [22] the FOS $D^{\alpha}x(t) = Ax(t)$, $1 \le \alpha < 2$, is stable if and only if there exists a matrix $P = P^* > 0$ satisfying

$$\mathrm{Sym}\left\{\begin{bmatrix} \cos\theta(A^TP) & \sin\theta(A^TP) \\ -\sin\theta(A^TP) & \cos\theta(A^TP) \end{bmatrix}\right\} < 0, \tag{5.7}$$

or

$$\mathrm{Sym}\left\{\begin{bmatrix} \cos\theta(A^TP) & -\sin\theta(A^TP) \\ \sin\theta(A^TP) & \cos\theta(A^TP) \end{bmatrix}\right\} < 0. \tag{5.8}$$

Applying Lemma 4.5, inequality (5.8) can be transformed into

$$\mathrm{Sym}\{e^{-j\theta}A^TP\} < 0.$$

This ends the proof. □

Lemma 5.4. *The FOSS $ED^\alpha x(t) = Ax(t)$, $0 < \alpha < 1$, is admissible if and only if there exist matrices $P, Q \in \mathbb{C}^{n\times n}$ satisfying*

$$E^TP = P^*E \geq 0, \tag{5.9}$$

$$E^TQ = Q^*E \geq 0, \tag{5.10}$$

$$\mathrm{Sym}\{A^T(e^{j\theta}P + e^{-j\theta}Q)\} < 0, \tag{5.11}$$

where $\theta = \frac{\pi}{2}(1-\alpha)$.

Proof. Sufficiency. Let $P = X$ and $Q = \bar{X}$. Then conditions (2.28a) and (2.28b) of Theorem 2.6 in Chapter 2 are satisfied, from which it follows that the FOSS $ED^\alpha x(t) = Ax(t)$ with $0 < \alpha < 1$ is admissible.

Necessity. Suppose that the FOSS $ED^\alpha x(t) = Ax(t)$ with $0 < \alpha < 1$ is admissible. Then there exist invertible matrices M and N satisfying

$$MEN = \begin{bmatrix} I & 0 \\ 0 & 0 \end{bmatrix}, \quad MAN = \begin{bmatrix} A_1 & 0 \\ 0 & I \end{bmatrix}.$$

By Lemma 5.2, there exist $P_1 = P_1^* > 0$ and $Q_1 = Q_1^* > 0$ satisfying

$$\mathrm{Sym}\{A_1^T(e^{j\theta}P_1 + e^{-j\theta}Q_1)\} < 0.$$

Let $P = M^T\left[\begin{smallmatrix} P_1 & 0 \\ 0 & -I \end{smallmatrix}\right]N^{-1}$ and $Q = M^T\left[\begin{smallmatrix} Q_1 & 0 \\ 0 & -I \end{smallmatrix}\right]N^{-1}$. Then

$$E^TP = P^*E = N^{-T}\begin{bmatrix} P_1 & 0 \\ 0 & 0 \end{bmatrix}N^{-1} \geq 0,$$

$$E^TQ = Q^*E = N^{-T}\begin{bmatrix} Q_1 & 0 \\ 0 & 0 \end{bmatrix}N^{-1} \geq 0,$$

$$\mathrm{Sym}\{A^T(e^{j\theta}P + e^{-j\theta}Q)\} = N^{-T}\begin{bmatrix} \mathrm{Sym}\{A_1^T(e^{j\theta}P_1 + e^{-j\theta}Q_1)\} & 0 \\ 0 & -4I \end{bmatrix}N^{-1} < 0.$$

This ends the proof. □

Based on these results, we give the following positive real lemma for FOSs.

Lemma 5.5. *Consider the FOS with transfer function $G(s) = C(s^\alpha I - A)^{-1}B + D$, $0 < \alpha < 1$. Then the system is extended strictly positive real if there exist matrices $P, Q \in \mathbb{H}_n$, $P > 0$, $Q > 0$, satisfying*

$$\begin{bmatrix} \mathrm{Sym}\{A^TX\} & X^*B - C^T \\ B^TX - C & -D - D^T \end{bmatrix} < 0, \tag{5.12}$$

where $X = e^{-j\theta}P + e^{j\theta}Q$, $\theta = \frac{\pi}{2}(1-\alpha)$.

Proof. The proof is composed of two steps. In Step 1, we prove the stability, and in Step 2, we prove the extended strictly positive realness.

Step 1. By inequality (5.12), there exist $P = P^* > 0$ and $Q = Q^* > 0$ satisfying

$$\mathrm{Sym}\{e^{-j\theta}A^T P + e^{j\theta}A^T Q\} < 0.$$

The system is stable by Lemma 5.2.

Step 2. Let $\lambda(w) = (jw)^\alpha = w^\alpha e^{j\frac{\pi}{2}\alpha}$, which denotes the ray in the complex plane as w ranges from 0 to $+\infty$. The ray can also be described by the set $\Lambda(\Psi, \Omega)$ defined in Lemma 4.8. For $\alpha \in (0, 1)$, $\Phi = \left[\begin{smallmatrix} 0 & e^{-j\theta} \\ e^{j\theta} & 0 \end{smallmatrix}\right]$ and $\Psi = \left[\begin{smallmatrix} 0 & e^{j\theta} \\ e^{-j\theta} & 0 \end{smallmatrix}\right]$. Then

$$\mathrm{Sym}\{G(jw)\} > 0 \tag{5.13}$$

can be written as

$$\begin{bmatrix} H(\lambda) \\ I \end{bmatrix}^* \Theta \begin{bmatrix} H(\lambda) \\ I \end{bmatrix} < 0 \quad \forall \lambda \in \Lambda, \tag{5.14}$$

where $H(\lambda) = (\lambda I_n - A)^{-1}B$ and $\Theta = \left[\begin{smallmatrix} 0 & -C^T \\ -C & -D-D^T \end{smallmatrix}\right]$. By Lemma 4.8, inequality (5.14) is equivalent to the fact that there exist $P, Q \in \mathbb{H}_n$, $Q > 0$, satisfying

$$\begin{bmatrix} A & B \\ I_n & 0 \end{bmatrix}^* \left(\begin{bmatrix} 0 & e^{-j\theta} \\ e^{j\theta} & 0 \end{bmatrix} \otimes P + \begin{bmatrix} 0 & e^{j\theta} \\ e^{-j\theta} & 0 \end{bmatrix} \otimes Q \right) \begin{bmatrix} A & B \\ I_n & 0 \end{bmatrix} + \begin{bmatrix} 0 & -C^T \\ -C & -D-D^T \end{bmatrix} < 0,$$

which can be rewritten as

$$\begin{bmatrix} \mathrm{Sym}\{A^T X\} & X^* B - C^T \\ B^T X - C & -D - D^T \end{bmatrix} < 0,$$

where $X = e^{-j\theta}P + e^{j\theta}Q$. Therefore inequality (5.13) can be derived from inequality (5.12), which indicates that for all $w \in [0, \infty)$,

$$\mathrm{Sym}\{G(jw)\} > 0$$

and, consequently,

$$\mathrm{Sym}\{G(j\infty)\} > 0.$$

This ends the proof. □

Lemma 5.6. *Given the FOS with transfer function $G(s) = C(s^\alpha I - A)^{-1}B + D$, $1 \le \alpha < 2$, the system is extended strictly positive real if and only if there exists a matrix $P \in \mathbb{H}_n$, $P > 0$, such that*

$$\begin{bmatrix} \mathrm{Sym}(A^T X) & X^* B - C^T \\ B^T X - C & -D - D^T \end{bmatrix} < 0, \tag{5.15}$$

where $X = e^{-j\theta} P$, $\theta = \frac{\pi}{2}(1 - \alpha)$.

Proof. Sufficiency. The sufficiency will be proved by two parts. Step 1 is the proof of the stability, and Step 2 is the proof of the extended strictly positive realness.

Step 1. By inequality (5.15) there exists a matrix $P \in \mathbb{H}_n$, $P > 0$, satisfying

$$\mathrm{Sym}(e^{-j\theta} A^T P) < 0.$$

By Lemma 5.3, the system is stable.

Step 2. Let $\lambda(w) = (jw)^\alpha = w^\alpha e^{j\frac{\pi}{2}\alpha}$, which denotes the ray in the complex plane as w ranges from 0 to $+\infty$. The ray can also be described by the set $\Lambda(\Psi, \Omega)$ defined in Lemma 4.8 with $\Phi = \left[\begin{smallmatrix} 0 & e^{-j\theta} \\ e^{j\theta} & 0 \end{smallmatrix}\right]$ and $\Psi = 0$ for $\alpha \in [1, 2)$. Then

$$\mathrm{Sym}\{G(jw)\} > 0 \tag{5.16}$$

can be written as

$$\begin{bmatrix} H(\lambda) \\ I_m \end{bmatrix}^* \Theta \begin{bmatrix} H(\lambda) \\ I_m \end{bmatrix} < 0 \quad \forall \lambda \in \Lambda, \tag{5.17}$$

where $H(\lambda) = (\lambda I_n - A)^{-1} B$ and $\Theta = \left[\begin{smallmatrix} 0 & -C^T \\ -C & -D-D^T \end{smallmatrix}\right]$. By Lemma 4.8, inequality (5.17) is equivalent to the fact that there exist $P, Q \in \mathbb{H}_n$, $Q > 0$, satisfying

$$\begin{bmatrix} A & B \\ I_n & 0 \end{bmatrix}^* (\Phi \otimes P + \Psi \otimes Q) \begin{bmatrix} A & B \\ I_n & 0 \end{bmatrix} + \begin{bmatrix} 0 & -C^T \\ -C & -D - D^T \end{bmatrix} < 0,$$

which can be rewritten as

$$\begin{bmatrix} \mathrm{Sym}\{A^T X\} & X^* B - C^T \\ B^T X - C & -D - D^T \end{bmatrix} < 0, \tag{5.18}$$

where $X = e^{-j\theta} P$. There exists $P \in \mathbb{H}_n$, $P > 0$, such that inequality (5.15) holds, which implies that inequality (5.18) also holds. Therefore inequality (5.16) holds. Hence the system is extended strictly positive real.

Necessity. Assume that the system is extended strictly positive real. Then inequality (5.16) holds, which is equivalent to the fact that there exist $P, Q \in \mathbb{H}_n$, $Q > 0$, such that inequality (5.18) holds by Lemma 4.8. It remains to prove that $P > 0$. From inequality (5.18), we have

$$e^{-j\theta} A^T P + e^{j\theta} P A < 0. \tag{5.19}$$

Since the system is stable, we have $\arg(spec(A^T)) > \frac{\pi\alpha}{2}$, indicating that $\arg(spec(e^{-j\theta}A^T)) > \frac{\pi}{2}$. Let

$$e^{-j\theta}A^T P + e^{j\theta}PA = -\Delta, \quad \Delta > 0.$$

Then

$$P = \int_0^{+\infty} e^{e^{-j\theta}A^T t}\Delta e^{(e^{-j\theta}A^T)^* t} > 0.$$

This ends the proof. □

Remark 5.2. The conditions for the positive realness of FOSs are proposed in [123], but they cannot guarantee the stability of FOSs. Lemmas 5.5 and 5.6 present novel positive real lemmas for FOSs with orders $\alpha \in (0,1)$ and $\alpha \in [1,2)$, respectively, which can both ensure the stability of FOSs. From Lemma 5.2 we can see that the term $\mathrm{Sym}\{A^T X\}$ in inequality (5.12) of Lemma 5.5 can determine the stability of FOSs of order $\alpha \in [1,2)$.

To show the superiority and effectiveness of the proposed results, in this section, we give a numerical example.

Example 5.1. Consider the FOS

$$\begin{aligned} D^{1.8}x(t) &= \begin{bmatrix} 14 & 4 \\ 4 & 14 \end{bmatrix} x(t) + \begin{bmatrix} -4 \\ 9 \end{bmatrix} w(t), \\ z(t) &= \begin{bmatrix} 6 & 3.5 \end{bmatrix} x(t) + 6w(t). \end{aligned} \tag{5.20}$$

The aim is to determine whether the FOS (5.20) satisfies the positive realness and extended strictly positive realness.

Using Theorem 1 in [123], we can find the following feasible solutions:

$$M = \begin{bmatrix} -0.2447 & -0.1969 \\ -0.1969 & -0.2324 \end{bmatrix}, \quad N = \begin{bmatrix} 0.7531 & 0.6061 \\ 0.6061 & 0.7152 \end{bmatrix}.$$

By Theorem 1 in [123], FOS (5.20) satisfies the positive realness.

Lemma 5.6 shows that no feasible solutions can be found. Because Lemma 5.6 gives sufficient and necessary conditions, FOS (5.20) does not satisfy the extended strictly positive realness.

From numerical simulation, it follows that the eigenvalues of the FOS (5.20) are 10 and 18, as shown in Figure 5.1, both located in the unstable areas. The eigenvalue curves of the system $\mathrm{Sym}\{G(jw)\}(w \geq 0)$ are shown in Figure 5.2, from which we can see that the eigenvalues of $\mathrm{Sym}\{G(jw)\}$ $(w \geq 0)$ are all greater than 0. According to Definition 5.1, FOS (5.20) does not satisfy the extended strictly positive realness because its transfer function is not always analytic on $\mathrm{Re}(s) \geq 0$.

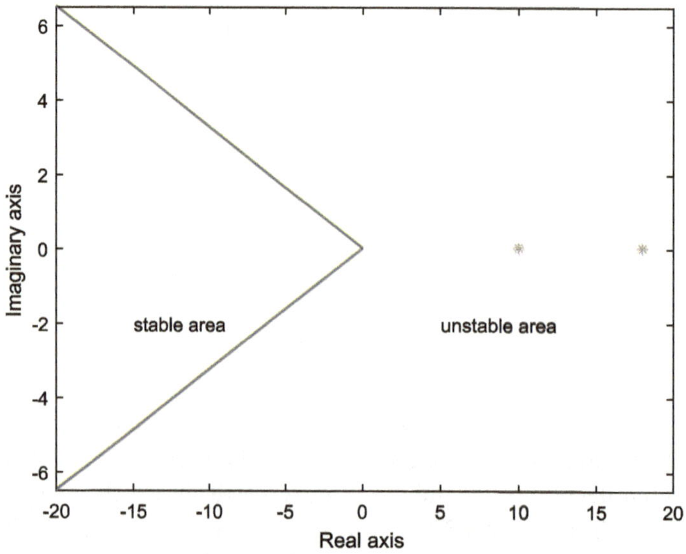

Figure 5.1: The location of the eigenvalues of FOS (5.20) in Example 5.1.

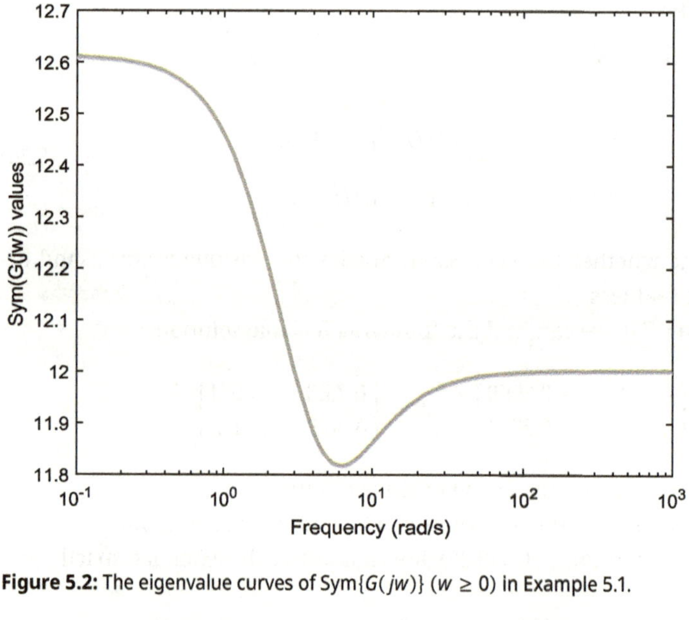

Figure 5.2: The eigenvalue curves of Sym$\{G(jw)\}$ ($w \geq 0$) in Example 5.1.

Therefore we can see that Theorem 1 in [123] only ensures the positive realness of the FOS, but it cannot ensure the stability of the system. In contrast to Theorem 1 in [123], Lemma 5.6 in this chapter not only ensures the extended strict positive realness of the system, but also ensures the stability of the system at the same time.

5.3 Positive real lemma

For the two cases of fractional order, $\alpha \in (0,1)$ and $\alpha \in [1,2)$, in this section, we present positive real lemmas for the nominal FOSSs (5.5).

5.3.1 Case $\alpha \in (1,2)$

Theorem 5.1. *Suppose that $D_0+D_0^T > 0$. The FOSS (5.5) is admissible and extended strictly positive real if and only if there exists $P \in \mathbb{C}^{n\times n}$ satisfying*

$$E^T P = P^* E \geq 0, \tag{5.21}$$

$$\begin{bmatrix} \mathrm{Sym}\{A_0^T X\} & X^* B_0 - C_0^T \\ B_0^T X - C_0 & -D_0 - D_0^T \end{bmatrix} < 0, \tag{5.22}$$

where $1 \leq \alpha < 2$, $\theta = \frac{\pi}{2}(1-\alpha)$, $X = e^{-j\theta} P$.

Proof. Sufficiency. By inequality (5.22), there exists P satisfying

$$e^{-j\theta} A_0^T P + e^{j\theta} P^* A_0 < 0. \tag{5.23}$$

By Lemma 4.5, inequality (5.23) can be transformed into

$$\mathrm{Sym}\left\{\begin{bmatrix} \cos\theta(A_0^T P) & -\sin\theta(A_0^T P) \\ \sin\theta(A_0^T P) & \cos\theta(A_0^T P) \end{bmatrix}\right\} < 0, \tag{5.24}$$

which can be rewritten as

$$\mathrm{Sym}\left\{\begin{bmatrix} \cos\theta & \sin\theta \\ -\sin\theta & \cos\theta \end{bmatrix} \otimes (A_0^T P)\right\} < 0. \tag{5.25}$$

By applying Theorem 1 in [63], the admissibility of FOSS (5.5) is proved with (5.21) and (5.25)

The next step is to prove that for all $w \in [0,\infty)$,

$$\mathrm{Sym}\{G(jw)\} > 0, \tag{5.26}$$

which can be written as

$$\mathrm{Sym}\{C_0(e^{j\frac{\pi}{2}\alpha} w^\alpha E - A_0)^{-1} B_0 + D_0\} > 0. \tag{5.27}$$

Let $\lambda(w) = (jw)^\alpha = w^\alpha e^{j\frac{\pi}{2}\alpha}$, which denotes the ray in the complex plane as w ranges from 0 to $+\infty$. The ray can also be described by the set $\Lambda(\Psi,\Omega)$. When $\alpha \in [1,2)$, $\Phi = \left[\begin{smallmatrix} 0 & e^{-j\theta} \\ e^{j\theta} & 0 \end{smallmatrix}\right]$ and $\Psi = 0$. Then inequality (5.27) can be rewritten as

$$\begin{bmatrix} H(\lambda) \\ I_m \end{bmatrix}^* \Theta \begin{bmatrix} H(\lambda) \\ I_m \end{bmatrix} < 0, \tag{5.28}$$

where $\Theta = \left[\begin{smallmatrix} 0 & -C^T \\ -C & -D-D^T \end{smallmatrix}\right]$ and $H(\lambda) = (\lambda E - A)^{-1}B$. By Lemma 4.8, inequality (5.28) is equivalent to the existence of $Q_1, Q_2 \in \mathbb{H}_n$, $Q_2 > 0$, satisfying

$$\begin{bmatrix} A_0 & B_0 \\ E & 0 \end{bmatrix}^* (\Phi \otimes Q_1 + \Psi \otimes Q_2) \begin{bmatrix} A_0 & B_0 \\ E & 0 \end{bmatrix} + \begin{bmatrix} 0 & -C_0^T \\ -C_0 & -D_0 - D_0^T \end{bmatrix} < 0,$$

which can be transformed into

$$\begin{bmatrix} \mathrm{Sym}\{A_0^T Y\} & Y^* B_0 - C_0^T \\ B_0^T Y - C_0 & -D_0 - D_0^T \end{bmatrix} < 0, \tag{5.29}$$

where $Y = e^{-j\theta} Q_1 E$. Let $P = Q_1 E$ in inequality (5.22). Then inequality (5.29) can be derived from inequality (5.22). Inequality (5.28) implies that for all $w \in [0, \infty)$,

$$\mathrm{Sym}\{G(jw)\} > 0$$

and, consequently, $\mathrm{Sym}\{G(j\infty)\} > 0$.

Necessity. Assume that FOSS (5.5) is admissible and extended strictly positive real. By Lemma 6 in [63], there exist invertible matrices N and M such that

$$A_0 = M \begin{bmatrix} A_1 & 0 \\ 0 & I \end{bmatrix} N, \quad E = M \begin{bmatrix} I & 0 \\ 0 & 0 \end{bmatrix} N, \tag{5.30}$$

where $A_1 \in \mathbb{R}^{r \times r}$. Denote

$$C_0 N^{-1} = [C_1 \quad C_2], \quad M^{-1} B_0 = \begin{bmatrix} B_1 \\ B_2 \end{bmatrix}. \tag{5.31}$$

Based on (5.30) and (5.31), the transfer function $G(s)$ can be described as

$$\begin{aligned} G(s) &= C_0 (s^\alpha E - A_0)^{-1} B_0 + D_0 \\ &= C_0 N^{-1} (s^\alpha M^{-1} E N^{-1} - M^{-1} A_0 N^{-1})^{-1} M^{-1} B_0 + D_0 \\ &= C_1 (s^\alpha I - A_1)^{-1} B_1 + D - C_2 B_2. \end{aligned}$$

By Lemma 5.6 there exists $P_1 \in \mathbb{H}_n$, $P_1 > 0$, satisfying

$$\begin{bmatrix} \mathrm{Sym}\{A_1^T X_1\} & X_1^* B_1 - C_1^T \\ B_1^T X_1 - C_1 & -D - C_2 B_2 - D^T - B_2^T C_2^T \end{bmatrix} < 0, \tag{5.32}$$

where $X_1 = e^{-j\theta} P_1$, $\theta = \frac{\pi}{2}(1 - \alpha)$, $1 \le \alpha < 2$. By Lemma 5.1, there exist matrices X_2 and X_3 satisfying

$$\Delta + \Gamma \tilde{X} \Pi + (\Gamma \tilde{X} \Pi)^T < 0, \tag{5.33}$$

where

$$\Delta = \begin{bmatrix} A_1^T X_1 + X_1^* A_1 & 0 & C_1^T - X_1^* B_1 \\ 0 & 0 & C_2^T \\ C_1 - B_1^T X_1 & C_2 & -D - D^T \end{bmatrix}, \quad \Gamma = \begin{bmatrix} 0 \\ I \\ -B_2^T \end{bmatrix},$$
$$\Pi = \begin{bmatrix} I & 0 & 0 \\ 0 & I & 0 \end{bmatrix}, \quad \tilde{X} = [X_2 \quad X_3], \tag{5.34}$$

which implies that

$$(\Pi^T)^{\perp} = [0 \quad 0 \quad I], \quad \Gamma^{\perp} = \begin{bmatrix} I & 0 & 0 \\ 0 & B_2^T & I \end{bmatrix}. \tag{5.35}$$

From inequalities (5.32) and $D_0 + D_0^T > 0$, it follows that

$$\Gamma^{\perp}\Delta(\Gamma^{\perp})^T = \begin{bmatrix} \mathrm{Sym}\{A_1^T X_1\} & X_1^* B_1 - C_1^T \\ B_1^T X_1 - C_1 & -D_0 - C_2 B_2 - D_0^T - B_2^T C_2^T \end{bmatrix} < 0,$$
$$(\Pi^T)^{\perp}\Delta((\Pi^T)^{\perp})^T = -D_0 - D_0^T < 0. \tag{5.36}$$

By Lemma 5.1, there exist matrices X_2 and X_3 such that (5.33) holds, and then it can be rewritten as

$$\begin{bmatrix} N^{-T}\mathrm{Sym}\{A_0^T X\}N^{-1} & N^{-T}(X^* B_0 - C_0^T) \\ (B_0^T X - C_0)N^{-1} & -D_0 - D_0^T \end{bmatrix} < 0, \tag{5.37}$$

which can be transformed into

$$\begin{bmatrix} \mathrm{Sym}\{A_0^T X\} & X^* B_0 - C_0^T \\ B_0^T X - C_0 & -D_0 - D_0^T \end{bmatrix} < 0, \tag{5.38}$$

where

$$X = M^{-T}\begin{bmatrix} X_1 & 0 \\ X_2 & X_3 \end{bmatrix} N = M^{-T}\begin{bmatrix} e^{-j\theta}P_1 & 0 \\ X_2 & X_3 \end{bmatrix} N$$
$$= e^{-j\theta} M^{-T}\begin{bmatrix} P_1 & 0 \\ e^{j\theta}X_2 & e^{j\theta}X_3 \end{bmatrix} N. \tag{5.39}$$

Let $P = M^{-T}\left[\begin{smallmatrix} P_1 & 0 \\ e^{j\theta}X_2 & e^{j\theta}X_3 \end{smallmatrix}\right]N$, which indicates that conditions (5.21) and (5.22) hold. This ends the proof. □

Theorem 5.1 contains equality constraints. From the perspective of numerical computation, it is necessary to provide results without equality constraints. When the order $\alpha \in [1, 2)$, the positive real lemma without equality constraints is as follows.

Theorem 5.2. *Suppose that $D_0 + D_0^T > 0$. System (5.5) is admissible and extended strictly positive real if and only if there exist matrices $P \in \mathbb{C}^{n\times n}$, $P > 0$, and $S \in \mathbb{R}^{(n-r)\times n}$ satisfying*

$$\begin{bmatrix} \mathrm{Sym}\{A_0^T X\} & X^* B_0 - C_0^T \\ B_0^T X - C_0 & -D_0 - D_0^T \end{bmatrix} < 0, \tag{5.40}$$

where $1 \le \alpha < 2$, $\theta = \frac{\pi}{2}(1-\alpha)$, and $X = e^{-j\theta}(PE + E_0 S)$ with E_0 any column-full rank matrix satisfying $E^T E_0 = 0$.

Proof. The proof of this theorem is similar to that of Theorem 5.1 and is omitted. □

Remark 5.3. When E in Theorems 5.3 and 5.1 is taken as the identity matrix I, Theorem 5.3 is equivalent to Lemma 5.5, and Theorem 5.1 is equivalent to Lemma 5.6. Therefore Theorem 5.3 is a generalization of Lemma 5.5, and Theorem 5.1 is a generalization of Lemma 5.6.

5.3.2 Case $\alpha \in (0, 1)$

Theorem 5.3. *Suppose that $D_0 + D_0^T > 0$. System (5.5) is admissible and extended strictly positive real if there exist matrices $P \in \mathbb{C}^{n\times n}$ and $Q \in \mathbb{C}^{n\times n}$ satisfying*

$$E^T Q = Q^* E \ge 0, \tag{5.41}$$

$$E^T P = P^* E \ge 0, \tag{5.42}$$

$$\begin{bmatrix} \mathrm{Sym}\{A_0^T X\} & X^* B_0 - C_0^T \\ B_0^T X - C_0 & -D_0 - D_0^T \end{bmatrix} < 0, \tag{5.43}$$

where $X = e^{-j\theta}P + e^{j\theta}Q$, $\theta = \frac{\pi}{2}(1-\alpha)$, $0 < \alpha < 1$.

Proof. By inequality (5.43), we have

$$\mathrm{Sym}\{A_0^T(e^{-j\theta}P + e^{j\theta}Q)\} < 0. \tag{5.44}$$

From (5.41), (5.42) and (5.44), it follows that system (5.5) is admissible by Lemma 5.4.

From inequality (5.43), by Lemma 4.7, we have

$$A_0^T X + X^* A_0 + (X^* B_0 - C_0^T)(D_0 + D_0^T)^{-1}(B_0^T X - C_0) < 0,$$

which implies that there exists a sufficiently small $Z > 0$ such that

$$A_0^T X + X^* A_0 + (X^* B_0 - C_0^T)(D_0 + D_0^T)^{-1}(B_0^T X - C_0) + Z < 0. \tag{5.45}$$

Let $W = (X^* B_0 - C_0^T)(D_0 + D_0^T)^{-1}(B_0^T X - C_0) + Z$. Then $W > 0$. Inequality (5.45) is equivalent to

$$-A_0^T X - X^* A_0 > W. \tag{5.46}$$

Furthermore, for $\alpha \in (0,1)$, we have

$$\begin{aligned}
&e^{-j\frac{\pi}{2}\alpha}w^{\alpha}E^{T}X + X^{*}e^{j\frac{\pi}{2}\alpha}w^{\alpha}E \\
&= w^{\alpha}e^{j(\frac{\pi}{2}-\pi\alpha)}E^{T}Q + w^{\alpha}e^{-j\frac{\pi}{2}}E^{T}P + w^{\alpha}e^{j(\pi\alpha-\frac{\pi}{2})}Q^{*}E + w^{\alpha}e^{j\frac{\pi}{2}}P^{*}E \\
&= w^{\alpha}e^{j(\frac{\pi}{2}-\pi\alpha)}E^{T}Q + w^{\alpha}e^{j(\pi\alpha-\frac{\pi}{2})}Q^{*}E \\
&= 2w^{\alpha}\sin(\pi\alpha)E^{T}Q \geq 0 \quad \forall w \in [0,\infty).
\end{aligned} \tag{5.47}$$

From (5.47), it follows that

$$\begin{aligned}
&\mathrm{Sym}\{G(jw)\} \\
&= C_0(e^{j\frac{\pi}{2}\alpha}w^{\alpha}E - A_0)^{-1}B_0 + B_0^T(e^{-j\frac{\pi}{2}\alpha}w^{\alpha}E - A_0)^{-T}C_0^T + D_0 + D_0^T \\
&= (C_0 - B_0^TX)(e^{j\frac{\pi}{2}\alpha}w^{\alpha}E - A_0)^{-1}B_0 + B_0^T(e^{-j\frac{\pi}{2}\alpha}w^{\alpha}E - A_0)^{-T}(C_0^T - X^{*}B_0) \\
&\quad + D_0 + D_0^T + B_0^TX(e^{j\frac{\pi}{2}\alpha}w^{\alpha}E - A_0)^{-1}B_0 + B_0^T(e^{-j\frac{\pi}{2}\alpha}w^{\alpha}E - A_0)^{-T}X^{*}B_0 \\
&= (C_0 - B_0^TX)(e^{j\frac{\pi}{2}\alpha}w^{\alpha}E - A_0)^{-1}B_0 + B_0^T(e^{-j\frac{\pi}{2}\alpha}w^{\alpha}E - A_0)^{-T}(C_0^T - X^{*}B_0) \\
&\quad + D_0 + D_0^T + B_0^T(e^{-j\frac{\pi}{2}\alpha}w^{\alpha}E - A_0)^{-T}((e^{-j\frac{\pi}{2}\alpha}w^{\alpha}E - A_0)^{T}X \\
&\quad + X^{*}(e^{j\frac{\pi}{2}\alpha}w^{\alpha}E - A_0))(e^{j\frac{\pi}{2}\alpha}w^{\alpha}E - A_0)^{-1}B_0.
\end{aligned}$$

From (5.46) and (5.47), we have

$$\begin{aligned}
&\mathrm{Sym}\{G(jw)\} \\
&> (C_0 - B_0^TX)(e^{j\frac{\pi}{2}\alpha}w^{\alpha}E - A_0)^{-1}B_0 + B_0^T(e^{-j\frac{\pi}{2}\alpha}w^{\alpha}E - A_0)^{-T}(C_0^T - X^{*}B_0) \\
&\quad + D_0 + D_0^T + B_0^T(e^{-j\frac{\pi}{2}\alpha}w^{\alpha}E - A_0)^{-T}W(e^{j\frac{\pi}{2}\alpha}w^{\alpha}E - A_0)^{-1}B_0 \\
&= [B_0^T(e^{-j\frac{\pi}{2}\alpha}w^{\alpha}E - A_0)^{-T} + (C_0 - B_0^TX)W^{-1}]W[(e^{j\frac{\pi}{2}\alpha}w^{\alpha}E - A_0)^{-1}B_0 \\
&\quad + W^{-1}(C_0^T - X^{*}B_0)] + D_0 + D_0^T - (C_0 - B_0^TX)W^{-1}(C_0^T - X^{*}B_0) \\
&\geq D_0 + D_0^T - (B_0^TX - C_0)W^{-1}(X^{*}B_0 - C_0^T).
\end{aligned}$$

Since

$$W - (X^{*}B_0 - C_0^T)(D_0 + D_0^T)^{-1}(B_0^TX - C_0) = Z > 0,$$

applying Lemma 4.7, we have

$$\begin{bmatrix} W & X^{*}B_0 - C_0^T \\ B_0^TX - C_0 & D_0 + D_0^T \end{bmatrix} > 0,$$

which is equivalent to

$$D_0 + D_0^T - (B_0^TX - C_0)W^{-1}(X^{*}B_0 - C_0^T) > 0.$$

Therefore

$$\mathrm{Sym}\{G(jw)\} > 0 \quad \forall w \in [0, \infty)$$

and, consequently,

$$\mathrm{Sym}\{G(j\infty)\} > 0.$$

This ends the proof. □

Remark 5.4. Theorem 5.3 provides conditions under which FOSSs satisfy the admissibility and extended strict positive realness. Theorem 5.3 includes two complex variables P and Q, and we can observe that $X = e^{-j\theta}P + e^{j\theta}Q$ is also a complex variable.

The following theorem presents an improvement to the above results, addressing the issue of complex variables.

Theorem 5.4. *Suppose that $D_0 + D_0^T > 0$. System (5.5) is admissible and extended strictly positive real if there exists a matrix $Y \in \mathbb{C}^{n\times n}$ satisfying*

$$E^T Y = Y^* E \geq 0, \tag{5.48}$$

$$\begin{bmatrix} \mathrm{Sym}\{A_0^T Z\} & Z^T B_0 - C_0^T \\ B_0^T Z - C_0 & -D_0 - D_0^T \end{bmatrix} < 0, \tag{5.49}$$

where $Z = e^{-j\theta}Y + e^{j\theta}\bar{Y}$, $\theta = \frac{\pi}{2}(1-\alpha)$, $0 < \alpha < 1$.

Proof. The proof can be conducted by showing that conditions (5.41)–(5.43) are equivalent to conditions (5.48)–(5.49). Suppose that

$$L_1(P) = \begin{bmatrix} \mathrm{Sym}\{e^{-j\theta}A_0^T P\} & e^{j\theta}P^* B_0 - C_0^T \\ e^{-j\theta}B_0^T P - C_0 & -D_0 - D_0^T \end{bmatrix},$$

$$L_2(Q) = \begin{bmatrix} \mathrm{Sym}\{e^{j\theta}A_0^T Q\} & e^{-j\theta}Q^* B_0 \\ e^{j\theta}B_0^T Q & 0 \end{bmatrix}.$$

Then conditions (5.41)–(5.43) are equivalent to

$$\begin{aligned} \exists P, Q \quad \text{s.t.} \quad & E^T Q = Q^* E \geq 0, \quad E^T P = P^* E \geq 0, \\ & L_1(P) + L_2(Q) < 0, \end{aligned} \tag{5.50}$$

and conditions (5.48)–(5.49) are equivalent to

$$\begin{aligned} \exists Y \quad \text{s.t.} \quad & E^T Y = Y^* E \geq 0, \\ & L_1(Y) + L_2(\bar{Y}) < 0. \end{aligned} \tag{5.51}$$

The purpose is to prove the equivalence of conditions (5.50) and (5.51).

(5.51) $\Rightarrow$ (5.50) can be proved by taking $Q = \bar{P}$.
(5.50) $\Rightarrow$ (5.51) can be proved as follows. Take $Y = P + \bar{Q}$. Then

$$E^T Y = E^T P + E^T \bar{Q} = P^* E + Q^* E = Y^* E \geq 0,$$

and

$$\begin{aligned} L_1(Y) + L_2(\bar{Y}) &= L_1(P) + L_1(\bar{Q}) + L_2(\bar{P}) + L_2(Q) \\ &= L_1(P) + L_2(Q) + \overline{L_1(P) + L_2(Q)} \\ &< 0. \end{aligned}$$

Therefore condition (5.50) is equivalent to condition (5.51). This ends the proof. □

Remark 5.5. As mentioned in Remark 5.4, the variable X in Theorem 5.3 is a complex variable. Although Y in Theorem 5.4 is a complex matrix, $Z = e^{-j\theta}Y + e^{j\theta}\bar{Y}$ is a real matrix, which can weaken some conservatism in the positive real controller design compared with Theorem 5.3.

Theorem 5.4 contains equality constraints. When the order $\alpha \in (0, 1)$, the positive real lemma without equality constraints is as follows.

Theorem 5.5. *Suppose that $D_0 + D_0^T > 0$. System (5.5) is admissible and extended strictly positive real if there exists matrices $Y \in \mathbb{C}^{n\times n}$, $Y > 0$, and $S \in \mathbb{R}^{(n-r)\times n}$ satisfying*

$$\begin{bmatrix} \mathrm{Sym}\{A_0^T Z\} & Z^T B_0 - C_0^T \\ B_0^T Z - C_0 & -D_0 - D_0^T \end{bmatrix} < 0, \tag{5.52}$$

where $Z = (e^{-j\theta}Y + e^{j\theta}\bar{Y})E + E_0 S$, $\theta = \frac{\pi}{2}(1 - \alpha)$, $0 < \alpha < 1$, with E_0 any column-full rank matrix satisfying $E^T E_0 = 0$.

Proof. The proof of this theorem is similar to that of Theorem 5.3 and is omitted. □

Example 5.2. Consider the fractional-order singular circuit model with system control output $z(t) = C_0 x(t) + D_0 w(t)$. Then the system can be described as

$$\begin{aligned} ED^{\alpha}x(t) &= A_0 x(t) + B_0 w(t), \\ z(t) &= C_0 x(t) + D_0 w(t), \end{aligned} \tag{5.53}$$

where

$$x(t) = \begin{bmatrix} i_1(t) \\ i_2(t) \\ i_3(t) \end{bmatrix}, \quad w(t) = \begin{bmatrix} e_1 \\ e_2 \end{bmatrix}, \quad E = \begin{bmatrix} L_1 & 0 & 0 \\ 0 & L_2 & 0 \\ 0 & 0 & 0 \end{bmatrix},$$

$$B_0 = \begin{bmatrix} 1 & 0 \\ 0 & 1 \\ 0 & 0 \end{bmatrix}, \quad A_0 = \begin{bmatrix} -R_1 & 0 & 0 \\ 0 & -R_2 & 0 \\ 1 & 1 & -1 \end{bmatrix}.$$

We can show that system (5.53) is the nominal FOSS (5.5) studied in this chapter. Take $L_1 = 10H$, $L_2 = 5H$, $R_1 = 1\Omega$, $R_2 = 2\Omega$, $C_0 = \left[\begin{smallmatrix} 0 & 1 & 0 \\ 1 & 0 & 0 \end{smallmatrix}\right]$ and $D_0 = \left[\begin{smallmatrix} 1 & 0 \\ 0 & 1 \end{smallmatrix}\right]$. Then the system parameters are

$$E = \begin{bmatrix} 10 & 0 & 0 \\ 0 & 5 & 0 \\ 0 & 0 & 0 \end{bmatrix}, \quad A_0 = \begin{bmatrix} -1 & 0 & 0 \\ 0 & -2 & 0 \\ 1 & 1 & -1 \end{bmatrix}, \quad B_0 = \begin{bmatrix} 1 & 0 \\ 0 & 1 \\ 0 & 0 \end{bmatrix}. \tag{5.54}$$

Consider the system order $\alpha = 0.8$. Our aim is to determine whether it satisfies the admissibility and extended strictly positive realness.

Using Theorem 5.3, we can find the following feasible solutions:

$$P = \begin{bmatrix} 1.3493 & 0.4691 - 0.0227\mathbf{i} & 0 \\ 0.9382 + 0.0454\mathbf{i} & 3.1761 & 0 \\ 29361.2464 + \mathbf{i} & 1 + 21332.0828\mathbf{i} & 1 - 21331.9210\mathbf{i} \end{bmatrix},$$

$$Q = \begin{bmatrix} 1.2746 & 0.4676 + 0.0391\mathbf{i} & 0 \\ 0.9353 - 0.0782\mathbf{i} & 3.121 & 0 \\ -36287.9754 & 9.6382 & 0 \end{bmatrix}.$$

Thus the FOSS (5.5) satisfies both the admissibility and extended strictly positive realness.

Using Theorem 5.4, we can find the following feasible solutions:

$$Y = \begin{bmatrix} 1.2123 & 0.4084 - 0.0499\mathbf{i} & 0 \\ 0.8167 + 0.0997\mathbf{i} & 2.8167 & 0 \\ 9096.7222\mathbf{i} & -2954.6251 & 2956.6710 \end{bmatrix}.$$

Thus the FOSS (5.5) satisfies both the admissibility and extended strictly positive realness.

Using Theorem 5.5, when $E_0 = \begin{bmatrix} 0 & 0 & 1 \end{bmatrix}^T$, we can find the following feasible solutions:

$$Y = \begin{bmatrix} 0.1366 & 0.0995 + 0.0829\mathbf{i} & 0 \\ 0.0995 - 0.0829\mathbf{i} & 0.6252 & 0 \\ 0 & 0 & 38771.8406 \end{bmatrix},$$

$$S = \begin{bmatrix} -6805.3009 & -6808.1821 & 6784.7608 \end{bmatrix}.$$

Thus the FOSS (5.5) satisfies both the admissibility and extended strictly positive realness.

By numerical simulation, the eigenvalues of the FOSS (5.54) with parameter (5.5) are $\{-0.4, -0.1\}$, and both eigenvalues are located in the stable areas as shown in Figure 5.3, i. e., the system is stable. Meanwhile, (i) $\det(s^\alpha E - A_0)$ is not constant 0, i. e., the system is regular; (ii) take the left and right maximal zeroed submatrices of the matrix E, $L_E = \begin{bmatrix} 0 & 0 & 1 \end{bmatrix}$ and $R_E^T = \begin{bmatrix} 0 & 0 & 1 \end{bmatrix}$; then $\operatorname{rank}(L_E A_0 R_E) = 1$, i. e., the system is impulse-free. Therefore the FOSS is admissible. The eigenvalue curves of $\operatorname{Sym}\{G(jw)\}$ $(w \geq 0)$ are shown in Figure 5.4, from which we can see that the eigenvalues of $\mathrm{Sym}\{G(jw)\}$ $(w \geq 0)$ are all greater than 0. So the system satisfies both the admissibility and extended strictly positive realness.

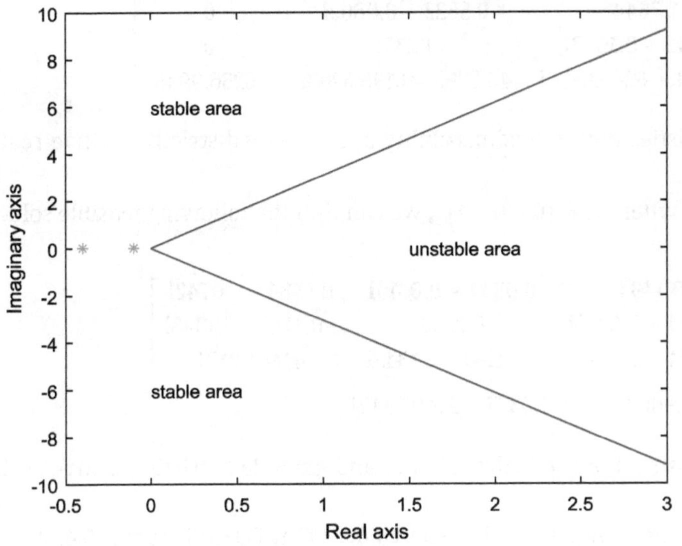

Figure 5.3: The location of the eigenvalues of FOSS (5.5) in Example 5.2.

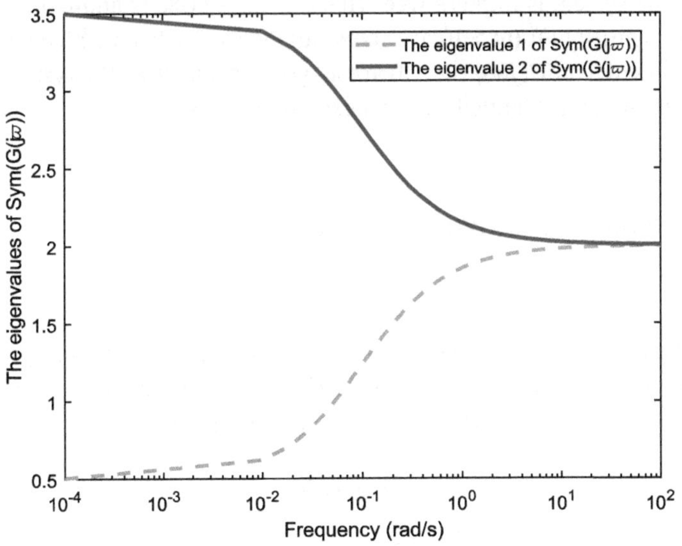

Figure 5.4: The eigenvalue curves of $\mathrm{Sym}\{G(jw)\}$ $(w \geq 0)$ in Example 5.2.

Example 5.3. Consider the same model and parameters as in Example 5.2, taking in this case the system order $\alpha = 1.5$. The aim is to determine whether it satisfies the admissibility and extended strictly positive realness.

Using Theorem 5.1, we can find the following feasible solutions:

$$P = \begin{bmatrix} 1.7644 & 0.5822 - 0.0802\mathbf{i} & 0 \\ 1.1645 + 0.1603\mathbf{i} & 4.2351 & 0 \\ 29.0724 + 6246.6797\mathbf{i} & 40.7395 + 6198.5783\mathbf{i} & -6256.0945\mathbf{i} \end{bmatrix}.$$

Thus the FOSS (5.5) satisfies both the admissibility and extended strictly positive real performance.

Using Theorem 5.2, when $E_0 = \begin{bmatrix} 0 & 0 & 1 \end{bmatrix}^T$, we can find the following feasible solutions:

$$P = \begin{bmatrix} 0.1494 & 0.0844 - 0.0319\mathbf{i} & 0.5584 - 1.0742\mathbf{i} \\ 0.0844 + 0.0319\mathbf{i} & 0.7245 & -0.1141 - 3.1343\mathbf{i} \\ 0.5584 + 1.0742\mathbf{i} & -0.1141 + 3.1343\mathbf{i} & 41441.0171 \end{bmatrix},$$

$$S = \begin{bmatrix} -9144.9067 & -9161.0375 & 9140.7349 \end{bmatrix}.$$

Thus the FOSS (5.5) satisfies both the admissibility and extended strictly positive real performance.

By numerical simulation, we get that the eigenvalues of the FOSS (5.5) are $\{-0.4, -0.1\}$, both located in the stable areas shown in Figure 5.5, i. e., the system is stable. Meanwhile, (i) $\det(s^\alpha E - A_0)$ is not constant 0, i. e., the system is regular; (ii) take the left and right maximal zeroed submatrices of the matrix E, $L_E = \begin{bmatrix} 0 & 0 & 1 \end{bmatrix}$ and $R_E^T = \begin{bmatrix} 0 & 0 & 1 \end{bmatrix}$; then $\operatorname{rank}(L_E A_0 R_E) = 1$, i. e., the system is impulse-free. Therefore the FOSS is admissible. The eigenvalue curves of $\operatorname{Sym}\{G(jw)\}$ $(w \geq 0)$ are shown in Figure 5.6, from which we can see that the eigenvalues of $\operatorname{Sym}\{G(jw)\}$ $(w \geq 0)$ are all greater than 0. So the system satisfies both the admissibility and extended strictly positive realness.

5.4 Robust state feedback positive real control

5.4.1 Case $\alpha \in (1, 2)$

Theorem 5.6. *Suppose that $D_0 + D_0^T > 0$. When the order $\alpha \in [1, 2)$, the closed-loop system (5.3) is admissible and extended strictly positive real if there exist a nonsingular matrix $Y \in \mathbb{R}^{n\times n}$, a matrix $L \in \mathbb{R}^{m\times n}$, and constants $\varepsilon_1 > 0$ and $\varepsilon_2 > 0$ satisfying*

$$EY = Y^* E^T \geq 0, \tag{5.55}$$

$$\begin{bmatrix} W_0 + \varepsilon_1 \Upsilon_1 \Upsilon_1^T + \varepsilon_2 \Upsilon_3 \Upsilon_3^T & \Upsilon_2^T & \Upsilon_4^T \\ \Upsilon_2 & -\varepsilon_1 I & 0 \\ \Upsilon_4 & 0 & -\varepsilon_2 I \end{bmatrix} < 0, \tag{5.56}$$

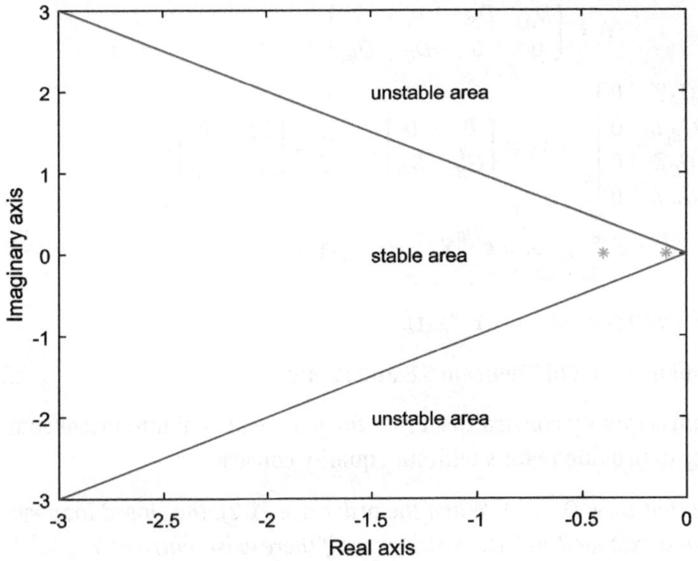

Figure 5.5: The location of the eigenvalues of FOSS (5.5) in Example 5.3.

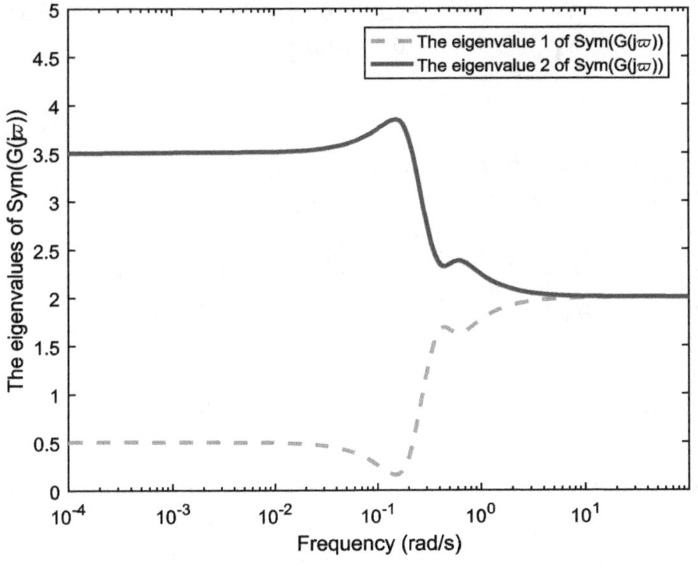

Figure 5.6: The eigenvalue curves of Sym$\{G(jw)\}$ ($w \geq 0$) in Example 5.3.

where

$$W_0 = \begin{bmatrix} \text{Sym}\{A_0 Z + B_{10}\hat{L}\} & (C_0 Z + B_{20}\hat{L} - B_0^T)^* \\ C_0 Z + B_{20}\hat{L} - B_0^T & -D_0 - D_0^T \end{bmatrix},$$

$$\Upsilon_1 = \begin{bmatrix} D_A & D_{B_1} & 0 & 0 \\ 0 & 0 & D_C & D_{B_2} \end{bmatrix},$$

$$\Upsilon_2 = \begin{bmatrix} G_A Z & 0 \\ G_{B_1}\hat{L} & 0 \\ G_C Z & 0 \\ G_{B_2}\hat{L} & 0 \end{bmatrix}, \quad \Upsilon_3 = \begin{bmatrix} 0 & 0 \\ G_B^T & D_D \end{bmatrix}, \quad \Upsilon_4 = \begin{bmatrix} D_B^T & 0 \\ 0 & G_D \end{bmatrix},$$

$$\hat{L} = e^{-j\theta} L, \quad Z = e^{-j\theta} Y, \quad \theta = \frac{\pi}{2}(1-\alpha).$$

Then the state feedback controller $u(t) = LY^{-1}x(t)$.

Proof. The proof is similar to that of Theorem 5.8 and is omitted. □

Theorem 5.6 contains equality constraints. From the perspective of numerical computation, it is necessary to provide results without equality constraints.

Theorem 5.7. *Suppose that $D_0 + D_0^T > 0$. When the order $\alpha \in [1, 2)$, the closed-loop system (5.3) is admissible and extended strictly positive real if there exist matrices $Y \in \mathbb{R}^{n\times n}$, $Y > 0$, $S \in \mathbb{R}^{(n-r)\times n}$, and $L \in \mathbb{R}^{m\times n}$ and constants $\varepsilon_1 > 0$ and $\varepsilon_2 > 0$ satisfying*

$$\begin{bmatrix} W_0 + \varepsilon_1 \Upsilon_1 \Upsilon_1^T + \varepsilon_2 \Upsilon_3 \Upsilon_3^T & \Upsilon_2^T & \Upsilon_4^T \\ \Upsilon_2 & -\varepsilon_1 I & 0 \\ \Upsilon_4 & 0 & -\varepsilon_2 I \end{bmatrix} < 0, \tag{5.57}$$

where

$$W_0 = \begin{bmatrix} \mathrm{Sym}\{A_0 Z + B_{10}\hat{L}\} & (C_0 Z + B_{20}\hat{L} - B_0^T)^* \\ C_0 Z + B_{20}\hat{L} - B_0^T & -D_0 - D_0^T \end{bmatrix},$$

$$\Upsilon_1 = \begin{bmatrix} D_A & D_{B_1} & 0 & 0 \\ 0 & 0 & D_C & D_{B_2} \end{bmatrix}, \quad \Upsilon_2 = \begin{bmatrix} G_A Z & 0 \\ G_{B_1}\hat{L} & 0 \\ G_C Z & 0 \\ G_{B_2}\hat{L} & 0 \end{bmatrix}, \quad \Upsilon_3 = \begin{bmatrix} 0 & 0 \\ G_B^T & D_D \end{bmatrix},$$

$$\Upsilon_4 = \begin{bmatrix} D_B^T & 0 \\ 0 & G_D \end{bmatrix}, \quad \hat{L} = e^{-j\theta} L, \quad Z = e^{-j\theta} J, \quad J = YE^T + E_0 S, \quad \theta = \frac{\pi}{2}(1-\alpha),$$

with E_0 a column-full rank matrix satisfying $EE_0 = 0$. (J is assumed to be nonsingular. Otherwise, we can choose any nonsingular real matrix $\hat{J}$ satisfying $E\hat{J} = \hat{J}^T E^T \geq 0$ and a small enough scalar $\xi \in (0, 1)$ that is not an eigenvalue of $-J\hat{J}^{-1}$ such that when J is replaced by $J + \xi\hat{J}$, inequality (5.57) still holds. Then $J + \xi\hat{J}$ is nonsingular.) Therefore we can obtain the state feedback controller $u(t) = LJ^{-1}x(t)$.

Proof. The proof of this theorem is similar to that of Theorem 5.3 and is omitted. □

5.4.2 Case $\alpha \in (0,1)$

Theorem 5.8. *Suppose that $D_0 + D_0^T > 0$. When the order $\alpha \in (0,1)$, the closed-loop system (5.3) is admissible and extended strictly positive real if there exist matrices $Y \in \mathbb{C}^{n\times n}$ and $L \in \mathbb{R}^{m\times n}$ and constants $\varepsilon_1 > 0$ and $\varepsilon_2 > 0$ satisfying*

$$EY = Y^*E^T \geq 0, \tag{5.58}$$

$$\begin{bmatrix} W_0 + \varepsilon_1 \Upsilon_1 \Upsilon_1^T + \varepsilon_2 \Upsilon_3 \Upsilon_3^T & \Upsilon_2^T & \Upsilon_4^T \\ \Upsilon_2 & -\varepsilon_1 I & 0 \\ \Upsilon_4 & 0 & -\varepsilon_2 I \end{bmatrix} < 0, \tag{5.59}$$

where

$$W_0 = \begin{bmatrix} \mathrm{Sym}\{A_0 Z + B_{10} L\} & (C_0 Z + B_{20} L - B_0^T)^T \\ C_0 Z + B_{20} L - B_0^T & -D_0 - D_0^T \end{bmatrix},$$

$$\Upsilon_1 = \begin{bmatrix} D_A & D_{B_1} & 0 & 0 \\ 0 & 0 & D_C & D_{B_2} \end{bmatrix},$$

$$\Upsilon_2 = \begin{bmatrix} G_A Z & 0 \\ G_{B_1} L & 0 \\ G_C Z & 0 \\ G_{B_2} L & 0 \end{bmatrix}, \quad \Upsilon_3 = \begin{bmatrix} 0 & 0 \\ G_B^T & D_D \end{bmatrix}, \quad \Upsilon_4 = \begin{bmatrix} D_B^T & 0 \\ 0 & G_D \end{bmatrix},$$

$$Z = e^{-j\theta} Y + e^{j\theta} \bar{Y}, \quad \theta = \frac{\pi}{2}(1-\alpha).$$

The Z here is assumed to be nonsingular. (If Z is singular, then we can choose any nonsingular real matrix $\hat{Z}$ satisfying $E\hat{Z} = \hat{Z}^T E^T \geq 0$ and a small enough scalar $\xi \in (0,1)$ that is not an eigenvalue of $-Z\hat{Z}^{-1}$ such that when Z is replaced by $Z + \xi\hat{Z}$, inequality (5.59) still holds. Then we can see that $Z + \xi\hat{Z}$ is nonsingular.) Therefore we can obtain the state feedback controller $u(t) = LZ^{-1}x(t)$.

Proof. By Lemma 4.7, inequality (5.59) is equivalent to

$$W_0 + \varepsilon_1 \Upsilon_1 \Upsilon_1^T + \varepsilon_2 \Upsilon_3 \Upsilon_3^T + \varepsilon_1^{-1} \Upsilon_2^T \Upsilon_2 + \varepsilon_2^{-1} \Upsilon_4^T \Upsilon_4 < 0. \tag{5.60}$$

Let $M_1 = \begin{bmatrix} F_A & 0 & 0 & 0 \\ 0 & F_{B_1} & 0 & 0 \\ 0 & 0 & F_C & 0 \\ 0 & 0 & 0 & F_{B_2} \end{bmatrix}$ and $M_2 = \begin{bmatrix} -F_B^T & 0 \\ 0 & -F_D \end{bmatrix}$. Then $M_1^T M_1 \leq I$ and $M_2^T M_2 \leq I$. By Lemma 4.6 there exist $\varepsilon_1 > 0$ and $\varepsilon_2 > 0$ such that

$$\begin{aligned} \mathrm{Sym}\{\Upsilon_1 M_1 \Upsilon_2\} &\leq \varepsilon_1 \Upsilon_1 \Upsilon_1^T + \varepsilon_1^{-1} \Upsilon_2^T M_1^T M_1 \Upsilon_2 \leq \varepsilon_1 \Upsilon_1 \Upsilon_1^T + \varepsilon_1^{-1} \Upsilon_2^T \Upsilon_2, \\ \mathrm{Sym}\{\Upsilon_3 M_1 \Upsilon_4\} &\leq \varepsilon_2 \Upsilon_3 \Upsilon_3^T + \varepsilon_2^{-1} \Upsilon_4^T M_2^T M_2 \Upsilon_4 \leq \varepsilon_2 \Upsilon_3 \Upsilon_3^T + \varepsilon_2^{-1} \Upsilon_4^T \Upsilon_4. \end{aligned} \tag{5.61}$$

From inequalities (5.60) and (5.61), we have

$$W_0 + \mathrm{Sym}\{\Upsilon_1 M_1 \Upsilon_2\} + \mathrm{Sym}\{\Upsilon_3 M_1 \Upsilon_4\} < 0. \tag{5.62}$$

Substituting Υ_1, Υ_2, Υ_3, Υ_4, M_1, and M_2 into (5.62), we have

$$W_0 + \mathrm{Sym}\left\{\begin{bmatrix} D_A F_A G_A Z + D_{B_1} F_{B_1} G_{B_1} L & 0 \\ D_C F_C G_C Z + D_{B_2} F_{B_2} G_{B_2} L - (D_B F_B G_B)^T & -D_D F_D G_D \end{bmatrix}\right\} < 0. \tag{5.63}$$

Note that $K = LZ^{-1}$, and thus $L = KZ$. Substituting W_0 and L into (5.63), we have

$$\begin{bmatrix} \mathrm{Sym}\{AZ + B_1 KZ\} & (CZ + B_2 KZ - B^T)^T \\ CZ + B_2 KZ - B^T & -D - D^T \end{bmatrix} < 0, \tag{5.64}$$

that is,

$$\begin{bmatrix} \mathrm{Sym}\{(A + B_1 K)Z\} & ((C + B_2 K)Z - B^T)^T \\ (C + B_2 K)Z - B^T & -D - D^T \end{bmatrix} < 0. \tag{5.65}$$

From (5.58) and (5.65) it follows that the system satisfies the admissibility and extended strictly positive realness. This ends the proof. □

Theorem 5.8 contains equality constraints. From the perspective of numerical computation, it is necessary to provide results without equality constraints.

Theorem 5.9. *Suppose that $D_0 + D_0^T > 0$. When the order $\alpha \in (0, 1)$, the closed-loop system (5.3) is admissible and extended strictly positive real if there exist matrices $Y \in \mathbb{C}^{n\times n}$, $Y > 0$, $S \in \mathbb{R}^{(n-r)\times n}$, and $L \in \mathbb{R}^{m\times n}$ and constants $\varepsilon_1 > 0$ and $\varepsilon_2 > 0$ satisfying*

$$\begin{bmatrix} W_0 + \varepsilon_1 \Upsilon_1 \Upsilon_1^T + \varepsilon_2 \Upsilon_3 \Upsilon_3^T & \Upsilon_2^T & \Upsilon_4^T \\ \Upsilon_2 & -\varepsilon_1 I & 0 \\ \Upsilon_4 & 0 & -\varepsilon_2 I \end{bmatrix} < 0, \tag{5.66}$$

where

$$W_0 = \begin{bmatrix} \mathrm{Sym}\{A_0 Z + B_{10} L\} & (C_0 Z + B_{20} L - B_0^T)^T \\ C_0 Z + B_{20} L - B_0^T & -D_0 - D_0^T \end{bmatrix},$$

$$\Upsilon_1 = \begin{bmatrix} D_A & D_{B_1} & 0 & 0 \\ 0 & 0 & D_C & D_{B_2} \end{bmatrix},$$

$$\Upsilon_2 = \begin{bmatrix} G_A Z & 0 \\ G_{B_1} L & 0 \\ G_C Z & 0 \\ G_{B_2} L & 0 \end{bmatrix}, \quad \Upsilon_3 = \begin{bmatrix} 0 & 0 \\ G_B^T & D_D \end{bmatrix}, \quad \Upsilon_4 = \begin{bmatrix} D_B^T & 0 \\ 0 & G_D \end{bmatrix},$$

$$Z = (e^{-j\theta} Y + e^{j\theta} \bar{Y}) E^T + E_0 S, \quad \theta = \frac{\pi}{2}(1 - \alpha),$$

with E_0 any column-full rank matrix satisfying $EE_0 = 0$. (Here Z is assumed to be nonsingular. Otherwise, we can choose any nonsingular real matrix $\hat{Z}$ satisfying $E\hat{Z} = \hat{Z}^T E^T \geq 0$ and a small enough scalar $\xi \in (0, 1)$ that is not an eigenvalue of $-Z\hat{Z}^{-1}$ such that when Z is *replaced by $Z + \xi\hat{Z}$, inequality (5.66) still holds. Then we can see that $Z + \xi\hat{Z}$ is nonsingular.) Therefore we can obtain the state feedback controller $u(t) = LZ^{-1}x(t)$.*

Proof. The proof of this theorem is similar to that of Theorem 5.8 and is omitted. □

Example 5.4. Example 5.2 analyzes the admissibility and extended strictly positive realness of a nominal FOSS of system order $\alpha = 0.8$, and in this example, we consider how to design controllers to make the closed-loop system (5.3) satisfy the robust admissibility and extended strictly positive realness when there are uncertainties in this model. Take the coefficient matrix and uncertain parameter matrix of the control input $u(t)$ as

$$\begin{gathered} B_{10} = \begin{bmatrix} 4 \\ 4 \\ 4 \end{bmatrix}, \quad B_{20} = \begin{bmatrix} 1 \\ 1 \end{bmatrix}, \quad D_{B1} = \begin{bmatrix} 0.1 \\ 0 \\ 0.1 \end{bmatrix}, \quad G_{B1} = 0.1, \quad D_{B2} = \begin{bmatrix} 0.1 \\ 0.1 \end{bmatrix}, \\ G_{B2} = 0.2, \quad D_A = \begin{bmatrix} 1.1 \\ 0.1 \\ 0 \end{bmatrix}, \quad G_A = \begin{bmatrix} 1 \\ 0.1 \\ 0 \end{bmatrix}^T, \quad D_B = \begin{bmatrix} 0.1 \\ 0.1 \\ 0.1 \end{bmatrix}, \quad G_B = \begin{bmatrix} 0.1 \\ 0.1 \end{bmatrix}^T, \\ D_C = \begin{bmatrix} 0.1 \\ 0.1 \end{bmatrix}, \quad G_C = \begin{bmatrix} 0.1 \\ 0.1 \\ 0.1 \end{bmatrix}, \quad D_D = \begin{bmatrix} 0.1 \\ 0.1 \end{bmatrix}, \quad G_D = \begin{bmatrix} 0.1 \\ 0.1 \end{bmatrix}^T. \end{gathered} \tag{5.67}$$

The open-loop system is first analyzed for admissibility and extended strictly positive realness. We can verify that the uncertain FOSS (5.1) fails to satisfy both the admissibility and extended strictly positive realness when the control input $u(t) = 0$. For example, when the uncertain parameters F_Γ, $\Gamma \in \{A, B, C, D\}$, are taken to be 1, the parameters A, B, C, D are as follows:

$$A = \begin{bmatrix} 0.1 & 0.11 & 0 \\ 0.1 & -1.99 & 0 \\ 1 & 1 & -1 \end{bmatrix}, \quad B = \begin{bmatrix} 1.01 & 0.01 \\ 0.01 & 1.01 \\ 0.01 & 0.01 \end{bmatrix},$$

$$C = \begin{bmatrix} 0.01 & 1.01 & 0.01 \\ 1.01 & 0.01 & 0.01 \end{bmatrix}, \quad D = \begin{bmatrix} 1.01 & 1.01 \\ 1.01 & 1.01 \end{bmatrix}.$$

At this point, the eigenvalues of the open-loop system are −0.3985 and 0.0105, as shown in Figure 5.7, and one eigenvalue is located in the unstable area. The $\mathrm{Sym}\{G(jw)\}$ $(w \geq 0)$ curve is shown in Figure 5.8, from which we can see that some of the values of $\mathrm{Sym}\{G(jw)\}$ $(w \geq 0)$ are less than 0. Therefore the open-loop system does not satisfy the admissibility and extended strictly positive realness.

According to Theorem 5.8, the following feasible solutions Y, L, ε_1, and ε_2 can be found:

$$Y = \begin{bmatrix} 3.505 & 4.1802 - 0.0246\mathbf{i} & 0 \\ 8.3605 + 0.0493\mathbf{i} & 13.1696 & 0 \\ -9.2299 & -14.2475 & 63.8898 \end{bmatrix},$$

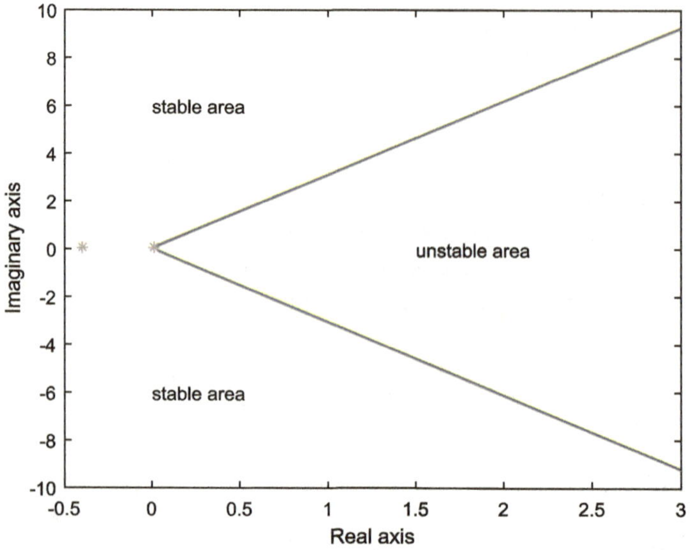

Figure 5.7: The eigenvalue location of the open-loop FOSS (5.1) in Example 5.4.

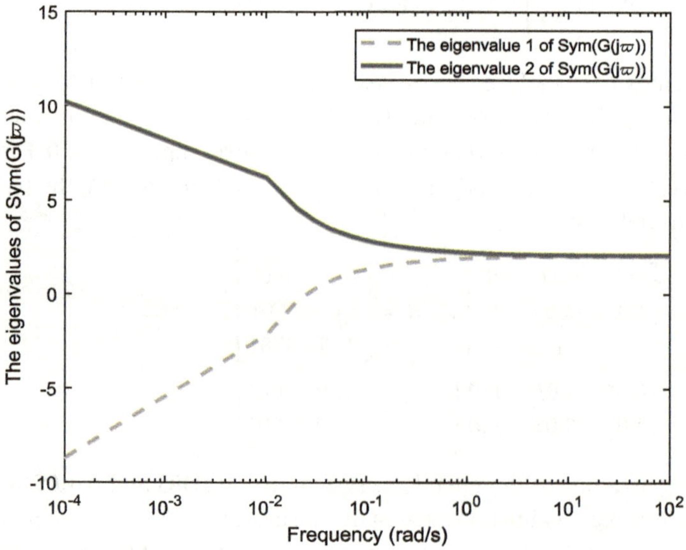

Figure 5.8: The eigenvalue curves of Sym$\{G(jw)\}$ ($w \geq 0$) of the open-loop FOSS (5.1) in Example 5.4.

$$L = \begin{bmatrix} -14.1024 & -19.0918 & -3.9412 \end{bmatrix}, \quad \varepsilon_1 = \varepsilon_2 = 1.$$

Therefore the designed state feedback controller is

$$u(t) = Kx(t) = \begin{bmatrix} -1.2162 & -0.4119 & -0.0324 \end{bmatrix} x(t).$$

According to Theorem 5.9, the following feasible solutions Y, L, ε_1, and ε_2 can be found:

$$Y = \begin{bmatrix} 0.2929 & 0.6748 + 0.0377\mathbf{i} & 0 \\ 0.6748 - 0.0377\mathbf{i} & 2.0801 & 0 \\ 0 & 0 & 80677.6566 \end{bmatrix},$$

$$S = \begin{bmatrix} -15.8997 & -23.364 & 97.44 \end{bmatrix},$$

$$L = \begin{bmatrix} -11.6996 & -15.5783 & -3.5737 \end{bmatrix}, \quad \varepsilon_1 = \varepsilon_2 = 1.$$

Therefore the designed state feedback controller is

$$u(t) = Kx(t) = \begin{bmatrix} -1.2869 & -0.4057 & -0.0367 \end{bmatrix} x(t).$$

To verify the effectiveness of the designed state-feedback positive real controller, a set of parameters is taken, and the result obtained by Theorem 5.8 is illustrated as an example. For example, when $F_A, F_B, F_C, F_D, F_{B1}, F_{B2}$ are all taken to be 1, the system parameters are as follows:

$$\begin{aligned} &A = \begin{bmatrix} 0.1 & 0.11 & 0 \\ 0.1 & -1.99 & 0 \\ 1 & 1 & -1 \end{bmatrix}, \quad B = \begin{bmatrix} 1.01 & 0.01 \\ 0.01 & 1.01 \\ 0.01 & 0.01 \end{bmatrix}, \quad C = \begin{bmatrix} 0.01 & 1.01 \\ 1.01 & 0.01 \\ 0.01 & 0.01 \end{bmatrix}^T, \\ &D = \begin{bmatrix} 1.01 & 1.01 \\ 1.01 & 1.01 \end{bmatrix}, \quad B_1 = \begin{bmatrix} 4.01 \\ 4 \\ 4.01 \end{bmatrix}, \quad B_2 = \begin{bmatrix} 1.02 \\ 1.02 \end{bmatrix}. \end{aligned} \tag{5.68}$$

The eigenvalues of the closed-loop system of uncertain FOSS (5.68) with parameter (5.3) under the action of the state-feedback controller derived by Theorem 5.8 are -0.9553 and -0.1904. As shown in Figure 5.9, the eigenvalues are all within the stability areas, i. e., the system is stable. Meanwhile, (i) $\det(s^\alpha E - A)$ is not constant 0, i. e., the system is regular; (ii) take the left and right maximal zeroed submatrices of the matrix E, $L_E = \begin{bmatrix} 0 & 0 & 1 \end{bmatrix}$ and $R_E^T = \begin{bmatrix} 0 & 0 & 1 \end{bmatrix}$; then $\operatorname{rank}(L_E A_0 R_E) = 1$, i. e., the system is impulse-free. Therefore the FOSS is admissible. The $\operatorname{Sym}\{G(jw)\}$ $(w \geq 0)$ curve of the closed-loop system is shown in Figure 5.10, from which we can see that the values of $\operatorname{Sym}\{G(jw)\}$ $(w \geq 0)$ are all greater than 0. Therefore the closed-loop system satisfies both the admissibility and extended strictly positive realness.

Remark 5.6. In Example 5.4, we consider an FOSS with norm-bounded uncertainties and verify that the system fails to satisfy both the admissibility and extended strictly positive realness when the uncertain parameter $F_\Gamma = 1$, $\Gamma \in \{A, B, C, D\}$. In fact, when $F_\Gamma \in (-1, 1)$, $\Gamma \in \{A, B, C, D\}$, i. e., when the uncertain parameter is distributed on $(-1, 1)$, the distribution of eigenvalues of the open-loop uncertain FOSS is shown in Figure 5.11, which shows that some of the eigenvalues are located in the unstable areas. At this

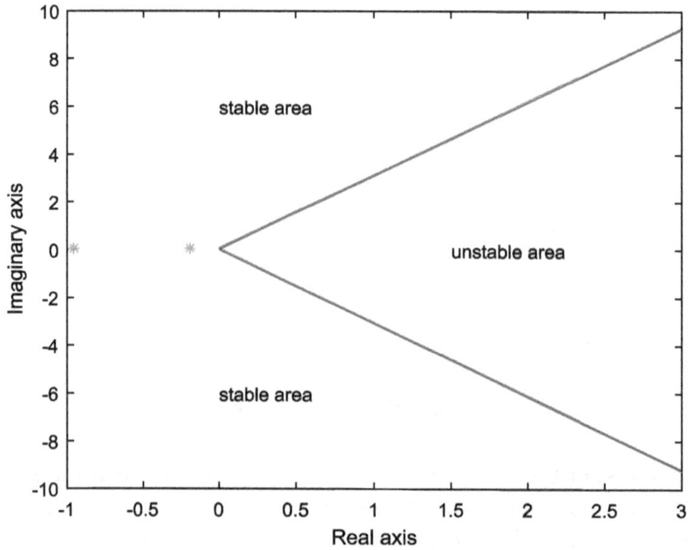

Figure 5.9: The eigenvalue location of the closed-loop FOSS (5.3) in Example 5.4.

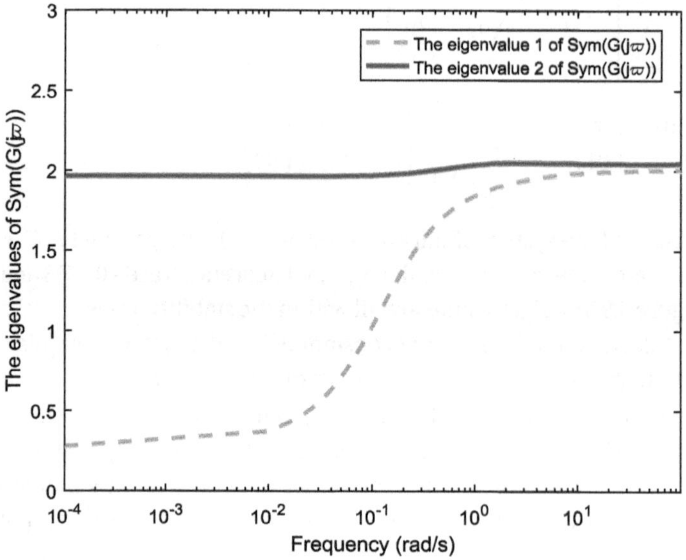

Figure 5.10: The curves of the closed-loop FOSS Sym$\{G(jw)\}$ ($w \geq 0$) in Example 5.4.

point, the distribution of eigenvalue curves of the open-loop uncertain FOSS Sym$\{G(jw)\}$ ($w \geq 0$) is shown in Figure 5.12, from which we can see that some of the eigenvalue curves of Sym$\{G(jw)\}$ ($w \geq 0$) have distributions less than 0. To make the system satisfy both the admissibility and extended strictly positive realness, we design state feedback

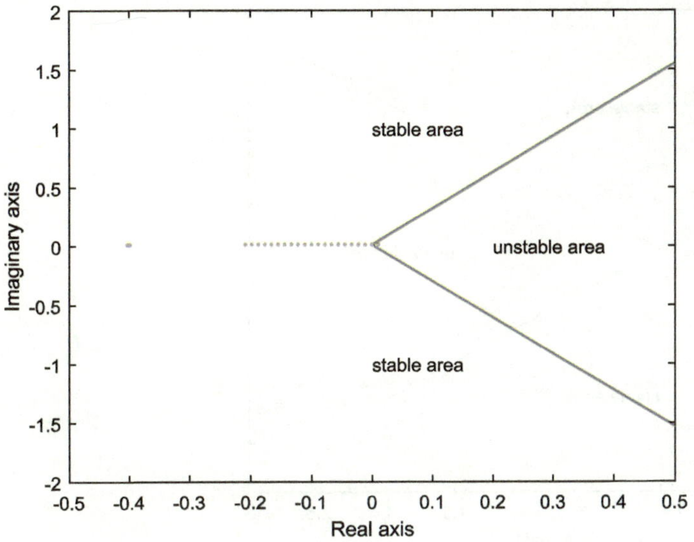

Figure 5.11: The distribution of the eigenvalues of the open-loop uncertain FOSS (5.1) in Example 5.4.

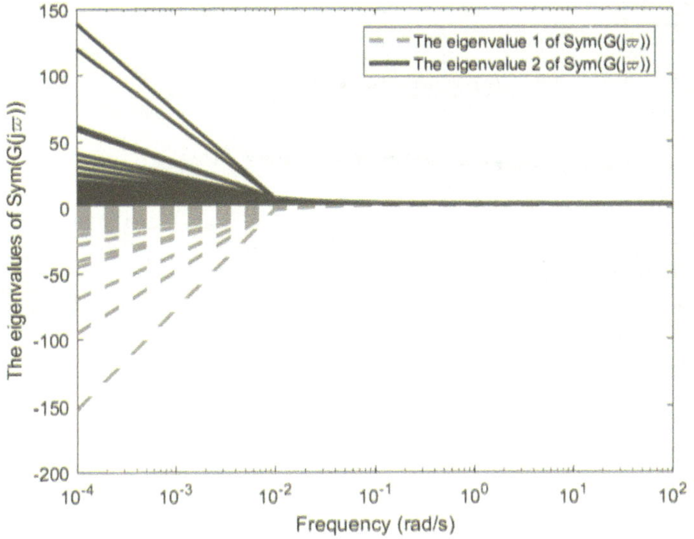

Figure 5.12: The distribution of the eigenvalue curves of Sym$\{G(jw)\}$ ($w \geq 0$) of the open-loop uncertain FOSS (5.1) in Example 5.4.

controllers based on Theorems 5.8 and 5.9 and verify that the system simultaneously satisfies the admissibility and extended strictly positive realness when the uncertain parameters $F_\Gamma = 1, \Gamma \in \{A, B, C, D, B_1, B_2\}$, are not determined. In fact, when $F_\Gamma \in (-1, 1)$, $\Gamma \in \{A, B, C, D, B_1, B_2\}$, i. e., when the uncertain parameter is distributed on $(-1, 1)$, the

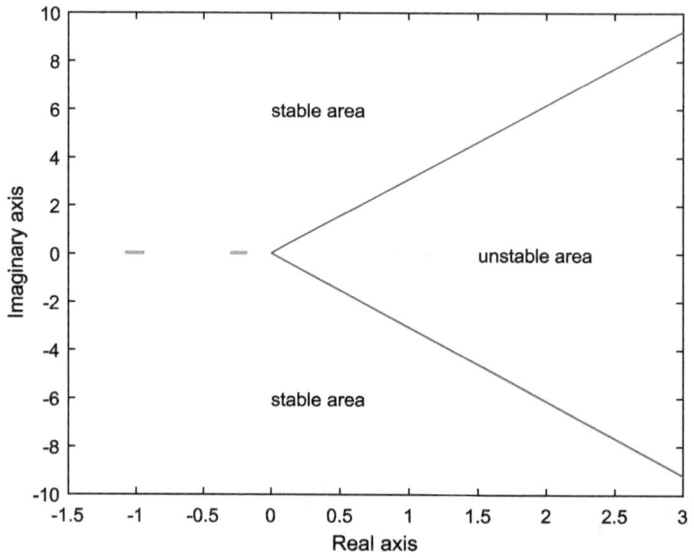

Figure 5.13: The distribution of the eigenvalues of the closed-loop uncertain FOSS (5.1) in Example 5.4.

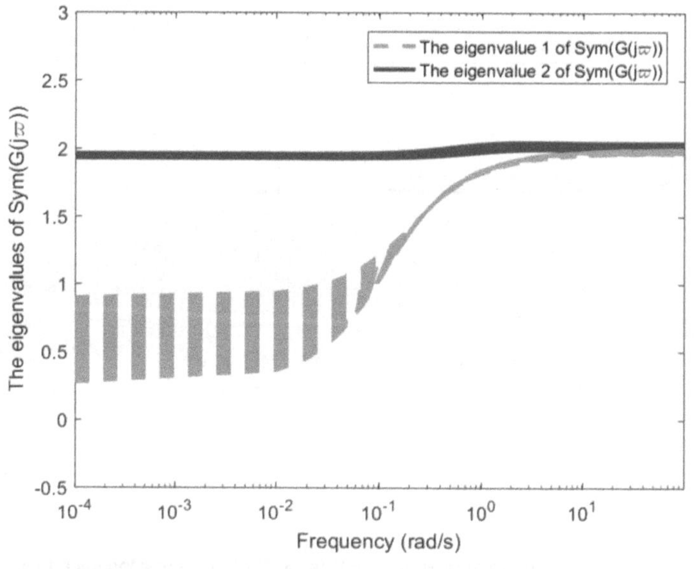

Figure 5.14: The distribution of the eigenvalue curves of Sym$\{G(jw)\}$ ($w \geq 0$) of the closed-loop uncertain FOSS (5.1) in Example 5.4.

eigenvalue distribution of the closed-loop uncertain FOSS is shown in Figure 5.13, and we can see that all eigenvalues are located in the stable areas. The eigenvalue curve distribution of the closed-loop uncertain FOSS Sym$\{G(jw)\}$ ($w \geq 0$) is shown in Figure 5.14, and we can see that all of them are greater than 0, that is, the system satisfies both the admissibility and extended strictly positive realness, indicating that the designed controller is effective.

5.5 Chapter summary

This chapter addresses the problem of robust positive real control of FOSSs with norm-bounded uncertainties, and the main contributions are as follows:

1) For nominal FOSs of orders $\alpha \in (0,1)$ and $\alpha \in [1,2)$, novel conditions have been proposed to ensure both the stability and extended strictly positive realness. This represents a significant improvement over the results in [123], which guarantee only the positive realness of the system, without addressing its stability.
2) For nominal FOSSs of order $\alpha \in (0,1)$, a positive real lemma involving two complex variables has been proposed. Subsequently, a positive real lemma with a single complex variable has been derived by leveraging the systems' linear superposition property, making it more advantageous for controller design. For nominal FOSSs of order $\alpha \in [1,2)$, nonconservative positive real lemmas have been introduced, meaning that the system satisfies both sufficient and necessary conditions for the admissibility and extended strictly positive realness.
3) For FOSSs of orders $\alpha \in (0,1)$ and $\alpha \in [1,2)$ with bounded uncertainty in the number of paradigms, a solution to the synthesis problem of the state-feedback positive real controllers has been given, and corresponding state-feedback controllers have been designed such that the closed-loop system satisfies both the admissibility and extended strictly positive realness. For FOSSs of orders $\alpha \in (0,1)$ and $\alpha \in [1,2)$ with norm-bounded uncertainties, solutions to the state-feedback controller synthesis problem have been provided. State feedback controllers have been designed to ensure that the closed-loop systems satisfy both the admissibility and extended strictly positive realness.

6 Fault detection

Abstract: In this chapter, we investigate the robust fault detection (FD) observer design problem in finite-frequency domains for fractional-order singular systems (FOSSs). We design an H_-/H_∞ fault detection observer for FOSSs to simultaneously satisfy the system admissibility, disturbance robustness, and fault sensitivity indices in finite-frequency domains. We give four theorems and four corollaries about the system admissibility and the performance index H_-/H_∞ in different frequency ranges. Then we obtain sufficient conditions in terms of linear matrix inequalities (LMIs). Finally, by two numerical examples, we verify the validity of the presented methods.

6.1 Introduction

Finding an effective way to diagnose faults has been a hot research topic. Among the existing research, the model-based FD methods are one of the most effective ways. Most FD methods are based on the robust fault detection observer. The observer is designed to produce a residual signal and a residual evaluation function with a threshold chosen to detect the faults. The main challenge here is to distinguish between faults and disturbances. Among many researches that have attempted to solve this problem, the H_- index has been proposed and studied to measure the fault sensitivity performance. The H_-/H_∞ FD observer is considered to be one of the most effective methods to simultaneously generate a residual signal with disturbance attenuation and fault sensitivity performance indices [51, 35, 109, 122].

Unfortunately, it has been pointed out in [95] that most full frequency H_-/H_∞ fault detection observers are valid only for failures that appear directly in the measurement equation. In practice, many faults and disturbances often occur within certain frequency ranges. The FD observer design problem over finite-frequency ranges for fractional-order system has been investigated in [43], but their method is not valid for singular systems and requires to select an arbitrary matrix and two scalars, which make their method problematic to apply in large-scale systems. Moreover, there is no definition and lemmas about the H_- indices in finite-frequency ranges for singular FOSs, and the existing H_∞ lemmas will be problematic to deal with the disturbances that effect on the sensor. As we know, there is no result about H_-/H_∞ FD observer design methods for FOSSs over finite-frequency ranges yet. So, how to design an H_-/H_∞ FD observer over finite-frequency ranges for FOSS is a valuable open problem.

With the above motivations, this chapter focuses on H_-/H_∞ FD observer design for FOSSs over finite-frequency ranges, and the main contributions can be summarized as follows: Firstly, an H_-/H_∞ FD observer is designed to simultaneously meet admissibility condition, disturbance robustness, and fault sensitivity indices in finite-frequency ranges. Four theorems and four corollaries about the system admissibility and the performance index H_-/H_∞ in different frequency ranges are given. Then, based on the gen-

https://doi.org/10.1515/9783111628646-006

eralized Kalman–Yakubovich–Popov (GKYP) lemma, the projection lemma, and some matrix decoupling techniques, the observer design conditions in different frequency domains are converted into solvable LMIs. Numerical examples are given to illustrate the validity of the proposed theorems and corollaries. Finally, the proposed methods are applied to a real electrical circuit to show their effectiveness.

6.2 Preliminaries

Consider the following FOSS:

$$\begin{cases} ED^{\alpha}x(t) = Ax(t) + Bu(t), \\ y(t) = Cx(t) + Du(t), \end{cases} \tag{6.1}$$

in which α stands for the fractional order, $y(t) \in \mathbb{R}^{n_y}$, $u(t) \in \mathbb{R}^{n_u}$ and $x(t) \in \mathbb{R}^{n}$ denote the measured output, control input, and state vector, $E, A \in \mathbb{R}^{n\times n}$, B, C, D are real constant matrices with $\operatorname{rank}(E) = r \leq n$, and D^{α} denotes the Caputo fractional-order derivative. When the pair (E, A) is regular, the transfer function matrix of system (6.1) is

$$G(s) = C(s^{\alpha}E - A)^{-1}B + D. \tag{6.2}$$

Definitions and lemmas required to obtain the main results are as follows.

Definition 6.1. The H_{-} norm of FOSS (6.1) is defined as

$$\|G\|_{-} := \inf_{\omega} \sigma_{\min}(G(j\omega)) \quad \forall \omega \in \Omega, \Omega \subseteq \mathbb{R}^{+},$$

where $\sigma_{\min}$ represents the minimum singular value.

Remark 6.1. Note that the H_{-} index is used for measuring the worst case of fault sensitivity. The H_{-} index in finite-frequency ranges for FOSs has been presented in [43]. It is extended to FOSSs in Definition 6.1.

Lemma 6.1 ([63]). *The FOSS* (6.1) *with* $\alpha \in [1, 2)$ *is admissible if and only if there exists a matrix* $Q \in \mathbb{R}^{n\times n}$ *such that*

$$\operatorname{Sym}\{\Theta \otimes (A^{T}Q)\} < 0, \quad E^{T}Q = Q^{T}E \geq 0, \tag{6.3}$$

where $\Theta = \begin{bmatrix} \sin\theta & -\cos\theta \\ \cos\theta & \sin\theta \end{bmatrix}$, $\theta = (1 - \alpha/2)\pi$.

Lemma 6.2 (GKYP lemma [45]). *Let* $A \in \mathbb{R}^{n\times n}$, $B \in \mathbb{R}^{n\times m}$, $\Theta \in \mathbb{H}_{n+m}$, *and* $\Phi, \Psi \in \mathbb{H}_2$. *The following statements are equivalent:*

(i) *For* $H(\lambda) = (\lambda E - A)^{-1}B$,

$$\begin{bmatrix} H(\lambda) \\ I \end{bmatrix}^{*} \Theta \begin{bmatrix} H(\lambda) \\ I \end{bmatrix} < 0 \quad \forall \lambda \in \bar{\Lambda}(\Phi, \Psi); \tag{6.4}$$

(ii) *There exist $P, Q \in \mathbb{H}_2$, $Q > 0$, such that*

$$\begin{bmatrix} A & B \\ E & 0 \end{bmatrix}^* (\Phi \otimes P + \Psi \otimes Q) \begin{bmatrix} A & B \\ E & 0 \end{bmatrix} + \Theta < 0, \tag{6.5}$$

where $\Phi = \left[\begin{smallmatrix} 0 & r \\ \bar{r} & 0 \end{smallmatrix}\right]$, $r = e^{j(1-\alpha)\pi/2}$, $\omega_c = jr(\omega_l^\alpha + \omega_h^\alpha)/2$, *and*

$$\Psi = \begin{cases} \begin{bmatrix} -1 & 0 \\ 0 & \omega_l^{2\alpha} \end{bmatrix}, & 0 \le \omega \le \omega_l, \\ \begin{bmatrix} -1 & \omega_c \\ \overline{\omega_c} & -\omega_l^\alpha \omega_h^\alpha \end{bmatrix}, & 0 \le \omega_l \le \omega \le \omega_h, \\ \begin{bmatrix} 1 & 0 \\ 0 & -\omega_h^{2\alpha} \end{bmatrix}, & 0 \le \omega_h \le \omega. \end{cases}$$

Lemma 6.3 (Projection lemma [86]). *Let $\mathcal{B} \in \mathbb{C}^{n\times m}$, $\mathcal{C} \in \mathbb{C}^{k\times n}$, and $\Xi \in \mathbb{H}_n$. Then the following statements are equivalent:*

(i) *There exists a matrix $\mathcal{W}$ satisfying*

$$\mathcal{B}\mathcal{W}\mathcal{C} + \mathcal{C}^*\mathcal{W}^*\mathcal{B}^* + \Xi < 0. \tag{6.6}$$

(ii) *The following two conditions hold:*

$$\begin{aligned} \mathcal{B}^{\perp}\Xi\mathcal{B}^{\perp *} &< 0 \quad \text{or } \mathcal{B}\mathcal{B}^* > 0. \\ \mathcal{C}^{*\perp}\Xi\mathcal{C}^{*\perp *} &< 0 \quad \text{or } \mathcal{C}^*\mathcal{C} > 0. \end{aligned} \tag{6.7}$$

Consider the following FOSS with disturbance and additive fault:

$$\begin{cases} ED^\alpha x(t) = Ax(t) + B_u u(t) + B_d d(t) + B_f f(t), \\ y(t) = Cx(t) + D_u u(t) + D_d d(t) + D_f f(t), \end{cases} \tag{6.8}$$

where $\alpha \in (0, 2)$ is the fractional order, $x(t) \in \mathbb{R}^n$, $d(t) \in \mathbb{R}^{n_d}$, $f(t) \in \mathbb{R}^{n_f}$, $u(t) \in \mathbb{R}^{n_u}$, and $y(t) \in \mathbb{R}^{n_y}$ stand for the pseudo-state, disturbance signal, fault signal, control input, and measured output, respectively, and E, A, B_u, B_f, B_d, C, D_u, D_f, D_d are known constant real system matrices. The disturbance and fault signals are assumed to be in certain frequency ranges: $\omega_d \in \Omega_d$ and $\omega_f \in \Omega_f$, respectively.

Remark 6.2. System (6.8) is assumed to be admissible and impulse observable. The observer design LMIs will be unsolvable if these two conditions are not met. The admissibility of the system can be judged by Lemmas 5.4 and 6.1. The impulse observability can be judged by the theorems proposed in [33].

To generate a residual signal $r(t)$ for fault detection, a fractional-order singular FD observer is designed as

$$\begin{cases} ED^{\alpha}\hat{x}(t) = A\hat{x}(t) + B_u u(t) + L(y(t) - \hat{y}(t)), \\ \hat{y}(t) = C\hat{x}(t) + D_u u(t), \\ r(t) = y(t) - \hat{y}(t), \end{cases} \tag{6.9}$$

where $r(t) \in \mathbb{R}^{n_y}$ denotes the residual signal, $\hat{y} \in \mathbb{R}^{n_y}$ stands for the output estimation signal, $\hat{x}(t) \in \mathbb{R}^n$ is the observer pseudo-state, and L is the gain matrix to be determined.

Denote $\bar{A} = A - LC$, $\overline{B_d} = B_d - LD_d$, and $\overline{B_f} = B_f - LD_f$. The observation error dynamics is as follows:

$$\begin{cases} ED^{\alpha}e(t) = \bar{A}e(t) + \overline{B_d}d(t) + \overline{B_f}f(t), \\ r(t) = Ce(t) + D_d d(t) + D_f f(t), \end{cases} \tag{6.10}$$

where $e(t) = x(t) - \hat{x}(t)$ is the estimation error.

The finite-frequency fault detection problem for FOSSs can be described as follows: For a admissible and observable system (6.8), a fault detection observer (6.9) should ensure that system (6.10) is admissible and satisfies the disturbance attenuation and fault sensitivity performance indices. The problem can be formulated as three constraints:

(i) For $G_{rd}(j\omega_d) = C((j\omega_d)^{\alpha}E - \bar{A})^{-1}\overline{B_d} + D_d$,

$$\|G_{rd}(j\omega_d)\|_{\infty} < \gamma \quad \forall \omega_d \in \Omega_d. \tag{6.11}$$

(ii) For $G_{rf}(j\omega_f) = C((j\omega_f)^{\alpha}E - \bar{A})^{-1}\overline{B_f} + D_f$,

$$\|G_{rf}(j\omega_f)\|_{-} > \beta \quad \forall \omega_f \in \Omega_f. \tag{6.12}$$

(iii) $(E, \bar{A})$ is regular, stable, and impulse-free.

6.3 Disturbance attenuation

LMI conditions for designing an observer gain matrix L in medium-frequency, low-frequency, and high-frequency ranges are given such that the error dynamics (6.10) satisfies the disturbance robustness condition (6.11).

Theorem 6.1 (Medium-frequency range). *Consider the error dynamics* (6.10) *with* $\alpha \in [0, 2)$. *The disturbance signal is supposed to be in the medium-frequency range* $\Omega_d = [\omega_l, \omega_h]$. *Let a scalar* ϵ_1 *be given. Then inequality* (6.11) *holds if there exists matrices* $P_d, Q_d \in \mathbb{H}_n$, $Q_d > 0$, *and a matrix variable* W *such that*

$$\begin{bmatrix} \Upsilon_{11} & \Upsilon_{12} & \Upsilon_{13} \\ \star & \Upsilon_{22} & \Upsilon_{23} \\ \star & \star & \Upsilon_{33} \end{bmatrix} < 0, \tag{6.13}$$

where $r = e^{j(1-\alpha)\pi/2}$, $\omega_c = jr(\omega_l^\alpha + \omega_h^\alpha)/2$, $W = \left[\begin{smallmatrix} W_1 & W_2 & W_3 \\ W_4 & W_5 & W_6 \end{smallmatrix}\right]$, *and*

$$\begin{aligned}
\Upsilon_{11} &= \begin{bmatrix} -Q_d & rP_d + \omega_c Q_d - W_1 \\ \star & -\omega_l^\alpha \omega_h^\alpha Q_d - \mathrm{Sym}\{W_4\} \end{bmatrix}, \\
\Upsilon_{12} &= \begin{bmatrix} -W_2 & -W_3 \\ -W_5 + W_4^* E^T + W_1^* A^T & -W_6 + W_1^* C^T \end{bmatrix}, \\
\Upsilon_{13} &= \begin{bmatrix} 0 & 0 \\ 0 & -W_1^* \end{bmatrix}, \\
\Upsilon_{22} &= \begin{bmatrix} \mathrm{Sym}\{AW_2 + EW_5\} & AW_3 + EW_6 + W_2^* C^T \\ \star & \mathrm{Sym}\{CW_3\} - \gamma^2 I \end{bmatrix}, \\
\Upsilon_{23} &= \begin{bmatrix} B_d - LD_d & -W_2^* + \epsilon_1 LC \\ D_d & -W_3^* \end{bmatrix}, \\
\Upsilon_{33} &= \begin{bmatrix} -I & 0 \\ \star & -\epsilon_1 I \end{bmatrix}.
\end{aligned}$$

Proof. Inequality (6.11) can be rewritten by some basic matrix calculations as follows:

$$\begin{bmatrix} G_{rd}^*(j\omega) \\ I \end{bmatrix}^* \begin{bmatrix} I & 0 \\ 0 & -\gamma^2 I \end{bmatrix} \begin{bmatrix} G_{rd}^*(j\omega) \\ I \end{bmatrix} < 0 \quad \forall \omega \in [\omega_l, \omega_h]. \tag{6.14}$$

By applying Lemma 6.2, we can show that (6.14) is equivalent to

$$\begin{bmatrix} \bar{A} & E \\ C & 0 \end{bmatrix} (\Phi \otimes P_d + \Psi \otimes Q_d) \begin{bmatrix} \bar{A} & E \\ C & 0 \end{bmatrix}^T + \begin{bmatrix} \overline{B_d} & 0 \\ D_d & I \end{bmatrix} \begin{bmatrix} I & 0 \\ 0 & -\gamma^2 I \end{bmatrix} \begin{bmatrix} \overline{B_d} & 0 \\ D_d & I \end{bmatrix}^T < 0. \tag{6.15}$$

Denote

$$\Xi_d = \begin{bmatrix} -Q_d & rP_d + \omega_c Q_d & 0 & 0 \\ \star & -\omega_l^\alpha \omega_h^\alpha Q_d & 0 & 0 \\ \star & \star & \overline{B_d}\overline{B_d}^T & \overline{B_d} D_d^T \\ \star & \star & \star & -\gamma^2 I + D_d D_d^T \end{bmatrix}. \tag{6.16}$$

Then (6.15) can be reformulated as

$$\begin{bmatrix} \bar{A}^T & C^T \\ E^T & 0 \\ I & 0 \\ 0 & I \end{bmatrix}^T \Xi_d \begin{bmatrix} \bar{A}^T & C^T \\ E^T & 0 \\ I & 0 \\ 0 & I \end{bmatrix} < 0. \tag{6.17}$$

Note that

$$[I \quad 0 \quad 0 \quad 0]\,\Xi_d\,[I \quad 0 \quad 0 \quad 0]^T = -Q_d < 0. \tag{6.18}$$

By Lemma 6.3, (6.17) and (6.18) hold if and only if there exists a matrix variable W of appropriate dimension such that

$$\Xi_d + \mathrm{Sym}\left(\begin{bmatrix} -I & 0 \\ 0 & -I \\ \bar{A} & E \\ C & 0 \end{bmatrix} W \begin{bmatrix} 0 & I & 0 & 0 \\ 0 & 0 & I & 0 \\ 0 & 0 & 0 & I \end{bmatrix}\right) < 0. \tag{6.19}$$

Letting $W = \left[\begin{smallmatrix} W_1 & W_2 & W_3 \\ W_4 & W_5 & W_6 \end{smallmatrix}\right]$, after some calculations, (6.19) can be rewritten as

$$\begin{bmatrix} \Upsilon_{11} & \bar{\Upsilon}_{12} \\ \star & \bar{\Upsilon}_{22} \end{bmatrix} + \begin{bmatrix} 0 \\ \mathcal{B} \end{bmatrix}\begin{bmatrix} 0 \\ \mathcal{B} \end{bmatrix}^T < 0, \quad \mathcal{B} = \begin{bmatrix} \bar{B}_d \\ D_d \end{bmatrix}, \tag{6.20}$$

where Υ_{11} is defined in (6.13), and

$$\bar{\Upsilon}_{12} = \begin{bmatrix} -W_2 & -W_3 \\ -W_5 + W_4^* E^T + W_1^* \bar{A}^T & -W_6 + W_1^* C^T \end{bmatrix},$$

$$\bar{\Upsilon}_{22} = \begin{bmatrix} \mathrm{Sym}\{\bar{A}W_2 + EW_5\} & \bar{A}W_3 + EW_6 + W_2^* C^T \\ \star & \mathrm{Sym}\{CW_3\} - \gamma^2 I \end{bmatrix}.$$

By the Schur complement lemma, (6.20) is equivalent to

$$\begin{bmatrix} \Upsilon_{11} & \bar{\Upsilon}_{12} & 0 \\ \star & \bar{\Upsilon}_{22} & \mathcal{B} \\ \star & \star & -I \end{bmatrix} < 0. \tag{6.21}$$

Let

$$\Theta_d = \begin{bmatrix} \Upsilon_{11} & \bar{\Upsilon}_{12} & 0 \\ \star & \bar{\Upsilon}_{22} & \mathcal{B} \\ \star & \star & -I \end{bmatrix}, \quad \mathcal{W} = [0 \quad -W_1 \quad -W_2 \quad -W_3 \quad 0].$$

Then condition (6.21) can be guaranteed by the following inequality with sufficiently large scalar ϵ_1:

$$\Pi_d = \begin{bmatrix} \Theta_d & \mathcal{W}^* \\ \star & -\epsilon_1 I \end{bmatrix} < 0. \tag{6.22}$$

Taking congruence transformation, this yields

$$\mathcal{T}_1 \Pi_d \mathcal{T}_1^T = \begin{bmatrix} \Upsilon_{11} & \Upsilon_{12} & \Upsilon_{13} \\ \star & \Upsilon_{22} & \Upsilon_{23} \\ \star & \star & \Upsilon_{33} \end{bmatrix} - \epsilon_1 \begin{bmatrix} 0 \\ \mathcal{C} \\ 0 \end{bmatrix} \begin{bmatrix} 0 \\ \mathcal{C} \\ 0 \end{bmatrix}^T < 0, \tag{6.23}$$

where

$$\mathcal{T}_1 = \begin{bmatrix} I & 0 & 0 \\ 0 & I & \mathcal{C} \\ 0 & 0 & I \end{bmatrix}, \quad \mathcal{C} = \begin{bmatrix} 0 & -LC \\ 0 & 0 \end{bmatrix}.$$

This condition is still nonlinear due to the bilinear item. Since $\epsilon_1 > 0$ and $LCC^T L^T \geq 0$, inequality (6.23) can be guaranteed by (6.13). This completes the proof. □

Corollary 6.1 (Low-frequency range). *Consider the error dynamics* (6.10) *with* $\alpha \in [0, 2)$. *The disturbance signal is supposed to be in the low-frequency range* $\Omega_d = (0, \omega_l]$. *Let a scalar* ϵ_1 *be given. Then inequality* (6.11) *holds if there exist matrices* $P_d, Q_d \in \mathbb{H}_n$, $Q_d > 0$, *and a matrix variable* W *such that*

$$\begin{bmatrix} \Upsilon_{11} & \Upsilon_{12} & \Upsilon_{13} \\ \star & \Upsilon_{22} & \Upsilon_{23} \\ \star & \star & \Upsilon_{33} \end{bmatrix} < 0, \tag{6.24}$$

where the matrices W, Υ_{12}, Υ_{13}, Υ_{22}, Υ_{23}, *and* Υ_{33} *are the same as in the case of* $\omega \in [\omega_l, \omega_h]$ (6.13), *and*

$$\Upsilon_{11} = \begin{bmatrix} -Q_d & rP_d - W_1 \\ \star & \omega_l^{2\alpha} Q_d - \mathrm{Sym}\{W_4\} \end{bmatrix}.$$

Proof. The proof of this corollary is the same as that of Theorem 6.1, so the details are omitted. □

Corollary 6.2 (High-frequency range). *Consider the error dynamics* (6.10) *with* $\alpha \in [0, 2)$. *The disturbance signal is supposed to be in the high-frequency range* $\Omega_d = [\omega_h, +\infty)$. *Let a scalar* ϵ_1 *be given. Then inequality* (6.11) *holds if there exist matrices* $P_d, Q_d \in \mathbb{H}_n$, $Q_d > 0$, *and a matrix variable* W *such that*

$$\begin{bmatrix} \Upsilon_{11} & \Upsilon_{12} & \Upsilon_{13} \\ \star & \Upsilon_{22} & \Upsilon_{23} \\ \star & \star & \Upsilon_{33} \end{bmatrix} < 0, \tag{6.25}$$

where the matrices W, Υ_{22}, Υ_{23}, *and* Υ_{33} *are the same as in the case of* $\omega \in [\omega_l, \omega_h]$ (6.13), *and*

$$\Upsilon_{11} = \begin{bmatrix} Q_d - \mathrm{Sym}\{W_1\} & rP_d - W_4^* \\ \star & -\omega_h^{2\alpha} Q_d \end{bmatrix},$$

$$\Upsilon_{12} = \begin{bmatrix} -W_2 + W_4^* E^T + W_1^* A^T & -W_3 + W_1^* C^T \\ -W_5 & -W_6 \end{bmatrix},$$

$$\Upsilon_{13} = \begin{bmatrix} 0 & -W_1^* \\ 0 & 0 \end{bmatrix}.$$

Proof. The proof of this corollary is the same as that of Theorem 6.1, so the details are omitted. □

6.4 Fault sensitivity

The aim of this part is to give conditions in the form of LMIs for fault sensitivity condition (6.12) in three different frequency domains.

Theorem 6.2 (Medium-frequency range). *Consider the error dynamics* (6.10) *with* $\alpha \in [0, 2)$. *The fault signal is supposed to be in the medium-frequency range* $\Omega_f = [\omega_l, \omega_h]$. *Let a scalar* ϵ_2 *be given. Then inequality* (6.12) *holds if there exists matrices* $P_f, Q_f \in \mathbb{H}_n$, $Q_f > 0$, *and matrix variables R and Y of appropriate dimensions such that*

$$\begin{bmatrix} \Sigma_{11} & \Sigma_{12} \\ \star & \Sigma_{22} \end{bmatrix} < 0, \tag{6.26}$$

where $r = e^{j(1-\alpha)\pi/2}$, $\omega_c = jr(\omega_l^\alpha + \omega_h^\alpha)/2$, $R = \left[\begin{smallmatrix} R_1 \\ R_2 \end{smallmatrix}\right]$, *and*

$$\Sigma_{11} = \begin{bmatrix} -\omega_l^\alpha \omega_h^\alpha E^T Q_f E - C^T C + \mathrm{Sym}\{R_1 A\} & -C^T D_f + R_1 B_f + A^T R_2^* \\ \star & \beta^2 I - D_f^T D_f + \mathrm{Sym}\{R_2 B_f\} \end{bmatrix},$$

$$\Sigma_{12} = \begin{bmatrix} -R_1 + rE^T P_f + \omega_c E^T Q_f & -R_1 + \epsilon_2 C^T L^T \\ -R_2 & -R_2 + \epsilon_2 D_f^T L^T \end{bmatrix},$$

$$\Sigma_{22} = \begin{bmatrix} -Q_f - \mathrm{Sym}\{Y\} & -Y^* \\ \star & -\epsilon_2 I \end{bmatrix}.$$

Proof. By some basic matrix calculations, we can easily show that condition (6.12) is equivalent to the following condition:

$$\begin{bmatrix} G_{rf}(j\omega) \\ I \end{bmatrix}^* \begin{bmatrix} -I & 0 \\ 0 & \beta^2 I \end{bmatrix} \begin{bmatrix} G_{rf}(j\omega) \\ I \end{bmatrix} < 0 \quad \forall \omega \in [\omega_l, \omega_h]. \tag{6.27}$$

By invoking Lemma 6.2, it follows that

$$\begin{bmatrix} \bar{A} & \overline{B_f} \\ E & 0 \end{bmatrix}^T (\Phi \otimes P_f + \Psi \otimes Q_f) \begin{bmatrix} \bar{A} & \overline{B_f} \\ E & 0 \end{bmatrix} + \begin{bmatrix} C & D_f \\ 0 & I \end{bmatrix}^T \begin{bmatrix} -I & 0 \\ 0 & \beta^2 I \end{bmatrix} \begin{bmatrix} C & D_f \\ 0 & I \end{bmatrix} < 0. \tag{6.28}$$

After a short calculation, (6.28) can be reformulated as

$$\mathcal{M} + \mathrm{Sym}\{\mathcal{P}\mathcal{A}\} - \mathcal{A}^T Q_f \mathcal{A} < 0, \tag{6.29}$$

where $\mathcal{A} = [\bar{A} \quad \overline{B_f}]$, and

$$\mathcal{M} = \begin{bmatrix} -\omega_l^a \omega_h^a E^T Q_f E - CC^T & -C^T D_f \\ \star & \beta^2 I - D_f^T D_f \end{bmatrix},$$

$$\mathcal{P} = \begin{bmatrix} E^T(rP_f + \omega_c Q_f) \\ 0 \end{bmatrix}.$$

Denote

$$\Xi_f = \begin{bmatrix} \mathcal{M} & \mathcal{P} \\ \star & -Q_f \end{bmatrix}. \tag{6.30}$$

Then (6.29) can be rewritten as follows:

$$[I \quad \mathcal{A}^T]\, \Xi_f \, [I \quad \mathcal{A}^T]^T < 0. \tag{6.31}$$

By Lemma 6.3, (6.31) is equivalent to

$$\Xi_f + \mathrm{Sym}\left(\begin{bmatrix} \mathcal{A}^T \\ -I \end{bmatrix} XI\right) < 0. \tag{6.32}$$

Denoting $X = [R^* \quad Y]$, (6.32) can be converted to

$$\Theta_f = \begin{bmatrix} \mathcal{M} + \mathrm{Sym}\{R\mathcal{A}\} & -R + \mathcal{P} + \mathcal{A}^T Y \\ \star & -Q_f - \mathrm{Sym}\{Y\} \end{bmatrix} < 0. \tag{6.33}$$

Condition (6.33) can be guaranteed by (6.34) with a sufficiently large scalar ϵ_2:

$$\Pi_f = \begin{bmatrix} \Theta_f & \mathcal{X} \\ \star & -\epsilon_2 I \end{bmatrix} < 0, \quad \mathcal{X} = \begin{bmatrix} -R \\ -Y^* \end{bmatrix}. \tag{6.34}$$

Taking congruence transformation with

$$\mathcal{T}_2 = \begin{bmatrix} I & 0 \\ \mathcal{L} & I \end{bmatrix}, \quad \mathcal{L} = \begin{bmatrix} 0 & 0 \\ -LC & -LD_f \end{bmatrix},$$

this yields

$$\mathcal{T}_2^T \Pi_f \mathcal{T}_2 = \begin{bmatrix} \Sigma_{11} & \Sigma_{12} \\ \star & \Sigma_{22} \end{bmatrix} - \epsilon_2 \begin{bmatrix} \mathcal{L}^T \\ 0 \end{bmatrix} \begin{bmatrix} \mathcal{L}^T \\ 0 \end{bmatrix}^T < 0. \tag{6.35}$$

Ignoring the semipositive bilinear item, the proof is completed. □

Corollary 6.3 (Low-frequency range). *Consider the error dynamics* (6.10) *with* $\alpha \in [0,2)$. *The fault signal is supposed to be in the low-frequency range* $\Omega_f = (0,\omega_l]$. *Let a scalar* ϵ_2 *be given. Then inequality* (6.12) *holds if there exists matrices* $P_f, Q_f \in \mathbb{H}_n$, $Q_f > 0$, *and matrix variables R and Y of appropriate dimensions such that*

$$\begin{bmatrix} \Sigma_{11} & \Sigma_{12} \\ \star & \Sigma_{22} \end{bmatrix} < 0, \tag{6.36}$$

where the matrices R and Σ_{22} *are the same as in the case of* $\omega \in [\omega_l, \omega_h]$ (6.26), *and*

$$\Sigma_{11} = \begin{bmatrix} \omega_l^{2\alpha} E^T Q_f E - C^T C + \mathrm{Sym}\{R_1 A\} & -C^T D_f + R_1 B_f + A^T R_2^* \\ \star & \beta^2 I - D_f^T D_f + \mathrm{Sym}\{R_2 B_f\} \end{bmatrix},$$

$$\Sigma_{12} = \begin{bmatrix} -R_1 + rE^T P_f & -R_1 + \epsilon_2 C^T L^T \\ -R_2 & -R_2 + \epsilon_2 D_f^T L^T \end{bmatrix}.$$

Proof. The details are omitted because the proof of this corollary is similar to that of Theorem 6.2. □

Corollary 6.4 (High-frequency range). *Consider the error dynamics* (6.10) *with* $\alpha \in [0,2)$. *The fault signal is supposed to be in the high-frequency range* $\Omega_f = [\omega_h, +\infty)$. *Let a scalar* ϵ_2 *be given. Then inequality* (6.12) *holds if there exists matrices* $P_f, Q_f \in \mathbb{H}_n$, $Q_f > 0$, *and matrix variables R and Y of appropriate dimensions such that*

$$\begin{bmatrix} \Sigma_{11} & \Sigma_{12} \\ \star & \Sigma_{22} \end{bmatrix} < 0, \tag{6.37}$$

where the matrix R is the same as in the case of $\omega \in [\omega_l, \omega_h]$ (6.26), *and*

$$\Sigma_{11} = \begin{bmatrix} -\omega_h^{2\alpha} E^T Q_f E - C^T C + \mathrm{Sym}\{R_1 A\} & -C^T D_f + R_1 B_f + A^T R_2^* \\ \star & \beta^2 I - D_f^T D_f + \mathrm{Sym}\{R_2 B_f\} \end{bmatrix},$$

$$\Sigma_{12} = \begin{bmatrix} -R_1 + rE^T P_f & -R_1 + \epsilon_2 C^T L^T \\ -R_2 & -R_2 + \epsilon_2 D_f^T L^T \end{bmatrix},$$

$$\Sigma_{22} = \begin{bmatrix} Q_f - \mathrm{Sym}\{Y\} & -Y^* \\ \star & -\epsilon_2 I \end{bmatrix}.$$

Proof. The details are omitted because the proof of this corollary is similar to that of Theorem 6.2. □

6.5 Fault detection

Some additional LMI conditions are given in this part to ensure the admissibility condition of the error dynamics (6.10) in two cases.

Theorem 6.3. *For system (6.10) with order $\alpha \in [0,1)$, let $r = e^{j(1-\alpha)\pi/2}$ and a scalar ϵ_3 be given. Then the system is admissible if there exists matrices $P_a, Q_a \in \mathbb{C}^{n\times n}$ and a matrix variable $X = [X_1 \quad X_2]$ of appropriate dimension such that*

$$E^T P_a = P_a^* E \geq 0, \quad E^T Q_a = Q_a^* E \geq 0, \tag{6.38}$$

and

$$\begin{bmatrix} \mathrm{Sym}\{A^T(rP_a + \bar{r}Q_a)\} & \bar{r}P_a^* + rQ_a^* + X_1^* & -X_1^* - \epsilon C^T L^T \\ \star & \mathrm{Sym}\{X_2\} & -X_2^* \\ \star & \star & -\epsilon_3 I \end{bmatrix} < 0. \tag{6.39}$$

Proof. By Lemma 5.4, system (6.10) is admissible if and only if there exist matrices $P_a, Q_a \in \mathbb{C}^{n\times n}$ such that $E^T P_a = P_a^* E \geq 0, E^T Q_a = Q_a^* E \geq 0$, and

$$\mathrm{Sym}\{\bar{A}^T(rP_a + \bar{r}Q_a)\} < 0. \tag{6.40}$$

Inequality (6.40) is nonconvex because of the coupling between matrices L and P_a, Q_a. To decouple them, reformulate the inequality as follows:

$$[I \quad -C^T L^T] \begin{bmatrix} \mathrm{Sym}\{A^T(rP_a + \bar{r}Q_a)\} & \bar{r}P_a^* + rQ_a^* \\ \star & 0 \end{bmatrix} \begin{bmatrix} I \\ -LC \end{bmatrix} < 0. \tag{6.41}$$

Then applying Lemma 6.3 to (6.41), it follows that

$$\begin{bmatrix} \mathrm{Sym}\{A^T(rP_a + \bar{r}Q_a)\} & \bar{r}P_a^* + rQ_a^* \\ \star & 0 \end{bmatrix} + \mathrm{Sym}\left(\begin{bmatrix} C^T L^T \\ I \end{bmatrix} XI\right) < 0. \tag{6.42}$$

Denote $X = [X_1 \quad X_2]$ and define

$$\Theta_a = \begin{bmatrix} \mathrm{Sym}\{A^T(rP_a + \bar{r}Q_a) + C^T L^T X_1\} & \bar{r}P_a^* + rQ_a^* + C^T L^T X_2 + X_1^* \\ \star & \mathrm{Sym}\{X_2\} \end{bmatrix}. \tag{6.43}$$

Then condition (6.42) can be guaranteed by the following LMI with a sufficiently large scalar ϵ_3:

$$\Pi_a = \begin{bmatrix} \Theta_a & -X^* \\ \star & -\epsilon_3 I \end{bmatrix} < 0. \tag{6.44}$$

Choose

$$\mathcal{T}_3 = \begin{bmatrix} I & 0 & LC \\ 0 & I & 0 \\ 0 & 0 & I \end{bmatrix}$$

and take the congruence transformation $\mathcal{T}_3^T \Pi_a \mathcal{T}_3$. Then, (6.44) is equivalent to the following condition:

$$\begin{bmatrix} \mathrm{Sym}\{A^T(rP_a + \bar{r}Q_a)\} & \bar{r}P_a^* + rQ_a^* + X_1^* & -X_1^* - \epsilon_3 C^T L^T \\ \star & \mathrm{Sym}\{X_2\} & -X_2^* \\ \star & \star & -\epsilon_3 I \end{bmatrix}$$

$$- \epsilon_3 \begin{bmatrix} LC & 0 & 0 \end{bmatrix}^T \begin{bmatrix} LC & 0 & 0 \end{bmatrix} < 0. \tag{6.45}$$

Ignoring the semipositive bilinear item, the proof is completed. □

Theorem 6.4. *For system (6.10) of order $\alpha \in [1, 2)$, let $r = e^{j(1-\alpha)\pi/2}$ and a scalar ϵ_3 be given. Then the system is admissible if there exist a matrix $P_a \in \mathbb{C}^{n\times n}$ and a matrix variable $X = [X_1 \quad X_2]$ of appropriate dimension such that*

$$E^T P_a = P_a^* E \geq 0 \tag{6.46}$$

and

$$\begin{bmatrix} \mathrm{Sym}\{rA^T P_a\} & \bar{r}P_a^* + X_1^* & -X_1^* - \epsilon_3 C^T L^T \\ \star & \mathrm{Sym}\{X_2\} & -X_2^* \\ \star & \star & -\epsilon_3 I \end{bmatrix} < 0. \tag{6.47}$$

Proof. By Theorem 4.3 system (6.10) with $\alpha \in [1, 2)$ is admissible if and only if there exists a matrix $P_a \in \mathbb{C}^{n\times n}$ such that

$$\mathrm{Sym}\{r\bar{A}^T P_a\} < 0, \quad E^T P_a = P_a^* E \geq 0, \quad r = e^{j(1-\alpha)\pi/2}. \tag{6.48}$$

Note that (6.48) has a similar form with (6.40), so they could be proved in the same way. Letting Q_a in (6.39) be 0, the proof is completed. □

To detect the fault signal, a widely used method is choosing a residual evaluation function $J(t)$ and a threshold J_{th} and using a logic unit [43]. Then the fault can be detected by

$$\begin{cases} J(t) > J_{th}, & \text{fault}, \\ J(t) \leq J_{th}, & \text{no fault}, \end{cases} \tag{6.49}$$

where $J(t)$ is defined as

$$J(t) = \sqrt{\frac{1}{t}\int_0^t r^T(\tau) r(\tau) d\tau}. \tag{6.50}$$

Noting that $\|G_{rd}(j\omega_d)\|_\infty < \gamma$, J_{th} can be chosen as

$$J_{th} = \gamma \cdot \sup \|d(t)\|_2. \tag{6.51}$$

In this section, we give two numerical examples to verify the validity of the fault detection observer design method. Example 6.1 tests the $0 < \alpha < 1$ case with the disturbance and fault both occurred in the low-frequency range, illustrates the validity of Corollary 6.1, Corollary 6.3, and Theorem 6.3, and interprets the advantages of finite-frequency observer design. In Example 6.2 the proposed method is applied to a electrical circuit system with fractional order $1 < \alpha < 2$, the disturbance is assumed to be in the medium-frequency range, and the fault is assumed to be a step function. Example 6.2 shows the effectiveness of Theorem 6.1, Corollary 6.3, and Theorem 6.4.

Example 6.1. Consider the singular FOS with the following system matrices:

$$E = \begin{bmatrix} 1 & 0 & 0 \\ 0 & 1 & 0 \\ 0 & 0 & 0 \end{bmatrix}, \quad A = \begin{bmatrix} -1 & 0.8 & 1.5 \\ -1 & -2 & -1 \\ -2 & -1 & -3 \end{bmatrix}, \quad B_d = \begin{bmatrix} -0.2 \\ 0 \\ 0.7 \end{bmatrix}, \quad B_f = \begin{bmatrix} 0.3 \\ 0.8 \\ 0.6 \end{bmatrix},$$
$$C = \begin{bmatrix} 1 & 1 & 0 \end{bmatrix}, \quad D_d = 0, \quad D_f = 1, \quad \alpha = 0.7.$$

We assume that both disturbance and fault occur within the low-frequency ranges $\Omega_d = (0, 1]$ and $\Omega_f = (0, 1]$. The disturbance signal is set up as $d(t) = 0.5e^{-0.05t}\sin(t)$, and the additive fault input is

$$f(t) = \begin{cases} 0, & t < 20 \text{ s}, \\ 1, & t \geq 20 \text{ s}. \end{cases}$$

Given $\gamma = 0.16$ and $\beta = 1.3$, we choose the scalars $\epsilon_1 = 4$, $\epsilon_2 = 4$, and $\epsilon_3 = 4$. Solve LMIs (6.24), (6.36), (6.38), and (6.39) and get the corresponding observer gain:

$$L = \begin{bmatrix} 0.0212 & -0.0067 & 0.0007 \end{bmatrix}^T.$$

Note that $\|d(t)\|_2 < 0.5$ and by (6.51) the detection threshold is $J_{th} = 0.08$.

Substitute L into the error dynamics (6.10). Then the eigenvalues of $A - LC$, the minimum singular values of G_{rf}, and the maximum singular values of G_{rd} are plotted in Figures 6.1, 6.2, and 6.3, respectively. We can see in the figures that the observer meets the design metrics.

Figure 6.2 also shows the advantages of our methods over the full frequency design methods. We can easily derive by the GKYP lemma that $\|G_{rf}\|_- < \beta$ in the full frequency range if and only if the following LMI is satisfied:

$$\begin{bmatrix} \text{Sym}\{r(A - LC)^T PE\} - C^T C & \star \\ r(B_f - LD_f)^T PE - D_f^T C & \beta^2 I - D_f^T D_f \end{bmatrix} < 0. \tag{6.52}$$

An upper bound of the β in this LMI is the minimum singular value of D_f, which is 1.0 in this example. In our method the value of β can be chosen greater than 1.0.

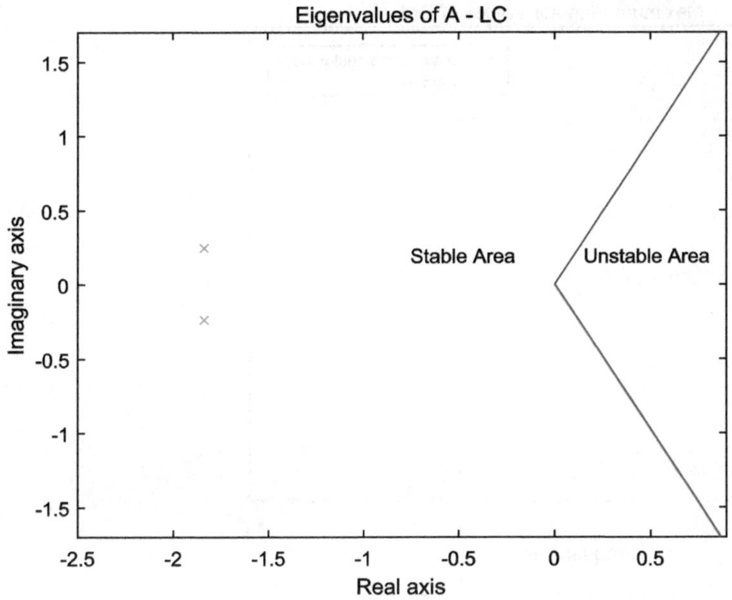

Figure 6.1: Eigenvalues of $A - LC$ in Example 6.1.

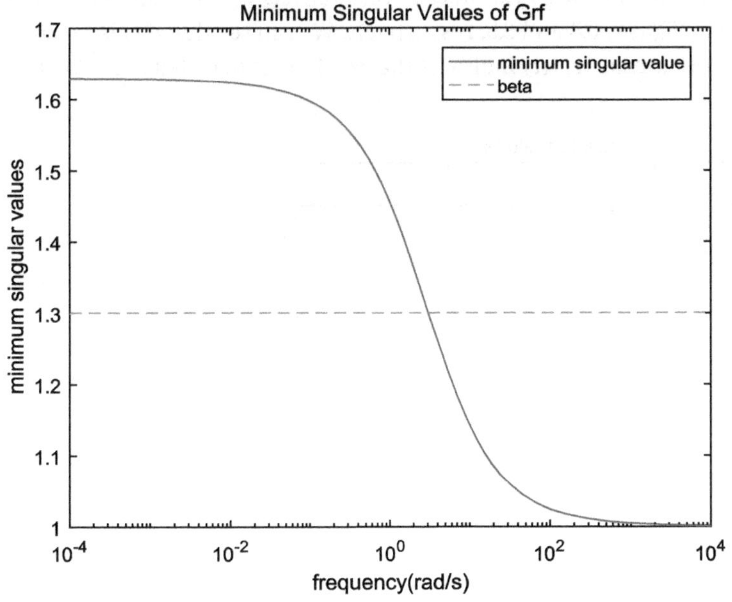

Figure 6.2: Minimum singular values of G_{rf} in Example 6.1.

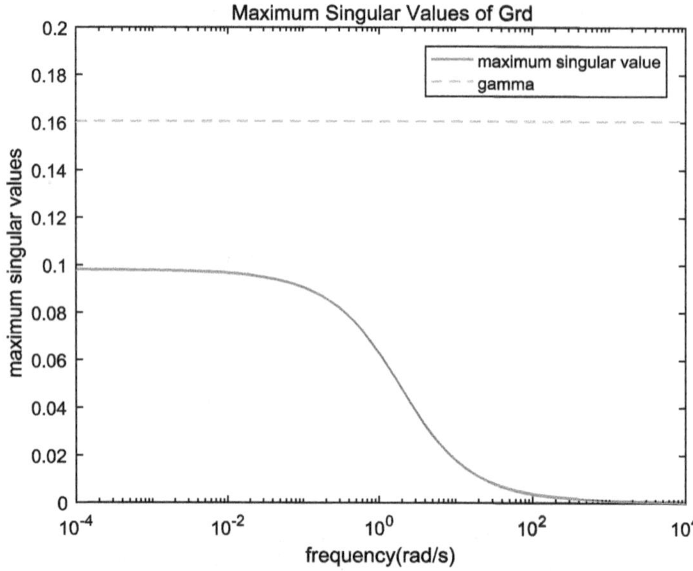

Figure 6.3: Maximum singular values of G_{rd} in Example 6.1.

The simulation results of the observer output residual signal and the evaluation function are plotted in Figures 6.4 and 6.5, respectively. We can see that $r(t)$ is sensitive to fault and robust to disturbance. $J(t)$ indicates that the fault is detected at $t \approx 20.08$ s.

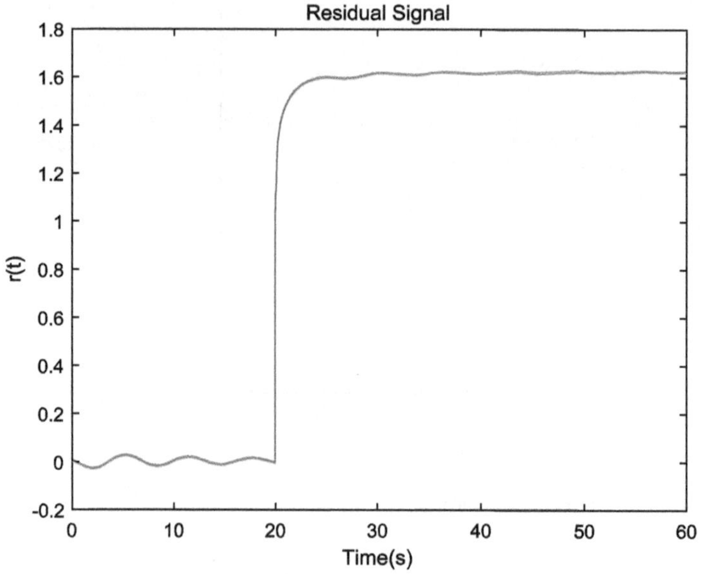

Figure 6.4: The residual signal in Example 6.1.

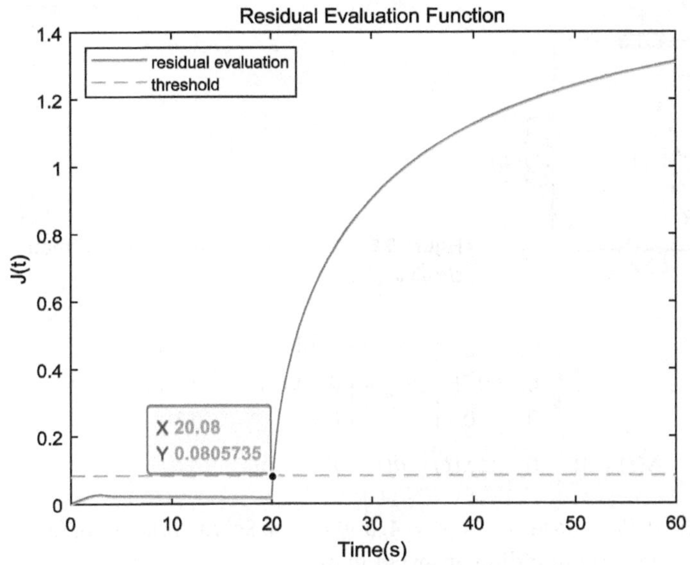

Figure 6.5: The residual evaluation function in Example 6.1.

Example 6.2. According to [41], the electrical circuit in Figure 6.6 can be described by

$$\begin{cases} e_1 = L_1 \frac{d^\alpha i_1}{dt^\alpha} + R_1 i_1 + L_3 \frac{d^\alpha i_3}{dt^\alpha} + R_3 i_3, \\ e_2 = L_2 \frac{d^\alpha i_1}{dt^\alpha} + R_2 i_2 + L_3 \frac{d^\alpha i_3}{dt^\alpha} + R_3 i_3, \\ i_3 = i_1 + i_2. \end{cases} \tag{6.53}$$

Consider the measured output $y(t) = i_3(t)$. Choosing the pseudo-state vector $x(t) = \begin{bmatrix} i_1(t) & i_2(t) & i_3(t) \end{bmatrix}^T$, the system is

$$\begin{aligned} \begin{bmatrix} L_1 & 0 & L_3 \\ 0 & L_2 & L_3 \\ 0 & 0 & 0 \end{bmatrix} D^\alpha x(t) &= -\begin{bmatrix} R_1 & 0 & R_3 \\ 0 & R_2 & R_3 \\ -1 & -1 & 1 \end{bmatrix} x(t) + \begin{bmatrix} 1 & 0 \\ 0 & 1 \\ 0 & 0 \end{bmatrix} \begin{bmatrix} e_1(t) \\ e_2(t) \end{bmatrix}, \\ y(t) &= \begin{bmatrix} 0 & 0 & 1 \end{bmatrix} x(t). \end{aligned} \tag{6.54}$$

When $\alpha = 1.35$, $L_1 = 5$, $L_2 = 4$, $L_3 = 0$, $R_1 = 4$, $R_2 = 3.2$, and $R_3 = 4$, system (6.54) is admissible and observable. Assume that system (6.54) has an additive unknown input disturbance $d(t)$ with $B_d = \begin{bmatrix} 2.0 & 2.4 & 0.7 \end{bmatrix}^T$, $D_d = -1$, $10 < \omega_d < 50$, $\|d(t)\|_2 \le 0.5$, and an additive fault signal $f(t)$ with $B_f = \begin{bmatrix} 2.5 & 3.2 & 0.6 \end{bmatrix}^T$, $D_f = 1$, $\omega_f < 1$. Then the system can be rewritten as

$$\begin{bmatrix} 1 & 0 & 0 \\ 0 & 1 & 0 \\ 0 & 0 & 0 \end{bmatrix} D^{1.35} x(t) = \begin{bmatrix} -0.8 & 0 & -0.8 \\ 0 & -0.8 & -1 \\ 1 & 1 & -1 \end{bmatrix} x(t)$$

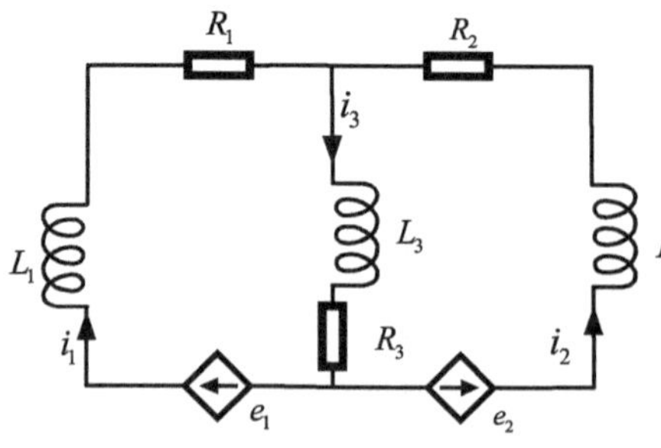

Figure 6.6: The fractional-order singular electrical circuit in [41].

$$+\begin{bmatrix} 0.2 & 0 \\ 0 & 0.25 \\ 0 & 0 \end{bmatrix} u(t) + \begin{bmatrix} 0.4 \\ 0.6 \\ 0.7 \end{bmatrix} d(t) + \begin{bmatrix} 0.5 \\ 0.8 \\ 0.6 \end{bmatrix} f(t),$$
$$y(t) = \begin{bmatrix} 0 & 0 & 1 \end{bmatrix} x(t) - d(t) + f(t).$$

Given $\gamma = 0.4$ and $\beta = 1.55$, choose $\epsilon_1 = 4$, $\epsilon_2 = 4$, and $\epsilon_3 = 4$. Solve LMIs (6.13), (6.36), (6.46), and (6.47) and get the corresponding observer gain:

$$L = \begin{bmatrix} -0.0621 & -0.0403 & 0.0393 \end{bmatrix}^T .$$

By (6.51), the detection threshold is $J_{th} = 0.2$. Substituting L and system matrices into the error dynamics (6.10), the singular values of G_{rf} and G_{rd} are plotted in Figures 6.7 and 6.8, respectively. The figures show that $\|G_{rf}(j\omega_f)\|_- > \beta \, \forall \omega_f < 1$ and $\|G_{rd}(j\omega_d)\|_\infty > \gamma$ $\forall \omega_d \in (10, 50)$. The eigenvalues of $A - LC$ are -0.8 and -2.4334, and Figure 6.9 demonstrates that the error dynamics (6.10) is stable.

Remark 6.3. As we can see in Figures 6.7 and 6.8, the H_- and H_∞ performance indices cannot meet the conditions in the whole frequency domain. However, in the given finite-frequency ranges, the design conditions were satisfied. Similarly to (6.52), we can easily obtain that $\gamma = 0.4$ is smaller than the lower bound in the full frequency case. This shows that by designing in finite-frequency ranges, greater β and smaller γ can be chosen, and thus the fault can be detected earlier.

The system is simulated by Matlab with

$$f(t) = \begin{cases} 0, & t < 20 \text{ s}, \\ 1, & t \geq 20 \text{ s}, \end{cases}$$
$$d(t) = 0.5e^{-0.04t} \sin(20t).$$

The results are plotted in Figures 6.10 and 6.11. We can see in Figure 6.10 that the residual signal $r(t)$ is sensitive to the fault signal (with the gain approximately 2.5) and robust to the disturbance (with the gain approximately 0.4). Figure 6.11 shows that the fault is detected timely at approximately $t = 20.27$ s.

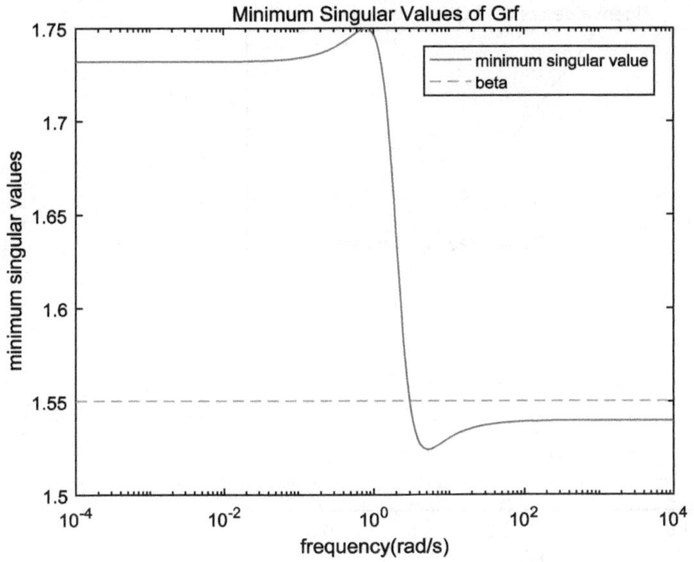

Figure 6.7: Minimum singular values of G_{rf} in Example 6.2.

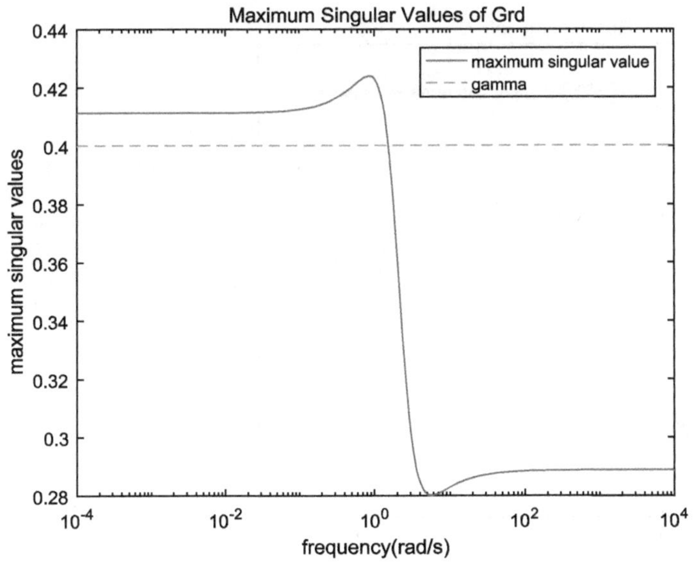

Figure 6.8: Maximum singular values of G_{rd} in Example 6.2.

6.6 Chapter summary

The robust fault detection observer design problem over finite-frequency ranges for FOSSs has been investigated in this chapter. The finite-frequency range fault detection

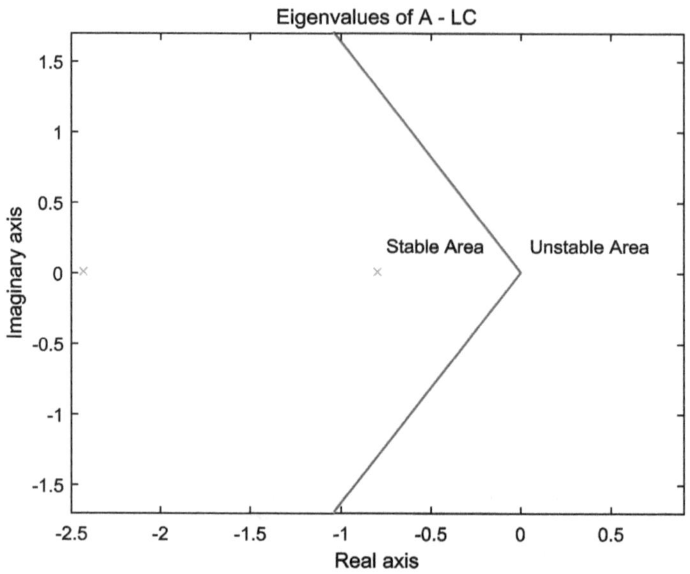

Figure 6.9: Eigenvalues of $A - LC$ in Example 6.2.

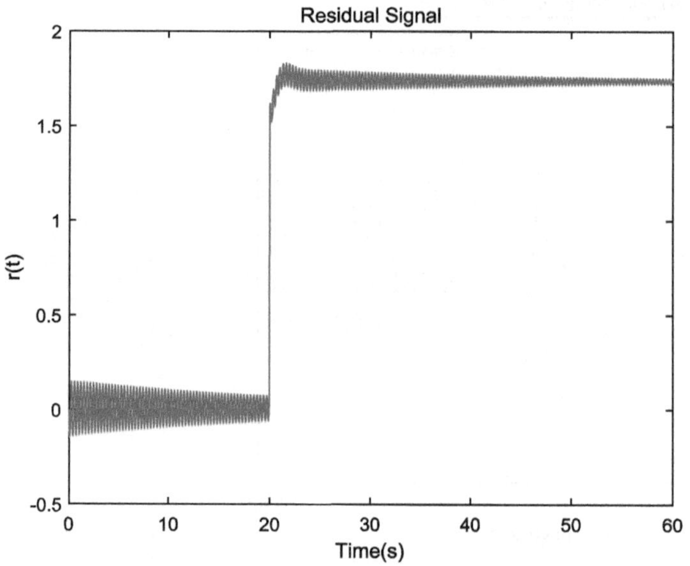

Figure 6.10: The output residual signal of the observer in Example 6.2.

problem has been formulated as designing a fault detection observer subject to admissibility constraint, disturbance robustness constraint, and fault sensitivity constraint. The H_- index of FOSS in finite-frequency ranges is defined, and the design method of the

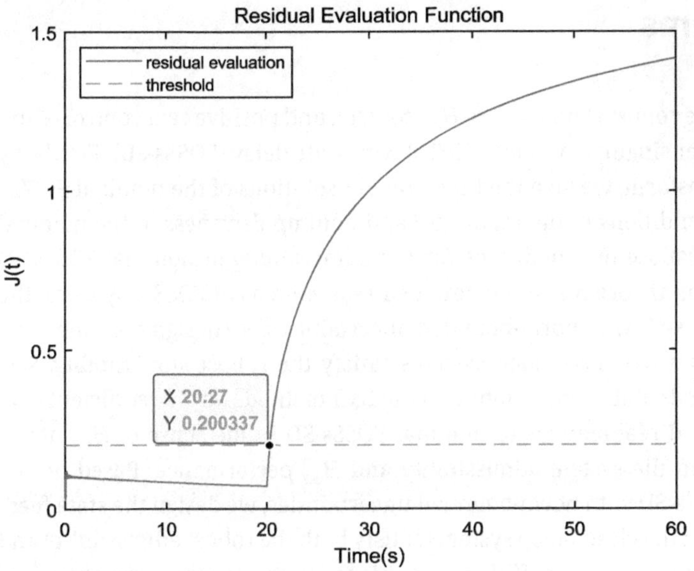

Figure 6.11: The residual evaluation function in Example 6.2.

H_-/H_∞ observer has been presented. Four theorems and four corollaries about the system admissibility and the performance index H_-/H_∞ in different frequency ranges are given. By invoking GKYP lemma, the projection lemma, and some matrices decoupling techniques the three design conditions have been obtained in terms of LMIs. Numerical simulations are given to validate the proposed theorems and corollaries. Finally, the proposed methods are applied to a real electrical circuit system, and the results show the effectiveness of the methods.

7 Delay systems

Abstract: We study the robust stabilization, H_∞ control, and positive real control of uncertain fractional-order singular systems (FOSSs) with state delay (FOSSs-SD). Firstly, by using the Laplace transform, we give the time-domain solutions of the nominal FOSSs-SD. We propose the conditions of the regularity and nonimpulsiveness of the nominal FOSSs-SD. Then we propose the conditions for the admissibility of nominal FOSSs-SD based on the small-gain theorem and the bounded real lemmas of FOSSs. By using the above results, for FOSSs-SD with norm-bounded uncertainties, we design the state feedback controllers so that the closed-loop systems satisfy the robust admissibility. Secondly, by using the admissibility conditions and the LMI method, we give sufficient conditions for the bounded real lemmas of nominal FOSSs-SD in the sense of H_∞ norm, which can ensure both the system admissibility and H_∞ performance. Based on the above results, for FOSSs-SD with norm-bounded uncertainties, we design the state feedback controllers so that the closed-loop systems satisfy both the robust admissibility and H_∞ performance. Thirdly, we give sufficient conditions for the positive real lemmas of nominal FOSSs-SD, which can ensure both the system admissibility and the extended strictly positive realness. Based on the above results, for FOSSs-SD with norm-bounded uncertainties, we design the state feedback controllers so that the closed-loop systems satisfy both the robust admissibility and extended strictly positive realness.

7.1 Introduction

In actual physical systems, from the occurrence of a signal to its reception, delay behavior inevitably occurs. This is reflected in the system model, where the past states of the system directly affect the current states. Such systems are referred to as systems with state delay. When fractional-order systems (FOSs) contain state delay, the stability of the systems may be destroyed. Unlike FOSs with state delay, FOSSs-SD require not only the study of the stability but also the consideration of regularity and nonimpulsiveness. Namely, the admissibility (regularity, absence of impulses, and stability) of the FOSSs-SD need to be investigated. The FOSSs-SD defined by the Riemann–Liouville fractional-order derivative are studied in [53], and sufficient conditions for the admissibility of the system are provided by using the Lyapunov method. However, the existence of matrix equation in [53] may lead to no solutions. In [98] the admissibility conditions of FOSSs-SD defined by the Caputo fractional derivative are given in LMI form by using the Lyapunov–Krasovskii method. However, note that the Caputo fractional derivative is different from the integer-order derivative. The method of directly adopting the integer-order Lyapunov–Krasovskii function for the Caputo fractional derivative must be improved. To study the admissibility of FOSDSs without using the Lyapunov method, the analysis of the Mittag-Leffler function and the scaling of norm inequality are employed

https://doi.org/10.1515/9783111628646-007

in [42], and sufficient admissibility conditions are obtained, but this method cannot be directly used to design the controllers.

From the above discussion it is clear that research on the admissibility of FOSSs-SD is still in its early stages, and many unresolved issues remain. Furthermore, the study of robust H_∞ control and positive real control for FOSSs-SD is still an open problem. Therefore this chapter conducts relevant research on the admissibility, robust stabilization, H_∞ control, and positive real control of FOSSs-SD, addressing the current challenges in this field.

7.2 Admissibility and robust stabilization

In this section, we consider the following uncertain FOSS-SD:

$$\begin{aligned} ED^{\alpha}x(t) &= Ax(t) + A_h x(t-h) + B_1 u(t), \\ x(t) &= \chi(t), \quad t \in (-h, 0], \end{aligned} \tag{7.1}$$

where $x(t) \in \mathbb{R}^n$ and $u(t) \in \mathbb{R}^p$ are the state vector and control input, $\chi(t)$ is the continuous vector initial value function, $E \in \mathbb{R}^{n\times n}$ is the constant matrix with $\operatorname{rank}(E) = r \le n$, $\alpha \in (0,1)$ is the fractional order, D^{α} is the Caputo fractional-order derivative, and A, A_h, and B_1 are matrices with uncertainties described as

$$\begin{aligned} A = A_0 + \Delta A, \quad A_h &= A_{h_0} + \Delta A_h, \quad B_1 = B_{10} + \Delta B_1, \\ \Delta\Gamma &= D_\Gamma F_\Gamma G_\Gamma, \quad \Gamma \in \{A, A_h, B_1\}, \end{aligned}$$

where $A_0 \in \mathbb{R}^{n\times n}$, $A_{h_0} \in \mathbb{R}^{n\times n}$, and $B_{10} \in \mathbb{R}^{m\times n}$ are known real matrices, and F_Γ are uncertain matrices satisfying $F_\Gamma^T F_\Gamma \le I$.

Consider the following state feedback controller:

$$u(t) = Kx(t), \tag{7.2}$$

with which FOSS-SD (7.1) can be transformed into

$$\begin{aligned} ED^{\alpha}x(t) &= (A + B_1 K)x(t) + A_h x(t-h), \\ x(t) &= \chi(t), \quad t \in (-h, 0]. \end{aligned} \tag{7.3}$$

Suppose that the uncertain matrices and the control input $u(t)$ are 0. Then system (7.1) can be described as the following nominal FOSS-SD:

$$\begin{aligned} ED^{\alpha}x(t) &= A_0 x(t) + A_{h_0} x(t-h), \\ x(t) &= \chi(t), \quad t \in (-h, 0]. \end{aligned} \tag{7.4}$$

The issues to be solved in this section are as follows:

(I) Establish the regularity and nonimpulsiveness conditions for nominal FOSS-SD (7.4) and derive the solutions of the system.
(II) Propose the admissibility conditions for nominal FOSS-SD (7.4).
(III) Design state feedback controllers for uncertain FOSS-SD (7.1) by using the admissibility conditions in (II) such that the closed-loop systems are robustly admissible.

To proceed, we introduce some lemmas and definitions. Similarly to integer-order singular systems (IOSSs) with state delay in [101], the regularity and nonimpulsiveness definitions of FOSS-SD (7.4) are as follows.

Definition 7.1. (I) For FOSS-SD (7.4), if $\det(\lambda E - A_0)$ ($\lambda = s^\alpha, s \in \mathbb{C}$) is not identically equal to 0, then the system is said to be regular.

(II) For FOSS-SD (7.4), if $\deg(\det(\lambda E - A_0)) = \operatorname{rank}(E)$ ($\lambda = s^\alpha, s \in \mathbb{C}$), then the system is said to be impulse-free.

Based on [46, 37], the H_∞ norm of a transfer function G is defined as $\|G\|_\infty = \sup_{\operatorname{Re}(s)\geq 0} \sigma_{\max}(G(s))$, and the stability definition of FOSS-SD (7.4) can be given.

Definition 7.2. For the FOSS-SD with transfer function G, if $\|G\|_\infty < +\infty$, then the system is said to be stable.

Definition 7.3. If FOSS-SD (7.4) is regular, impulse-free, and stable, then the system is said to be admissible.

According to Theorem 3.1 in [103], we can state the following lemma.

Lemma 7.1. *Consider the linear fractional-order neutral delay system $D^\alpha x(t) - CD^\alpha x(t - h) = Ax(t) + Bx(t-h)$, $x(t) = \chi(t)$, $t \in (-h, 0]$, $\alpha \in (0,1)$. The system has a unique solution*

$$x(t) = X_\alpha(t)\chi(0) - X_\alpha(t)C\chi(-h) + B\int_{-h}^{0} X(t-\tau-h)\chi(\tau)d\tau + C\int_{-h}^{0} X_{\alpha+1}(t-\tau-h)\chi(\tau)d\tau$$

with $X(t) = L^{-1}\{H^{-1}(s)\}$, $X_\alpha(t) = L^{-1}\{s^{\alpha-1}H^{-1}(s)\}$, *and* $H^{-1}(s) = (s^\alpha I - A - Cs^\alpha e^{-sh} - Be^{-sh})^{-1}$.

Lemma 7.2 ([37]). *Consider interconnected systems S_1 and S_2 with transfer functions G_{S_1} and G_{S_2}. Suppose that S_1 and S_2 are stable, i. e., for* $\operatorname{Re}(s) \geq 0$, $\|G_{S_1}\|_\infty < +\infty$ *and* $\|G_{S_1}\|_\infty < +\infty$. *The stability condition for this interconnected system is* $\|G_{S_1}\|_\infty \|G_{S_1}\|_\infty < 1$.

Theorem 7.1. *FOSS-SD (7.4) is regular if and only if there exist matrices Ψ and Φ such that*

$$\Phi E \Psi = \operatorname{diag}(I_r, N) \quad \text{and} \quad \Phi A_0 \Psi = \operatorname{diag}(A_1, I_{n-r}),$$

where $A_1 \in \mathbb{R}^{r\times r}$, and $N \in \mathbb{R}^{(n-r)\times(n-r)}$ is a nilpotent matrix.

Proof. The theorem can be proved by using the matrix decomposition method of Lemma 2 in [106], and thus the proof is omitted for brevity. □

Followed by Theorem 7.1, we can obtain the nonimpulsiveness conditions for FOSS-SDs.

Theorem 7.2. *If the matrix N in Theorem 7.1 is zero, then FOSS-SD (7.4) is impulse-free.*

Proof. By Theorem 7.1 there exist matrices Ψ and Φ such that

$$\Phi E \Psi = \operatorname{diag}(I_r, N) \quad \text{and} \quad \Phi A_0 \Psi = \operatorname{diag}(A_1, I_{n-r}).$$

Let $\Phi A_{h_0} \Psi = \begin{bmatrix} A_{h_1} & A_{h_2} \\ A_{h_3} & A_{h_4} \end{bmatrix}$. FOSS-SD (7.4) can be transformed into

$$\begin{bmatrix} I_r & 0 \\ 0 & N \end{bmatrix} \begin{bmatrix} D^\alpha x_1(t) \\ D^\alpha x_2(t) \end{bmatrix} = \begin{bmatrix} A_1 & 0 \\ 0 & I_{n-r} \end{bmatrix} \begin{bmatrix} x_1(t) \\ x_2(t) \end{bmatrix} + \begin{bmatrix} A_{h_1} & A_{h_2} \\ A_{h_3} & A_{h_4} \end{bmatrix} \begin{bmatrix} x_1(t-h) \\ x_2(t-h) \end{bmatrix}, \tag{7.5}$$

where

$$\begin{bmatrix} x_1(t) \\ x_2(t) \end{bmatrix} = \Psi^{-1} x(t), \quad \begin{bmatrix} x_1(t-h) \\ x_2(t-h) \end{bmatrix} = \Psi^{-1} x(t-h).$$

If $N = 0$, then (7.5) becomes

$$D^\alpha x_1(t) = A_1 x_1(t) + A_{h_1} x_1(t-h) + A_{h_2} x_2(t-h), \tag{7.6}$$

$$0 = x_2(t) + A_{h_3} x_1(t-h) + A_{h_4} x_2(t-h). \tag{7.7}$$

Taking fractional-order derivatives on (7.7) yields

$$0 = D^\alpha x_2(t) + A_{h_3} D^\alpha x_1(t-h) + A_{h_4} D^\alpha x_2(t-h). \tag{7.8}$$

Combining (7.7) and (7.8) gives

$$\begin{aligned} D^\alpha x_2(t) = {} & x_2(t) + A_{h_3} x_1(t-h) + A_{h_4} x_2(t-h) \\ & - A_{h_3} D^\alpha x_1(t-h) - A_{h_4} D^\alpha x_2(t-h). \end{aligned} \tag{7.9}$$

By (7.6) and (7.9), (7.5) is equivalent to

$$\begin{aligned} D^\alpha \eta(t) - \hat{C} D^\alpha \eta(t-h) &= \hat{A} \eta(t) + \hat{A}_{h_0} \eta(t-h), \\ \eta(t) &= \Psi^{-1} \chi(t), \end{aligned} \tag{7.10}$$

where

$$\eta(t) = \begin{bmatrix} x_1(t) \\ x_2(t) \end{bmatrix}, \quad \hat{A} = \begin{bmatrix} A_1 & 0 \\ 0 & I_{n-r} \end{bmatrix}, \quad \hat{A}_{h_0} = \begin{bmatrix} A_{h_1} & A_{h_2} \\ A_{h_3} & A_{h_4} \end{bmatrix}, \quad \hat{C} = \begin{bmatrix} 0 & 0 \\ -A_{h_3} & -A_{h_4} \end{bmatrix}.$$

We can see that system (7.10) is a linear fractional-order neutral delay system. By Lemma 7.1 the system has a unique solution

$$\eta(t) = X_\alpha(t)\Psi^{-1}\chi(0) - X_\alpha(t)C\Psi^{-1}\chi(-h) + \hat{A}_{h_0}\int_{-h}^{0} X(t-\tau-h)\Psi^{-1}\chi(\tau)d\tau$$
$$+ \hat{C}\int_{-h}^{0} X_{\alpha+1}(t-\tau-h)\Psi^{-1}\chi(\tau)d\tau, \tag{7.11}$$

where

$$X(t) = L^{-1}\{H^{-1}(s)\}, \quad X_\alpha(t) = L^{-1}\{s^{\alpha-1}H^{-1}(s)\},$$
$$H^{-1}(s) = (s^\alpha I - \hat{A} - \hat{C}s^\alpha e^{-sh} - \hat{A}_{h_0}e^{-sh})^{-1}.$$

From (7.11) we can see that there are no impulsive terms in the time domain solutions $x(t) = \Psi\eta(t)$ of system (7.4). This completes the proof. □

By Theorem 7.2, FOSS-SD (7.4) is impulse-free if $N = 0$. The solutions of the system on $t \geq 0$ are given by the following theorem.

Theorem 7.3. *The solutions of FOSS-SD* (7.4) *are as follows:*

$$x(t) = \Psi X_\alpha(t)\Psi^{-1}\chi(0) - \Psi X_\alpha(t)C\Psi^{-1}\chi(-h) + \Psi B\int_{-h}^{0} X(t-\tau-h)\Psi^{-1}\chi(\tau)d\tau$$
$$+ \Psi C\int_{-h}^{0} X_{\alpha+1}(t-\tau-h)\Psi^{-1}\chi(\tau)d\tau,$$

where

$$X(t) = L^{-1}\{H^{-1}(s)\}, \quad X_\alpha(t) = L^{-1}\{s^{\alpha-1}H^{-1}(s)\},$$
$$H^{-1}(s) = (s^\alpha I - \hat{A} - \hat{C}s^\alpha e^{-sh} - \hat{A}_{h_0}e^{-sh})^{-1},$$
$$\hat{A} = \begin{bmatrix} A_1 & 0 \\ 0 & I_{n-r} \end{bmatrix}, \quad \hat{A}_{h_0} = \begin{bmatrix} A_{h_1} & A_{h_2} \\ A_{h_3} & A_{h_4} \end{bmatrix}, \quad \hat{C} = \begin{bmatrix} 0 & 0 \\ -A_{h_3} & -A_{h_4} \end{bmatrix},$$

and the matrix Ψ *is given in Theorem* 7.1.

Theorem 7.4. *FOSS-SD* (7.4) *is admissible if there exist matrices* $P \in \mathbb{C}^{n\times n}$ *and* $Q \in \mathbb{C}^{n\times n}$, $Q > 0$, *such that*

$$E^T P = P^* E \geq 0, \tag{7.12}$$
$$\begin{bmatrix} \mathrm{Sym}\{A_0^T X\} + Q & X^* A_{h_0} \\ A_{h_0}^T X & -Q \end{bmatrix} < 0, \tag{7.13}$$

where $X = e^{j\theta}P + e^{-j\theta}\bar{P}$, $\theta = \frac{\pi}{2}(1-\alpha)$, $0 < \alpha < 1$.

Proof. The proof consists of two parts. In Part I, we prove that the system is regular and impulse-free, and in Part II, we prove that the system is stable.

Part I. Note that $\operatorname{rank}(E) = r \leq n$. There exist nonsingular real matrices Π_1 and Π_2 such that

$$\tilde{E} = \Pi_1 E \Pi_2 = \begin{bmatrix} I_r & 0 \\ 0 & 0 \end{bmatrix}.$$

Let

$$\tilde{A} = \Pi_1 A_0 \Pi_2 = \begin{bmatrix} A_1 & A_2 \\ A_3 & A_4 \end{bmatrix}, \quad \tilde{P} = \Pi_1^{-T} P \Pi_2 = \begin{bmatrix} P_1 & P_2 \\ P_3 & P_4 \end{bmatrix},$$

$$\tilde{X} = \Pi_1^{-T} X \Pi_2 = \begin{bmatrix} X_1 & X_2 \\ X_3 & X_4 \end{bmatrix} = \begin{bmatrix} e^{j\theta} P_1 + e^{-j\theta} \bar{P}_1 & e^{j\theta} P_2 + e^{-j\theta} \bar{P}_2 \\ e^{j\theta} P_3 + e^{-j\theta} \bar{P}_3 & e^{j\theta} P_4 + e^{-j\theta} \bar{P}_4 \end{bmatrix}.$$

Premultiplying and postmultiplying (7.12) by Π_2^T and Π_2 give

$$\tilde{E}^T \tilde{P} = \tilde{P}^* \tilde{E} \geq 0. \tag{7.14}$$

Substituting $\tilde{E}$ and $\tilde{P}$ into (7.14) yields

$$\begin{bmatrix} P_1 & P_2 \\ 0 & 0 \end{bmatrix} = \begin{bmatrix} P_1^* & 0 \\ P_2^* & 0 \end{bmatrix} \geq 0.$$

Therefore $P_1 = P_1^*$, $P_2 = 0$. Since $Q > 0$, by (7.13) we have

$$\operatorname{Sym}\{A_0^T X\} < 0. \tag{7.15}$$

Premultiplying and postmultiplying (7.15) by Π_2^T and Π_2 give

$$\operatorname{Sym}\{\tilde{A}^T \tilde{X}\} < 0, \tag{7.16}$$

i. e.,

$$\operatorname{Sym}\left\{ \begin{bmatrix} A_1 & A_2 \\ A_3 & A_4 \end{bmatrix}^T \begin{bmatrix} e^{j\theta} P_1 + e^{-j\theta} \bar{P}_1 & 0 \\ e^{j\theta} P_3 + e^{-j\theta} \bar{P}_3 & e^{j\theta} P_4 + e^{-j\theta} \bar{P}_4 \end{bmatrix} \right\} < 0. \tag{7.17}$$

From (7.17), we have

$$\operatorname{Sym}\left\{ \begin{bmatrix} A_1^T (e^{j\theta} P_1 + e^{-j\theta} \bar{P}_1) + A_3^T (e^{j\theta} P_3 + e^{-j\theta} \bar{P}_3) & A_3^T (e^{j\theta} P_4 + e^{-j\theta} \bar{P}_4) \\ A_2^T (e^{j\theta} P_1 + e^{-j\theta} \bar{P}_1) + A_4^T (e^{j\theta} P_3 + e^{-j\theta} \bar{P}_3) & A_4^T (e^{j\theta} P_4 + e^{-j\theta} \bar{P}_4) \end{bmatrix} \right\} < 0. \tag{7.18}$$

Therefore

$$\operatorname{Sym}\{A_4^T (e^{j\theta} P_4 + e^{-j\theta} \bar{P}_4)\} < 0. \tag{7.19}$$

By Lemma 2.10, A_4 is nonsingular. Let $\lambda = s^\alpha$. Then

$$\begin{aligned}\det(\lambda E - A) &= \det(\Pi_1^{-1})\det(\lambda\tilde{E} - \tilde{A})\det(\Pi_2^{-1})\\ &= \det(\Pi_1^{-1})\det(-A_4^{-1})\det(\lambda I - (A_1 - A_2A_4^{-1}A_3))\det(\Pi_2^{-1}).\end{aligned} \tag{7.20}$$

Note that Π_1, Π_2, and A_4 are nonsingular and $\operatorname{rank}(A_1) = r$. From (7.20), we can see that $\det(\lambda E - A)$ is not identically equal to 0 for different λ, jointly with $\deg(\det(\lambda E - A)) = r = \operatorname{rank}(E)$. The system is regular and impulse-free by Definition 7.1.

Part II. Let the output of system (7.4) be $y(t) = x(t)$. The original system is equivalent to the following two interconnected subsystems S_1 and S_2:

$$S_1 : \begin{cases} ED^\alpha x(t) = A_0x(t) + A_{h_0}w(t), \\ y(t) = x(t), \end{cases} \tag{7.21}$$

$$S_2 : w(t) = x(t - h). \tag{7.22}$$

The transfer functions of systems S_1 and S_2 are

$$G_{S_1}(s) = (s^\alpha E - A_0)^{-1}A_{h_0}, \tag{7.23}$$

$$G_{S_2}(s) = e^{-sh}. \tag{7.24}$$

For $\operatorname{Re}(s) \geq 0$, we have $\|G_{S_2}\|_\infty \leq 1$. By Lemma 7.2 the interconnected subsystems S_1 and S_2 are stable if $\|G_{S_1}\|_\infty < 1$. Let $\Sigma = \Sigma^*$ and suppose Σ is nonsingular. Then $\|\Sigma^{-1}G_{S_2}\Sigma\|_\infty \leq 1$. The interconnected subsystems S_1 and S_2 are stable if $\|\Sigma^{-1}G_{S_1}\Sigma\|_\infty < 1$, i. e., $\|\Sigma^{-1}((s^\alpha E - A)^{-1}A_{h_0})\Sigma\|_\infty < 1$. By Theorem 4.1, if

$$\begin{bmatrix} \operatorname{Sym}\{A_0^T(e^{j\theta}P + e^{-j\theta}\bar{P})\} & (e^{j\theta}P + e^{-j\theta}\bar{P})^*(A_{h_0}\Sigma) & \Sigma^{-T} \\ (A_{h_0}\Sigma)^T(e^{j\theta}P + e^{-j\theta}\bar{P}) & -I & 0 \\ \Sigma^{-1} & 0 & -I \end{bmatrix} < 0, \tag{7.25}$$

then $\|\Sigma^{-1}((s^\alpha E - A_0)^{-1}A_{h_0})\Sigma\|_\infty < 1$. Inequality (7.25) is equivalent to

$$\operatorname{Sym}\{A_0^T(e^{j\theta}P + e^{-j\theta}\bar{P})\} + (e^{j\theta}P + e^{-j\theta}\bar{P})^*A_{h_0}\Sigma\Sigma^TA_{h_0}^T(e^{j\theta}P + e^{-j\theta}\bar{P}) + \Sigma^{-T}\Sigma^{-1} < 0. \tag{7.26}$$

Let $Q = \Sigma^{-T}\Sigma^{-1} > 0$. Then inequality (7.26) is equivalent to

$$\operatorname{Sym}\{A_0^T(e^{j\theta}P + e^{-j\theta}\bar{P})\} + (e^{j\theta}P + e^{-j\theta}\bar{P})^*A_{h_0}Q^{-1}A_{h_0}^T(e^{j\theta}P + e^{-j\theta}\bar{P}) + Q < 0. \tag{7.27}$$

Inequality (7.27) can be written as

$$\begin{bmatrix} \operatorname{Sym}\{A_0^T(e^{j\theta}P + e^{-j\theta}\bar{P})\} + Q & (e^{j\theta}P + e^{-j\theta}\bar{P})^*A_{h_0} \\ A_{h_0}^T(e^{j\theta}P + e^{-j\theta}\bar{P}) & -Q \end{bmatrix} < 0. \tag{7.28}$$

This completes the proof. □

Based on Theorem 7.4 and the duality of the system, we can obtain the following corollary.

Corollary 7.1. *FOSS-SD (7.4) is admissible if there exist matrices $P \in \mathbb{C}^{n\times n}$ and $Q \in \mathbb{C}^{n\times n}$, $Q > 0$, such that*

$$EP = P^* E^T \geq 0, \tag{7.29}$$

$$\begin{bmatrix} \mathrm{Sym}\{A_0 X\} + Q & A_{h_0} X \\ X^* A_{h_0}^T & -Q \end{bmatrix} < 0, \tag{7.30}$$

where $X = e^{j\theta} P + e^{-j\theta} \bar{P}$, $\theta = \frac{\pi}{2}(1-\alpha)$, $0 < \alpha < 1$.

Now we provide a design method of the controller (7.2) such that the closed-loop system (7.3) is robustly admissible.

Theorem 7.5. *Given the uncertain FOSS-SD (7.1), there exists a state feedback controller (7.2) such that (7.3) is robustly admissible if there exist matrices $P \in \mathbb{C}^{n\times n}$, $Q \in \mathbb{C}^{n\times n}$, $Q > 0$, and $Y \in \mathbb{R}^{m\times n}$ and a constant $\varepsilon > 0$ satisfying*

$$EP = P^* E^T \geq 0, \tag{7.31}$$

$$\begin{bmatrix} W_0 + \varepsilon \Upsilon_1 \Upsilon_1^T & \Upsilon_2^T \\ \Upsilon_2 & -\varepsilon I \end{bmatrix} < 0, \tag{7.32}$$

where

$$W_0 = \begin{bmatrix} \mathrm{Sym}\{A_0 X + B_{10} Y\} + Q & A_{h_0} X \\ X^* A_{h_0}^T & -Q \end{bmatrix}, \quad \Upsilon_1 = \begin{bmatrix} D_A & D_{B_1} & D_{A_h} \\ 0 & 0 & 0 \end{bmatrix},$$

$$\Upsilon_2 = \begin{bmatrix} G_A X & 0 \\ G_{B_1} Y & 0 \\ 0 & G_{A_h} X \end{bmatrix}, \quad X = e^{j\theta} P + e^{-j\theta} \bar{P}, \quad \theta = \frac{\pi}{2}(1-\alpha), \quad 0 < \alpha < 1.$$

(X can be considered as a nonsingular matrix. Otherwise, we could choose a nonsingular real matrix $\hat{X}$ satisfying $E\hat{X} = \hat{X}^T E^T \geq 0$ and a constant $\xi \in (0,1)$ such that (i) ξ is not an eigenvalue of $-X\hat{X}^{-1}$ and (ii) replacing X in inequality (7.32) by $X + \xi\hat{X}$, the inequality still holds. Now $X + \xi\hat{X}$ is nonsingular.) The controller gain in (7.2) can be given as $K = YX^{-1}$.

Proof. Inequality (7.32) is equivalent to

$$W_0 + \varepsilon \Upsilon_1 \Upsilon_1^T + \varepsilon^{-1} \Upsilon_2^T \Upsilon_2 < 0. \tag{7.33}$$

Let $M_1 = \begin{bmatrix} F_A & 0 & 0 \\ 0 & F_{B_1} & 0 \\ 0 & 0 & F_{A_h} \end{bmatrix}$, for which $M_1^T M_1 \leq I$. By Lemma 4.6, for $\varepsilon > 0$, we have

$$\mathrm{Sym}\{\Upsilon_1 M_1 \Upsilon_2\} \leq \varepsilon \Upsilon_1 \Upsilon_1^T + \varepsilon^{-1} \Upsilon_2^T M_1^T M_1 \Upsilon_2 \leq \varepsilon \Upsilon_1 \Upsilon_1^T + \varepsilon^{-1} \Upsilon_2^T \Upsilon_2. \tag{7.34}$$

Combining inequalities (7.33) and (7.34) gives

$$W_0 + \mathrm{Sym}\{\Upsilon_1 M_1 \Upsilon_2\} < 0. \tag{7.35}$$

Substituting W_0, Υ_1, M_1, and Υ_2 yields

$$\begin{bmatrix} \mathrm{Sym}\{AX + B_1 Y\} + Q & A_h X \\ X^* A_h^T & -Q \end{bmatrix} < 0. \tag{7.36}$$

Note that $u(t) = Kx(t) = YX^{-1}x(t)$, from which we have $Y = KX$. Substituting Y into inequality (7.36) yields

$$\begin{bmatrix} \mathrm{Sym}\{AX + B_1 KX\} + Q & A_h X \\ X^* A_h^T & -Q \end{bmatrix} < 0, \tag{7.37}$$

i. e.,

$$\begin{bmatrix} \mathrm{Sym}\{(A + B_1 K)X\} + Q & A_h X \\ X^* A_h^T & -Q \end{bmatrix} < 0. \tag{7.38}$$

By Corollary 7.1, the closed-loop system (7.3) is robustly admissible by (7.31) and (7.38). This completes the proof. □

7.3 Robust H_∞ control

Consider the FOSS-SD

$$\begin{aligned} ED^\alpha x(t) &= Ax(t) + A_h x(t-h) + B_1 u(t) + Bw(t), \\ z(t) &= Cx(t) + B_2 u(t), \\ x(t) &= \chi(t), \quad t \in (-h, 0], \end{aligned} \tag{7.39}$$

where $x(t) \in \mathbb{R}^n$, $z(t) \in \mathbb{R}^q$, $w(t) \in \mathbb{R}^p$, and $u(t) \in \mathbb{R}^m$ are the state vector, measured output, disturbance input, and control input, $\chi(t)$ is the continuous vector initial value function, $E \in \mathbb{R}^{n\times n}$ is the constant matrix with $\mathrm{rank}(E) = r \le n$, $\alpha \in (0,1)$ is the fractional order, D^α is the Caputo fractional-order derivative, and A, A_h, B, C, B_1, and B_2 are matrices with uncertainties described as

$$\begin{gathered} A = A_0 + \Delta A, \quad A_h = A_{h_0} + \Delta A_h, \quad B = B_0 + \Delta B, \quad B_1 = B_{10} + \Delta B_1 \\ C = C_0 + \Delta C, \quad B_2 = B_{20} + \Delta B_2, \quad \Delta\Gamma = D_\Gamma F_\Gamma G_\Gamma, \quad \Gamma \in \{A, A_h, B, C, B_1, B_2\}, \end{gathered}$$

where $A_0, A_{h_0}, B_0, C_0, B_{10}, B_{20}, D_\Gamma$, and G_Γ are known real matrices, and F_Γ are uncertain matrices satisfying $F_\Gamma^T F_\Gamma \le I$.

Consider the state feedback controller

$$u(t) = Kx(t), \tag{7.40}$$

with which, FOSS-SD (7.39) can be transformed into

$$\begin{aligned} ED^{\alpha}x(t) &= (A + B_1K)x(t) + A_h x(t-h) + Bw(t), \\ z(t) &= (C + B_2K)x(t), \\ x(t) &= \chi(t), \quad t \in (-h, 0]. \end{aligned} \tag{7.41}$$

The transfer function of system (7.41) is represented by

$$G_K(s) = (C + B_2K)(s^{\alpha}E - (A + B_1K) - A_h e^{-sh})^{-1}B. \tag{7.42}$$

Suppose that the uncertain matrices and control input $u(t)$ are 0. Then the system can be described as the following nominal FOSS-SD:

$$\begin{aligned} ED^{\alpha}x(t) &= A_0x(t) + A_{h_0}x(t-h) + B_0w(t), \\ z(t) &= C_0x(t), \\ x(t) &= \chi(t), \quad t \in (-h, 0], \end{aligned} \tag{7.43}$$

with transfer function

$$G_N(s) = C_0(s^{\alpha}E - A_0 - A_{h_0}e^{-sh})^{-1}B_0. \tag{7.44}$$

The purposes of this section are to propose the bounded real lemmas in the sense of the H_∞ norm for FOSS-SDs and to design the state feedback controller for FOSS-SDs with norm-bounded uncertainties.

Based on the admissibility conditions for the FOSS-SDs provided in Section 7.2, in this section, we present the bounded real lemmas for the FOSS-SD (7.43).

Theorem 7.6. *For a given positive scalar γ, the FOSS-SD (7.43) is admissible and its transfer function satisfies $\|G_N\|_\infty < \gamma$ if there exist matrices $P \in \mathbb{C}^{n\times n}$ and $Q \in \mathbb{C}^{n\times n}$, $Q > 0$, such that*

$$E^T P = P^* E \geq 0, \tag{7.45}$$

$$\begin{bmatrix} \mathrm{Sym}\{A_0^T X\} + Q & X^T A_{h_0} & C_0^T & X^T B_0 \\ A_{h_0}^T X & -Q & 0 & 0 \\ C_0 & 0 & -\gamma I & 0 \\ B_0^T X & 0 & 0 & -\gamma I \end{bmatrix} < 0, \tag{7.46}$$

where $X = e^{j\theta}P + e^{-j\theta}\bar{P}$, $\theta = \frac{\pi}{2}(1-\alpha)$, $\alpha \in (0,1)$.

Proof. We divide the proof into two parts. In Step I, we prove the admissibility of the system, and in Step II, we prove that $\|G_N\|_\infty < \gamma$.

Step I. According to inequality (7.46), there exist matrices $P \in \mathbb{C}^{n\times n}$ and $Q \in \mathbb{C}^{n\times n}$ such that

$$\begin{bmatrix} \mathrm{Sym}\{A_0^T X\} + Q & X^T A_{h_0} \\ A_{h_0}^T X & -Q \end{bmatrix} < 0. \tag{7.47}$$

From (7.45) and (7.46), according to Theorem 7.4, the FOSS-SD (7.43) is admissible.

Step II. According to the Schur complement lemma, inequality (7.46) is equivalent to

$$\mathrm{Sym}\{A_0^T X\} + Q + X^T A_{h_0} Q^{-1} A_{h_0}^T X + \gamma^{-1} C_0^T C_0 + \gamma^{-1} X^T B_0 B_0^T X < 0. \tag{7.48}$$

Let

$$\begin{aligned} \Omega &= -(\mathrm{Sym}\{A_0^T X\} + Q + X^T A_{h_0} Q^{-1} A_{h_0}^T X + \gamma^{-1} C_0^T C_0 + \gamma^{-1} X^T B_0 B_0^T X), \\ \Phi &= e^{-j\frac{\pi}{2}\alpha} w^\alpha E^T X + e^{j\frac{\pi}{2}\alpha} w^\alpha X^* E. \end{aligned} \tag{7.49}$$

Then

$$\Omega > 0. \tag{7.50}$$

Noticing that $E^T P = P^* E \geq 0$, we have

$$E^T \bar{P} = \bar{P}^* E \geq 0. \tag{7.51}$$

Because $X = e^{j\theta} P + e^{-j\theta} \bar{P}$, we have $X^T = X^*$. Therefore, for $w \geq 0$,

$$\begin{aligned} \Phi &= e^{-j\frac{\pi}{2}\alpha} w^\alpha E^T X + e^{j\frac{\pi}{2}\alpha} w^\alpha X^* E \\ &= e^{j(\frac{\pi}{2} - \pi\alpha)} w^\alpha E^T P + e^{-j\frac{\pi}{2}} w^\alpha E^T \bar{P} + e^{j(-\frac{\pi}{2} + \pi\alpha)} w^\alpha P^* E + e^{j\frac{\pi}{2}} w^\alpha \bar{P}^* E^T \\ &= 2\sin(\pi\alpha) w^\alpha E^T P \geq 0. \end{aligned} \tag{7.52}$$

From (7.48) and (7.49) we have

$$\begin{aligned} &\mathrm{Sym}\{(e^{-j\frac{\pi}{2}\alpha} w^\alpha E^T - A_0^T - A_{h_0}^T e^{jwh}) X\} - X^T A_{h_0} Q^{-1} A_{h_0}^T X - \gamma^{-1} C_0^T C_0 \\ &\quad - \gamma^{-1} X^T B_0 B_0^T X - Q - \Omega + e^{jwh} A_{h_0}^T X + e^{-jwh} X^T A_{h_0} - \Phi = 0. \end{aligned} \tag{7.53}$$

Let

$$\begin{aligned} \Delta &= e^{j\frac{\pi}{2}\alpha} w^\alpha E - A_0 - A_{h_0} e^{-jwh}, \\ W &= Q + X^T A_{h_0} Q^{-1} A_{h_0}^T X - e^{jwh} A_{h_0}^T X - e^{-jwh} X^T A_{h_0}. \end{aligned}$$

Then we have

$$\begin{aligned} W &= Q + X^T A_{h_0} Q^{-1} A_{h_0}^T X - e^{jwh} A_{h_0}^T X - e^{-jwh} X^T A_{h_0} \\ &= (e^{-jwh} X^T A_{h_0} - Q) Q^{-1} (e^{jwh} A_{h_0}^T X - Q) \geq 0. \end{aligned} \tag{7.54}$$

Equality (7.53) can be transformed into

$$\Delta^* X + X^T \Delta - W - \gamma^{-1} C_0^T C_0 - \gamma^{-1} X^T B_0 B_0^T X - \Omega - \Phi = 0. \tag{7.55}$$

From (7.50), (7.52), and (7.54), we have $\Omega > 0$, $\Phi \geq 0$, and $W \geq 0$. Therefore

$$\Delta^* X + X^T \Delta = W + \gamma^{-1} C_0^T C_0 + \gamma^{-1} X^T B_0 B_0^T X + \Omega + \Phi > 0. \tag{7.56}$$

By (7.56), Δ is nonsingular. Pre- and postmultiplying (7.55) by Δ^{-*} and Δ^{-1}, respectively, we obtain that

$$\begin{aligned} &X\Delta^{-1} + \Delta^{-*} X^T - \gamma^{-1} \Delta^{-*} X^T B_0 B_0^T X \Delta^{-1} \\ &\quad - \Delta^{-*} (W + \Omega + \Phi) \Delta^{-1} - \gamma^{-1} \Delta^{-*} C_0^T C_0 \Delta^{-1} = 0. \end{aligned} \tag{7.57}$$

Pre- and postmultiplying (7.57) by B_0^T and B_0, respectively, we obtain that

$$\begin{aligned} &\text{Sym}\{B_0^T X \Delta^{-1} B_0\} - \gamma^{-1} B_0^T \Delta^{-*} X^T B_0 B_0^T X \Delta^{-1} B_0^T \\ &\quad - B_0^T \Delta^{-*} (W + \Omega + \Phi) \Delta^{-1} B_0 - \gamma^{-1} B_0^T \Delta^{-*} C_0^T C_0 \Delta^{-1} B_0 = 0. \end{aligned} \tag{7.58}$$

Equality (7.58) can be transformed into

$$\Gamma_1 + \Gamma_2 + \Gamma_3 = 0, \tag{7.59}$$

where

$$\begin{aligned} \Gamma_1 &= -\gamma^{-1} (B_0^T \Delta^{-*} X^T B_0 - \gamma I)(B_0^T \Delta^{-*} X^T B_0 - \gamma I)^* \leq 0, \\ \Gamma_2 &= -B_0^T \Delta^{-*} (W + \Omega + \Phi) \Delta^{-1} B_0 < 0, \\ \Gamma_3 &= \gamma I - \gamma^{-1} B_0^T \Delta^{-*} C_0^T C_0 \Delta^{-1} B_0. \end{aligned}$$

From (7.59) we have

$$\Gamma_3 > 0, \tag{7.60}$$

that is, $B_0^T \Delta^{-*} C_0^T C_0 \Delta^{-1} B_0 < \gamma^2 I$. The proof is ended. □

Based on Corollary 7.1, we can obtain the following corollary.

Corollary 7.2. *For a given positive scalar γ, the FOSS-SD* (7.43) *is admissible and its transfer function satisfies $\|G_N\|_\infty < \gamma$ if there exist matrices $P \in \mathbb{C}^{n\times n}$ and $Q \in \mathbb{C}^{n\times n}$, $Q > 0$, such that*

$$EP = P^* E^T \geq 0, \tag{7.61}$$

$$\begin{bmatrix} \mathrm{Sym}\{A_0 X\} + Q & A_{h_0} X & B_0 & X^T C_0^T \\ X^* A_{h_0}^T & -Q & 0 & 0 \\ B_0^T & 0 & -\gamma I & 0 \\ C_0 X & 0 & 0 & -\gamma I \end{bmatrix} < 0, \tag{7.62}$$

where $X = e^{j\theta} P + e^{-j\theta} \bar{P}$, $\theta = \frac{\pi}{2}(1 - \alpha)$, $\alpha \in (0, 1)$.

Remark 7.1. This section presents the bounded real lemmas for FOSS-SDs based on the LMI method, i. e., the conditions under which the system satisfies both the admissibility and H_∞ performance. When the time-delay term is zero, the results obtained in this section are consistent with those obtained in Chapter 4, indicating that the bounded real lemmas for FOSS-SDs are an extension of the bounded real lemmas for FOSSs.

Remark 7.2. From (7.52) in the proof of Theorem 7.6 we can see that the results obtained in this section are applicable to systems with $\alpha \in (0, 1)$ because $\sin(\pi\alpha) > 0$ for $\alpha \in (0, 1)$. For $\alpha \in (1, 2)$, $\sin(\pi\alpha) < 0$. Therefore, for the bounded real lemmas of FOSS-SDs with $\alpha \in (1, 2)$, new methods need to be adopted. This will be one of the future research directions of this work.

Based on the bounded real lemmas FOSS-SDs, we give the solution to the H_∞ controller synthesis problem for the closed-loop system (7.41) and the design of the corresponding state feedback H_∞ controller as follows.

Theorem 7.7. *For a given positive scalar γ, there exists a state feedback controller* (7.40) *such that the closed-loop system* (7.41) *is admissible and its transfer function satisfies $\|G_K\|_\infty < \gamma$ if there exist a real constant $\varepsilon_1 > 0$ and matrices $P \in \mathbb{C}^{n \times n}$, $Q \in \mathbb{C}^{n \times n}$, $Q > 0$, and $Y \in \mathbb{R}^{m \times n}$ such that*

$$EP = P^* E^T \geq 0, \tag{7.63}$$

$$\begin{bmatrix} W_0 + \varepsilon_1 Y_1 Y_1^T & Y_2^T \\ Y_2 & -\varepsilon_1 I \end{bmatrix} < 0, \tag{7.64}$$

where

$$Y_1 = \begin{bmatrix} D_A & D_{B_1} & D_{A_h} & D_B & 0 & 0 \\ 0 & 0 & 0 & 0 & 0 & 0 \\ 0 & 0 & 0 & 0 & 0 & 0 \\ 0 & 0 & 0 & 0 & D_C & D_{B_2} \end{bmatrix}, \quad Y_2 = \begin{bmatrix} G_A X & 0 & 0 & 0 \\ G_{B_1} Y & 0 & 0 & 0 \\ 0 & G_{A_h} X & 0 & 0 \\ 0 & 0 & G_B & 0 \\ G_C X & 0 & 0 & 0 \\ G_{B_2} Y & 0 & 0 & 0 \end{bmatrix},$$

$$W_0 = \begin{bmatrix} \mathrm{Sym}\{A_0X + B_{10}Y\} + Q & A_{h_0}X & B_0 & (C_0X + B_{20}Y)^T \\ X^* A_{h_0}^T & -Q & 0 & 0 \\ B_0^T & 0 & -\gamma I & 0 \\ C_0X + B_{20}Y & 0 & 0 & -\gamma I \end{bmatrix},$$

$$X = e^{j\theta}P + e^{-j\theta}\bar{P}, \quad \theta = \frac{\pi}{2}(1-\alpha), \quad \alpha \in (0,1).$$

(Here X can be considered nonsingular. Otherwise, we can choose any nonsingular real matrix $\hat{X}$ such that $E\hat{X} = \hat{X}^T E^T \geq 0$ and any real constant $\xi \in (0,1)$ such that (i) it is not an eigenvalue of the matrix $-X\hat{X}^{-1}$ and (ii) when X in (7.64) is replaced with $X + \xi\hat{X}$, (7.64) still holds. In this case, we can see that $X + \xi\hat{X}$ meets the conditions and is nonsingular.)

At this point, the state feedback controller (7.40) can be designed as $u(t) = Kx(t) = YX^{-1}x(t)$.

Proof. According to the Schur complement, (7.64) is equivalent to

$$W_0 + \varepsilon_1 \Upsilon_1 \Upsilon_1^T + \varepsilon_1^{-1} \Upsilon_2^T \Upsilon_2 < 0. \tag{7.65}$$

Let

$$M_1 = \begin{bmatrix} F_A & 0 & 0 & 0 & 0 & 0 \\ 0 & F_{B_1} & 0 & 0 & 0 & 0 \\ 0 & 0 & F_{A_h} & 0 & 0 & 0 \\ 0 & 0 & 0 & F_B & 0 & 0 \\ 0 & 0 & 0 & 0 & F_C & 0 \\ 0 & 0 & 0 & 0 & 0 & F_{B_2} \end{bmatrix}.$$

Since $F_\Gamma^T F_\Gamma \leq I, \Gamma \in \{A, A_h, B, C, B_1, B_2\}$, we have

$$M_1^T M_1 \leq I.$$

According to Lemma 4.6, for any real constant $\varepsilon_1 > 0$, we have

$$\mathrm{Sym}\{\Upsilon_1 M_1 \Upsilon_2\} \leq \varepsilon_1 \Upsilon_1 \Upsilon_1^T + \varepsilon_1^{-1} \Upsilon_2^T M_1^T M_1 \Upsilon_2 \leq \varepsilon_1 \Upsilon_1 \Upsilon_1^T + \varepsilon_1^{-1} \Upsilon_2^T \Upsilon_2. \tag{7.66}$$

Inequalities (7.65) and (7.66) give

$$W_0 + \mathrm{Sym}\{\Upsilon_1 M_1 \Upsilon_2\} < 0. \tag{7.67}$$

Take W_0, Υ_1, M_1 and Υ_2 in (7.67). Then we have the inequality

$$\begin{bmatrix} \mathrm{Sym}\{AX + B_1Y\} + Q & A_hX & B & X^TC^T + Y^TB_2^T \\ X^* A_h^T & -Q & 0 & 0 \\ B^T & 0 & -\gamma I & 0 \\ CX + B_2Y & 0 & 0 & -\gamma I \end{bmatrix} < 0. \tag{7.68}$$

Note that the state feedback controller is $u(t) = Kx(t) = YX^{-1}x(t)$. Substituting $Y = KX$ into (7.68) yields

$$\begin{bmatrix} \mathrm{Sym}\{(A + B_1K)X\} + Q & A_hX & B & X^T(C + B_2K)^T \\ X^*A_h^T & -Q & 0 & 0 \\ B^T & 0 & -\gamma I & 0 \\ (C + B_2K)X & 0 & 0 & -\gamma I \end{bmatrix} < 0. \tag{7.69}$$

By Corollary 7.2 and equations (7.63) and (7.69), the theorem is proved. □

Remark 7.3. This section presents the bounded real lemmas and robust H_∞ controller design methods for FOSS-SDs. Here are two points of explanation:

1. In all the results of this section, when the state delay is zero, the results are consistent with those in Chapter 4. This indicates that the H_∞ control results for FOSS-SDs in this section are more general than the H_∞ control results for FOSSs in Chapter 4. It is important to emphasize that although the results of this section include the H_∞ control results for FOSSs with $\alpha \in (0, 1)$ in Chapter 4, the results of Chapter 2 still have their necessity because 1) the methods used in this section and Chapter 2 are different, and 2) the bounded real lemmas for FOSS-SDs studied in this section require the system to both satisfy the H_∞ performance and admissibility, where the admissibility results are given using the small gain theorem and the bounded real lemmas for FOSSs. Therefore the results of Chapter 4 are also the foundation for the research in this section.
2. When the fractional-order derivative coefficient matrix E is set to the identity matrix, the results of this chapter are applicable to FOSs with state delay. This fills the gap in the research field of H_∞ control for FOSs with state delay.

Now we present two numerical examples to show the effectiveness of our results.

Example 7.1. Consider the nominal FOSS-SD

$$\begin{aligned} ED^\alpha x(t) &= A_0x(t) + A_{h_0}x(t-h) + B_0w(t), \\ z(t) &= C_0x(t), \\ x(t) &= \chi(t), \quad t \in (-h, 0], \end{aligned} \tag{7.70}$$

where

$$E = \begin{bmatrix} 10 & 0 & 0 \\ 0 & 5 & 0 \\ 0 & 0 & 0 \end{bmatrix}, \quad A_0 = \begin{bmatrix} -1 & 0 & 0 \\ 0 & -2 & 0 \\ 1 & 1 & -1 \end{bmatrix}, \quad A_{h_0} = \begin{bmatrix} -0.4 & 0 & 0 \\ 0 & -0.3 & 0 \\ 0 & 0 & 0 \end{bmatrix},$$

$$B_0 = \begin{bmatrix} 1 & 0 \\ 0 & 1 \\ 0 & 0 \end{bmatrix}, \quad C_0 = \begin{bmatrix} 0 & 1 & 0 \\ 1 & 0 & 0 \end{bmatrix}, \quad h = 0.1, \quad \alpha = 0.8, \quad \gamma = 2.$$

We can see that system (7.70) is the FOSS-SD (7.43) studied in this section. The goal is to determine whether it satisfies the admissibility and H_∞ performance.

According to Theorem 7.6, we can find the following feasible solutions:

$$P = \begin{bmatrix} 0.7694 & 0.0218 & 0 \\ 0.0436 & 2.4854 & 0 \\ -3861.5377 & 11872.7986\mathbf{i} & 1 - 11863.5535\mathbf{i} \end{bmatrix},$$

$$Q = \begin{bmatrix} 7059.5513 & 7053.0005 & -7049.6853 \\ 7053.0005 & 7051.8274 & -7043.8707 \\ -7049.6853 & -7043.8707 & 7040.6285 \end{bmatrix}.$$

Therefore the FOSS-SD (7.70) is admissible, and its transfer function satisfies $\|G_N\|_\infty < \gamma = 2$.

From the perspective of numerical simulation, when the initial condition is chosen as $\chi(t) = \begin{bmatrix} 2 & -4 & -2 \end{bmatrix}^T$, the state responses of the system are shown in Figure 7.1. As we can seen from the figure, all states tend to zero, indicating that the system is stable. Additionally, let $\lambda = s^\alpha$. Then (i) $\det(\lambda E - A_0)$ is not identically zero, indicating that the system is regular, and (ii) $\deg(\det(\lambda E - A_0)) = \operatorname{rank}(E)$, indicating that the system is impulse-free. Therefore the system is admissible. The maximum singular values of the system transfer function are shown in Figure 7.2. As we can see from the figure, the peak value of the maximum singular values is $0.7133 < 2$. Hence the system is admissible, and its transfer function satisfies $\|G_N\|_\infty < \gamma = 2$.

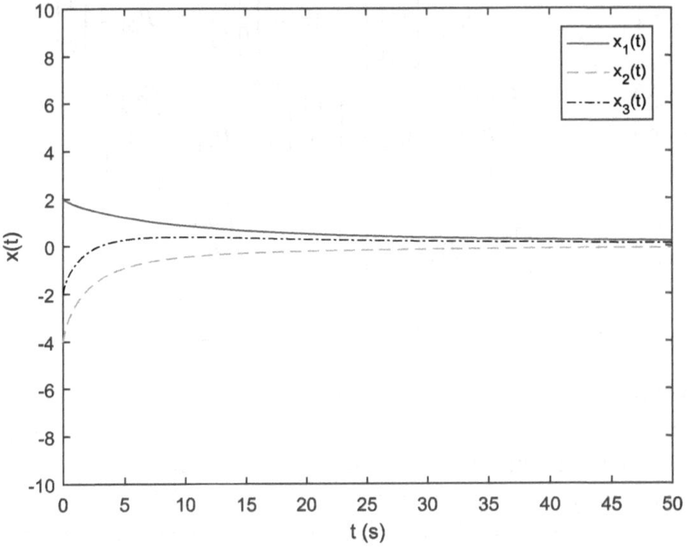

Figure 7.1: The time responses the FOSS-SD (7.43) in Example 7.1.

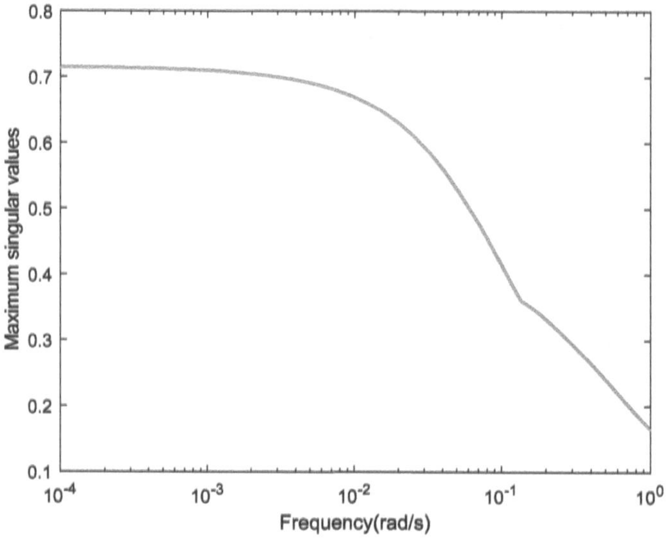

Figure 7.2: Maximum singular values of the FOSS-SD (7.43) in Example 7.1.

Example 7.2. Consider the FOSS-SD with norm-bounded uncertainties

$$E = \begin{bmatrix} 10 & 0 & 0 \\ 0 & 5 & 0 \\ 0 & 0 & 0 \end{bmatrix}, \quad A = \begin{bmatrix} -1 & 0 & 0 \\ 0 & -2 & 0 \\ 1 & 1 & -1 \end{bmatrix}, \quad D_A = \begin{bmatrix} 1.1 \\ 0.1 \\ 0 \end{bmatrix}, \quad G_A = \begin{bmatrix} 1 \\ 0.1 \\ 0 \end{bmatrix}^T,$$

$$B_0 = \begin{bmatrix} 1 & 0 \\ 0 & 1 \\ 0 & 0 \end{bmatrix}, \quad C_0 = \begin{bmatrix} 0 & 1 & 0 \\ 1 & 0 & 0 \end{bmatrix}, \quad D_B = \begin{bmatrix} 0.1 \\ 0.1 \\ 0.1 \end{bmatrix}, \quad G_B = \begin{bmatrix} 0.1 \\ 0.1 \end{bmatrix}^T, \quad B_{20} = \begin{bmatrix} 1 \\ 1 \end{bmatrix},$$

$$A_{h_0} = \begin{bmatrix} 5 & 0 & 0 \\ 0 & -3 & 0 \\ 0 & 0 & 0 \end{bmatrix}, \quad D_{A_h} = \begin{bmatrix} 0.1 \\ 0 \\ 0 \end{bmatrix}, \quad G_{A_h} = \begin{bmatrix} 0.1 \\ 0 \\ 0 \end{bmatrix}^T, \quad D_{B1} = \begin{bmatrix} 0.1 \\ 0 \\ 0.1 \end{bmatrix},$$

$$B_{10} = \begin{bmatrix} 4 \\ 4 \\ 4 \end{bmatrix}, \quad D_C = \begin{bmatrix} 0.1 \\ 0.1 \end{bmatrix}, \quad D_{B2} = \begin{bmatrix} 0.1 \\ 0.1 \end{bmatrix}, \quad G_C = \begin{bmatrix} 0.1 \\ 0.1 \\ 0.1 \end{bmatrix},$$

$$G_{B1} = 0.1, \quad G_{B2} = 0.2, \quad h = 0.1, \quad \alpha = 0.8, \quad \gamma = 2.$$

The goal is to design a state feedback controller (7.40) such that the closed-loop system (7.41) satisfies the robust admissibility and H_∞ performance. First, analyze the robust admissibility and H_∞ performance of the open-loop system. When $u(t) = 0$, the system is inadmissible (some states diverge, and the system is not stable), meaning that the system has poles located in the right half-plane. This also causes the H_∞ norm of the system to approach infinity, indicating that the system does not satisfy the admissibility and H_∞ performance.

According to Theorem 7.7, we can find the following feasible solutions:

$$P = \begin{bmatrix} 0.2427 & -0.1332 + 0.0577\mathbf{i} & 0 \\ -0.2663 - 0.1154\mathbf{i} & 0.821 & 0 \\ 0.3441 & -0.4912 & 53.9927 \end{bmatrix},$$

$$Q = \begin{bmatrix} 6.028 & 6.4947 & 9.7347 \\ 6.4947 & 10.2106 & 9.6676 \\ 9.7347 & 9.6676 & 76.1834 \end{bmatrix}, \quad Y = \begin{bmatrix} -3.7483 & -4.3482 & -3.7407 \end{bmatrix}.$$

Therefore the corresponding state feedback controller gain is obtained as $K = \begin{bmatrix} -12.9803 & -5.2081 & -0.0364 \end{bmatrix}$.

To verify the effectiveness of the designed state feedback H_∞ controller, we chose a set of parameters from the uncertain system. For instance, when $F_A, F_{A_h}, F_B, F_C, F_{B1}$, and F_{B2} are all set to 1, the system parameters are

$$\begin{aligned} A &= \begin{bmatrix} 0.1 & 0.11 & 0 \\ 0.1 & -1.99 & 0 \\ 1 & 1 & -1 \end{bmatrix}, \quad B = \begin{bmatrix} 1.01 & 0.01 \\ 0.01 & 1.01 \\ 0.01 & 0.01 \end{bmatrix}, \quad C = \begin{bmatrix} 0.01 & 1.01 \\ 1.01 & 0.01 \\ 0.01 & 0.01 \end{bmatrix}^T, \\ A_h &= \begin{bmatrix} 5.01 & 0 & 0 \\ 0 & -3 & 0 \\ 0 & 0 & 0 \end{bmatrix}, \quad B_1 = \begin{bmatrix} 4.01 \\ 4 \\ 4.01 \end{bmatrix}, \quad B_2 = \begin{bmatrix} 1.02 \\ 1.02 \end{bmatrix}. \end{aligned} \tag{7.71}$$

When the initial condition of the system is $\chi(t) = \begin{bmatrix} 1 & -2.5 & -1.0482 \end{bmatrix}^T$, under the action of the controller obtained from Theorem 7.7, the state responses of the closed-loop system are shown in Figure 7.3. As we can see from the figure, all states asymptotically approach zero, indicating that the system is stable. Additionally, let $\lambda = s^\alpha$. Then (i) $\det(\lambda E - A)$ is not identically zero, indicating that the system is regular, and (ii) $\deg(\det(\lambda E - A)) = \operatorname{rank}(E)$, indicating that the system is impulse-free. Therefore the system is robustly admissible. The maximum singular values of the systems transfer function are shown in Figure 7.4. As we can see from the figure, the peak value of the maximum singular values is $1.1170 < 2$. Hence the system is robustly admissible and satisfies the H_∞ performance.

7.4 Robust positive real control

Consider the FOSS-SD with norm-bounded uncertainties

$$\begin{aligned} ED^\alpha x(t) &= Ax(t) + A_h x(t-h) + B_1 u(t) + Bw(t), \\ z(t) &= Cx(t) + B_2 u(t) + Dw(t), \\ x(t) &= \chi(t), \quad t \in (-h, 0], \end{aligned} \tag{7.72}$$

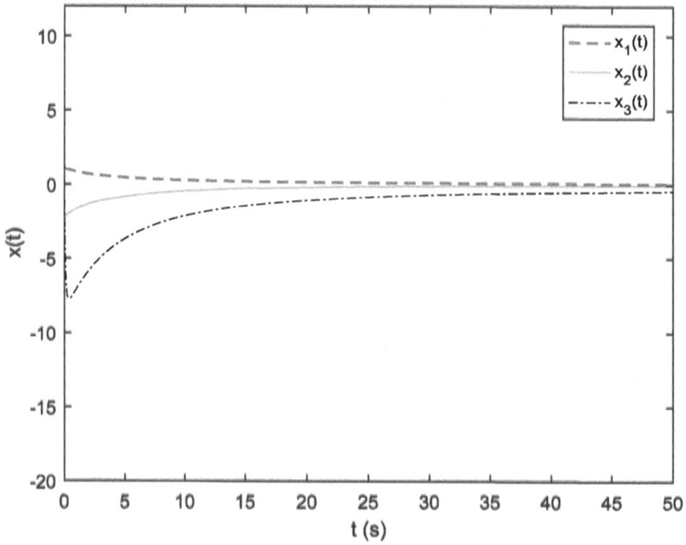

Figure 7.3: The time responses the FOSS-SD (7.41) in Example 7.2.

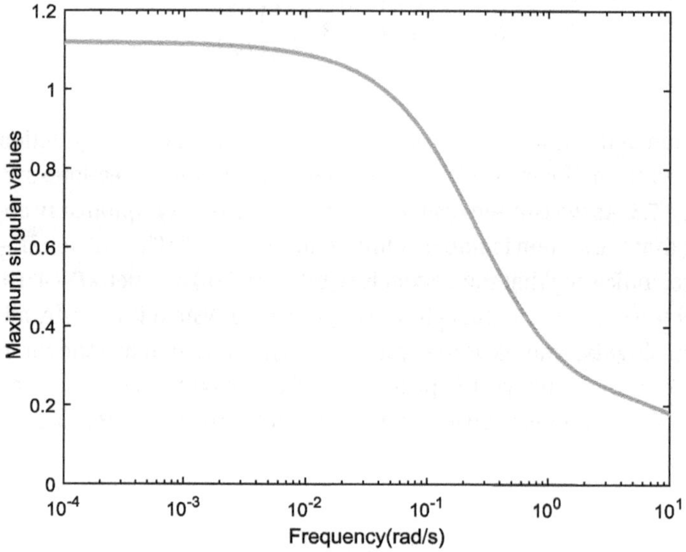

Figure 7.4: Maximum singular values of the FOSS-SD (7.41) in Example 7.2.

where $x(t) \in \mathbb{R}^n$, $z(t) \in \mathbb{R}^q$, $w(t) \in \mathbb{R}^p$, and $u(t) \in \mathbb{R}^m$ are the state vector, control output, disturbance input, and control input, respectively, $\chi(t)$ is the initial continuous vector function, $\alpha \in (0,1)$ is the system order, $h > 0$ is a real constant representing the delay, and $E \in \mathbb{R}^{n\times n}$ is a known real matrix with $\operatorname{rank}(E) = r \le n$, D^α is the Caputo fractional derivative, and A, A_h, B, C, B_1, B_2, and D are uncertain matrices satisfying

$$\begin{aligned} A = A_0 + \Delta A, \quad A_h &= A_{h_0} + \Delta A_h, \quad B = B_0 + \Delta B, \\ B_1 &= B_{10} + \Delta B_1, \quad D = D_0 + \Delta D, \\ C &= C_0 + \Delta C, \quad B_2 = B_{20} + \Delta B_2, \\ \Gamma &= D_\Gamma F_\Gamma G_\Gamma, \quad \Gamma \in \{A, A_h, B, C, B_1, B_2\}, \end{aligned}$$

where $A_0, A_{h_0}, B_0, C_0, D_0, B_{10}, B_{20}, D_\Gamma$, and G_Γ are known real matrices, and F_Γ is a matrix with norm-bounded uncertainties satisfying $F_\Gamma^T F_\Gamma \le I$.

Consider the state feedback controller

$$u(t) = Kx(t). \tag{7.73}$$

Applying the state feedback controller (7.73) to control system (7.72), we have

$$\begin{aligned} ED^\alpha x(t) &= (A + B_1K)x(t) + A_h x(t-h) + Bw(t), \\ z(t) &= (C + B_2K)x(t) + Dw(t), \\ x(t) &= \chi(t), \quad t \in (-h, 0], \end{aligned} \tag{7.74}$$

with transfer function

$$G_K(s) = (C + B_2K)(s^\alpha E - (A + B_1K) - A_h e^{-sh})^{-1} B + D. \tag{7.75}$$

When the uncertain parameters and the control input $u(t)$ in system (7.72) are both zero, the system can be rewritten as the nominal FOSS-SD

$$\begin{aligned} ED^\alpha x(t) &= A_0 x(t) + A_{h_0} x(t-h) + B_0 w(t), \\ z(t) &= C_0 x(t) + D_0 w(t), \\ x(t) &= \chi(t), \quad t \in (-h, 0]. \end{aligned} \tag{7.76}$$

The transfer function of system (7.76) is

$$G_N(s) = C_0(s^\alpha E - A_0 - A_{h_0} e^{-sh})^{-1} B_0 + D_0. \tag{7.77}$$

We refer to Definition 5.1 in Chapter 5 and consider the following definitions.

Definition 7.4. (1) If the transfer function (7.77) is analytic in $\operatorname{Re}(s) > 0$ and satisfies $\operatorname{Sym}\{G(s)\} > 0$ when $\operatorname{Re}(s) > 0$, then system (7.76) is said to be positive real. (2) If the transfer function (7.77) is analytic in $\operatorname{Re}(s) \ge 0$ and satisfies $\operatorname{Sym}\{G(jw)\} > 0$ when $w \ge 0$, then system (7.76) is said to be strictly positive real. (3) If system (7.76) is strictly positive real and its transfer function satisfies $\operatorname{Sym}\{G(j\infty)\} > 0$, then system (7.76) is said to be extended strictly positive real.

The objectives of this section are:

1) To provide the positive real lemma for the FOSS-SD (7.76), that is, the conditions under which system (7.76) satisfies both admissibility and extended strict positive realness.
2) For the FOSS-SD (7.72) with norm-bounded uncertainties, to design a state feedback controller (7.73) such that the closed-loop system (7.74) satisfies both robust admissibility and extended strict positive realness.

Based on the admissibility conditions of FOSS-SDs in Section 7.2, in this section, we present the positive real lemma for the FOSS-SD (7.76).

Theorem 7.8. *Assume that $D_0 + D_0^T > 0$. The FOSS-SD* (7.76) *is admissible and extended strictly positive real if there exist $P \in \mathbb{C}^{n\times n}$ and $Q \in \mathbb{C}^{n\times n}$, $Q > 0$, such that*

$$E^T P = P^* E \geq 0, \tag{7.78}$$

$$\begin{bmatrix} \mathrm{Sym}\{A_0^T X\} + Q & X^* A_{h_0} & X^* B_0 - C_0^T \\ A_{h_0}^T X & -Q & 0 \\ B_0^T X - C_0 & 0 & -D_0 - D_0^T \end{bmatrix} < 0, \tag{7.79}$$

where $X = e^{-j\theta} P + e^{j\theta} \bar{P}$, $\theta = \frac{\pi}{2}(1 - \alpha)$, $0 < \alpha < 1$.

Proof. From (7.79) we have

$$\begin{bmatrix} \mathrm{Sym}\{A_0^T X\} + Q & X^* A_{h_0} \\ A_{h_0}^T X & -Q \end{bmatrix} < 0. \tag{7.80}$$

From equations (7.78) and (7.80), by Theorem 7.4 the FOSS-SD (7.76) is admissible.

From (7.79), we have the following inequality:

$$\begin{aligned} &A_0^T X + X^* A_0 + Q + X^* A_{h_0} Q^{-1} A_{h_0}^T X \\ &\quad + (X^* B_0 - C_0^T)(D_0 + D_0^T)^{-1}(B_0^T X - C_0) < 0. \end{aligned}$$

There exists a constant matrix $Z > 0$ such that

$$\begin{aligned} &A_0^T X + X^* A_0 + Q + X^* A_{h_0} Q^{-1} A_{h_0}^T X \\ &\quad + (X^* B_0 - C_0^T)(D_0 + D_0^T)^{-1}(B_0^T X - C_0) + Z < 0. \end{aligned} \tag{7.81}$$

Let $W = (X^* B_0 - C_0^T)(D_0 + D_0^T)^{-1}(B_0^T X - C_0) + Z$. Then $W > 0$. Note that

$$\begin{aligned} &Q + X^T A_{h_0} Q^{-1} A_{h_0}^T X - e^{jwh} A_{h_0}^T X - e^{-jwh} X^T A_{h_0} \\ &= (e^{-jwh} X^T A_{h_0} - Q) Q^{-1} (e^{jwh} A_{h_0}^T X - Q) \geq 0. \end{aligned} \tag{7.82}$$

Thus

$$-e^{jwh}A_{h_0}^T X - e^{-jwh}X^T A_{h_0} \geq -Q - X^T A_{h_0} Q^{-1} A_{h_0}^T X. \tag{7.83}$$

Furthermore,

$$\begin{aligned}
&- A_0^T X - X^* A_0 - e^{jwh}A_{h_0}^T X - e^{-jwh}X^T A_{h_0} \\
&\quad \geq -A_0^T X - X^* A_0 - Q - X^T A_{h_0} Q^{-1} A_{h_0}^T X > W > 0.
\end{aligned} \tag{7.84}$$

According to (7.78), we also have

$$E^T \bar{P} = \bar{P}^* E \geq 0.$$

Meanwhile, for $\alpha \in (0,1)$, we have

$$\begin{aligned}
&e^{-j\frac{\pi}{2}\alpha} w^\alpha E^T X + X^* e^{j\frac{\pi}{2}\alpha} w^\alpha E \\
&\quad = w^\alpha e^{j(\frac{\pi}{2}-\pi\alpha)} E^T \bar{P} + w^\alpha e^{-j\frac{\pi}{2}} E^T P + w^\alpha e^{j(\pi\alpha-\frac{\pi}{2})} \bar{P}^* E + w^\alpha e^{j\frac{\pi}{2}} P^* E \\
&\quad = w^\alpha e^{j(\frac{\pi}{2}-\pi\alpha)} E^T \bar{P} + w^\alpha e^{j(\pi\alpha-\frac{\pi}{2})} \bar{P}^* E \\
&\quad = 2w^\alpha \sin(\pi\alpha) E^T \bar{P} \geq 0 \quad \forall w \in [0,\infty).
\end{aligned} \tag{7.85}$$

Based on (7.85), we can obtain the following equality:

$$\begin{aligned}
&\mathrm{Sym}\{G(jw)\} \\
&\quad = C_0(e^{j\frac{\pi}{2}\alpha} w^\alpha E - A_0 - A_{h_0} e^{-jwh})^{-1} B_0 + B_0^T (e^{-j\frac{\pi}{2}\alpha} w^\alpha E - A_0 - A_{h_0} e^{jwh})^{-T} C_0^T + D_0 + D_0^T \\
&\quad = (C_0 - B_0^T X)(e^{j\frac{\pi}{2}\alpha} w^\alpha E - A_0 - A_{h_0} e^{-jwh})^{-1} B_0 + B_0^T (e^{-j\frac{\pi}{2}\alpha} w^\alpha E \\
&\qquad - A_0 - A_{h_0} e^{jwh})^{-T} (C_0^T - X^* B_0) \\
&\qquad + D_0 + D_0^T + B_0^T X (e^{j\frac{\pi}{2}\alpha} w^\alpha E - A_0 - A_{h_0} e^{-jwh})^{-1} B_0 \\
&\qquad + B_0^T (e^{-j\frac{\pi}{2}\alpha} w^\alpha E - A_0 - A_{h_0} e^{jwh})^{-T} X^* B_0 \\
&\quad = (C_0 - B_0^T X)(e^{j\frac{\pi}{2}\alpha} w^\alpha E - A_0 - A_{h_0} e^{-jwh})^{-1} B_0 \\
&\qquad + B_0^T (e^{-j\frac{\pi}{2}\alpha} w^\alpha E - A_0 - A_{h_0} e^{jwh})^{-T} (C_0^T - X^* B_0) \\
&\qquad + D_0 + D_0^T + B_0^T (e^{-j\frac{\pi}{2}\alpha} w^\alpha E - A_0 - A_{h_0} e^{jwh})^{-T} ((e^{-j\frac{\pi}{2}\alpha} w^\alpha E - A_0 - A_{h_0} e^{jwh})^T X \\
&\qquad + X^* (e^{j\frac{\pi}{2}\alpha} w^\alpha E - A_0 - A_{h_0} e^{-jwh}))(e^{j\frac{\pi}{2}\alpha} w^\alpha E - A_0 - A_{h_0} e^{-jwh})^{-1} B_0.
\end{aligned}$$

From (7.84) and (7.85) we have

$$\begin{aligned}
&\mathrm{Sym}\{G(jw)\} \\
&\quad > (C_0 - B_0^T X)(e^{j\frac{\pi}{2}\alpha} w^\alpha E - A_0 - A_{h_0} e^{-jwh})^{-1} B_0 \\
&\qquad + B_0^T (e^{-j\frac{\pi}{2}\alpha} w^\alpha E - A_0 - A_{h_0} e^{jwh})^{-T} (C_0^T - X^* B_0) \\
&\qquad + D_0 + D_0^T + B_0^T (e^{-j\frac{\pi}{2}\alpha} w^\alpha E - A_0 - A_{h_0} e^{jwh})^{-T} W (e^{j\frac{\pi}{2}\alpha} w^\alpha E - A_0 - A_{h_0} e^{-jwh})^{-1} B_0
\end{aligned}$$

$$
\begin{aligned}
&= [B_0^T(e^{-j\frac{\pi}{2}\alpha}w^\alpha E - A_0 - A_{h_0}e^{jwh})^{-T} + (C_0 - B_0^T X)W^{-1}]W[(e^{j\frac{\pi}{2}\alpha}w^\alpha E - A_0 - A_{h_0}e^{-jwh})^{-1}B_0 \\
&\quad + W^{-1}(C_0^T - X^* B_0)] + D_0 + D_0^T - (C_0 - B_0^T X)W^{-1}(C_0^T - X^* B_0) \\
&\geq D_0 + D_0^T - (B_0^T X - C_0)W^{-1}(X^* B_0 - C_0^T).
\end{aligned}
$$

Note that

$$
W - (X^* B_0 - C_0^T)(D_0 + D_0^T)^{-1}(B_0^T X - C_0) = Z > 0.
$$

Using Lemma 4.7, we have

$$
\begin{bmatrix} W & X^* B_0 - C_0^T \\ B_0^T X - C_0 & D_0 + D_0^T \end{bmatrix} > 0,
$$

which is equivalent to

$$
D_0 + D_0^T - (B_0^T X - C_0)W^{-1}(X^* B_0 - C_0^T) > 0.
$$

Therefore,

$$
\mathrm{Sym}\{G(jw)\} > 0 \quad \forall w \in [0, \infty),
$$

and hence

$$
\mathrm{Sym}\{G(j\infty)\} > 0.
$$

Proof is completed. □

From Corollary 7.1 we can obtain the following corollary.

Corollary 7.3. *Assume that $D_0 + D_0^T > 0$. The FOSS-SD (7.76) is admissible and extended strictly positive real if there exist $P \in \mathbb{C}^{n\times n}$ and $Q \in \mathbb{C}^{n\times n}$, $Q > 0$, such that*

$$
EP = P^* E^T \geq 0, \tag{7.86}
$$

$$
\begin{bmatrix} \mathrm{Sym}\{A_0 X\} + Q & A_{h_0} X & X^* C_0^T - B_0 \\ X^T A_{h_0}^T & -Q & 0 \\ C_0 X - B_0^T & 0 & -D_0 - D_0^T \end{bmatrix} < 0, \tag{7.87}
$$

where $X = e^{-j\theta}P + e^{j\theta}\bar{P}$, $\theta = \frac{\pi}{2}(1 - \alpha)$, $0 < \alpha < 1$.

Remark 7.4. This section provides the positive real lemmas for FOSS-SDs based on the LMI method, that is, the conditions under which the system satisfies both admissibility and extended strict positive realness. When the state delay is zero, the results obtained in this section are consistent with those obtained in Chapter 5, indicating that the positive real lemma for FOSS-SDs is a generalization of the positive real lemma for FOSSs.

Remark 7.5. From inequality (7.85) in the proof of Theorem 7.8 we can see that the results obtained in this chapter are applicable to systems with order $\alpha \in (0,1)$ because $\sin(\pi\alpha) > 0$ for $\alpha \in (0,1)$. When $\alpha \in (1,2)$, $\sin(\pi\alpha) < 0$. Therefore new methods are needed for the bounded real lemma for FOSS-SDs with order $\alpha \in (1,2)$, which is one of the future research directions of this paper.

Based on the positive real lemma for FOSS-SDs, in this section, we present the solution to the positive real controller synthesis problem for the closed-loop system (7.74) and design the corresponding state feedback positive real controller. The results are as follows.

Theorem 7.9. *Assume that $D_0 + D_0^T > 0$. When $\alpha \in (0,1)$, there exists a state feedback controller* (7.73) *such that the closed-loop system* (7.74) *is admissible and extended strictly positive real if and only if there exist a real constant $\varepsilon_1 > 0$ and matrices $P \in \mathbb{C}^{n\times n}$, $Q \in \mathbb{C}^{n\times n}$, $Q > 0$, and $Y \in \mathbb{R}^{m\times n}$ such that*

$$EP = P^* E^T \geq 0, \tag{7.88}$$

$$\begin{bmatrix} W_0 + \varepsilon_1 \Upsilon_1 \Upsilon_1^T & \Upsilon_2^T \\ \Upsilon_2 & -\varepsilon_1 I \end{bmatrix} < 0, \tag{7.89}$$

where

$$W_0 = \begin{bmatrix} \mathrm{Sym}\{A_0 X + B_{10} Y\} + Q & A_{h_0} X & (C_0 X + B_{20} Y - B_0^T)^T \\ X^T A_{h_0}^T & -Q & 0 \\ C_0 X + B_{20} Y - B_0^T & 0 & -D_0 - D_0^T \end{bmatrix},$$

$$\Upsilon_1 = \begin{bmatrix} D_A & D_{B_1} & D_{A_h} & -D_B & 0 & 0 & 0 \\ 0 & 0 & 0 & 0 & 0 & 0 & 0 \\ 0 & 0 & 0 & 0 & D_C & D_{B_2} & -D_D \end{bmatrix},$$

$$\Upsilon_2 = \begin{bmatrix} G_A X & 0 & 0 \\ G_{B_1} Y & 0 & 0 \\ 0 & G_{A_h} X & 0 \\ 0 & 0 & G_B \\ G_C X & 0 & 0 \\ G_{B_2} Y & 0 & 0 \\ 0 & 0 & G_D \end{bmatrix},$$

$$X = e^{-j\theta} P + e^{j\theta} \bar{P}, \quad \theta = \frac{\pi}{2}(1-\alpha).$$

(Here X can be considered nonsingular. Otherwise, we can choose a nonsingular real matrix $\hat{X}$ such that $E\hat{X} = \hat{X}^T E^T \geq 0$ and then choose a real constant $\xi \in (0,1)$ such that (i) *it is not an eigenvalue of the matrix $-X\hat{X}^{-1}$ and* (ii) *inequality* (7.89) *still holds when X is replaced with $X + \xi\hat{X}$. We can seen that then resulting $X + \xi\hat{X}$ is nonsingular.)*

In this case the state feedback controller (7.73) *can be designed as* $u(t) = Kx(t) = YX^{-1}x(t)$.

Proof. By Lemma 4.7, (7.89) is equivalent to

$$W_0 + \varepsilon_1 \Upsilon_1 \Upsilon_1^T + \varepsilon_1^{-1} \Upsilon_2^T \Upsilon_2 < 0. \tag{7.90}$$

Let

$$M_1 = \begin{bmatrix} F_A & 0 & 0 & 0 & 0 & 0 & 0 \\ 0 & F_{B_1} & 0 & 0 & 0 & 0 & 0 \\ 0 & 0 & F_{A_h} & 0 & 0 & 0 & 0 \\ 0 & 0 & 0 & F_B & 0 & 0 & 0 \\ 0 & 0 & 0 & 0 & F_C & 0 & 0 \\ 0 & 0 & 0 & 0 & 0 & F_{B_2} & 0 \\ 0 & 0 & 0 & 0 & 0 & 0 & F_D \end{bmatrix}.$$

From $F_\Gamma^T F_\Gamma \leq I, \Gamma \in \{A, A_h, B, C, B_1, B_2, D\}$, we have

$$M_1^T M_1 \leq I.$$

According to Lemma 4.6, for any constant $\varepsilon_1 > 0$, we have

$$\text{Sym}\{\Upsilon_1 M_1 \Upsilon_2\} \leq \varepsilon_1 \Upsilon_1 \Upsilon_1^T + \varepsilon_1^{-1} \Upsilon_2^T M_1^T M_1 \Upsilon_2 \leq \varepsilon_1 \Upsilon_1 \Upsilon_1^T + \varepsilon_1^{-1} \Upsilon_2^T \Upsilon_2. \tag{7.91}$$

From (7.90) and (7.91) we have

$$W_0 + \text{Sym}\{\Upsilon_1 M_1 \Upsilon_2\} < 0. \tag{7.92}$$

Substituting W_0, Υ_1, M_1, and Υ_2 into (7.92), we have

$$\begin{bmatrix} \text{Sym}\{AX + B_1 Y\} + Q & A_h X & X^*(CX + B_2 Y)^T - B \\ X^T A_h^T & -Q & 0 \\ CX + B_2 Y - B^T & 0 & -D - D^T \end{bmatrix} < 0. \tag{7.93}$$

Since the state feedback controller is $u(t) = Kx(t) = YX^{-1}x(t)$, we have $Y = KX$. Substituting Y into (7.93), we have

$$\begin{bmatrix} \text{Sym}\{(A + B_1 K)X\} + Q & A_h X & X^*((C + B_2 K)X)^T - B \\ X^T A_h^T & -Q & 0 \\ (C + B_2 K)X - B^T & 0 & -D - D^T \end{bmatrix} < 0. \tag{7.94}$$

By Corollary 7.3, (7.88), and (7.94) the theorem is proved. □

Remark 7.6. This chapter provides the positive real lemma and robust positive real controller design methods for FOSS-SDs. Here are two points of explanation:

- In all the results of this chapter, when the delay term coefficient is zero, the results of this chapter are consistent with those of Chapter 5, indicating that the positive real control results of FOSS-SDs in this chapter are a generalization of the positive real control results of FOSSs in Chapter 5. It should be emphasized that although the results of this chapter include the positive real control results for FOSSs with $\alpha \in (0,1)$ in Chapter 5, the related results in Chapter 5 are still necessary. Although both this chapter and Chapter 5 use the LMI method for $\alpha \in (0,1)$, the presence of delay terms in this chapter makes the treatment more complex. The derivation process of the results in Chapter 5 provides ideas for the analysis of this chapter and forms the basis for the research in this chapter.
- When the fractional derivative coefficient matrix E is taken as the identity matrix, the results of this chapter apply to fractional-order time-delay systems, indicating the strong applicability of the results of this chapter.

We provide two numerical examples to verify the effectiveness of our results.

Example 7.3. Consider the nominal FOSS in Example 5.2 of Chapter 5. When the system has a time-delay term $\begin{bmatrix} -0.1 & 0 & 0 \\ 0 & -0.3 & 0 \\ 0 & 0 & 0 \end{bmatrix} x(t-0.1)$, it can be described as

$$\begin{aligned} ED^{\alpha}x(t) &= A_0x(t) + A_{h_0}x(t-h) + B_0w(t), \\ z(t) &= C_0x(t) + D_0w(t), \\ x(t) &= \chi(t), \quad t \in (-h, 0], \end{aligned} \tag{7.95}$$

where

$$E = \begin{bmatrix} 10 & 0 & 0 \\ 0 & 5 & 0 \\ 0 & 0 & 0 \end{bmatrix}, \quad A_0 = \begin{bmatrix} -1 & 0 & 0 \\ 0 & -2 & 0 \\ 1 & 1 & -1 \end{bmatrix}, \quad A_{h_0} = \begin{bmatrix} -0.1 & 0 & 0 \\ 0 & -0.3 & 0 \\ 0 & 0 & 0 \end{bmatrix},$$

$$B_0 = \begin{bmatrix} 1 & 0 \\ 0 & 1 \\ 0 & 0 \end{bmatrix}, \quad C_0 = \begin{bmatrix} 0 & 1 & 0 \\ 1 & 0 & 0 \end{bmatrix}, \quad D_0 = \begin{bmatrix} 1 & 0 \\ 0 & 1 \end{bmatrix}, \quad h = 0.1, \quad \alpha = 0.8.$$

We can see that system (7.95) is the FOSS-SD (7.76) studied in this chapter. The problem is to determine whether it satisfies admissibility and extended strict positive realness.

According to Theorem 7.8, we can find the following feasible solution:

$$P = \begin{bmatrix} 1.0191 & 0.4151 + 0.0832i & 0 \\ 0.8302 - 0.1665i & 2.4426 & 0 \\ -3296.1573 & -3298.2281 & 3297.6693 \end{bmatrix},$$

$$Q = \begin{bmatrix} 6230.7113 & 6233.3870 & -6231.1838 \\ 6233.3870 & 6238.3950 & -6233.2833 \\ -6231.1838 & -6233.2833 & 6232.8573 \end{bmatrix}.$$

Therefore the FOSS-SD (7.76) is admissible and extended strictly positive real.

From the perspective of numerical simulation, when the initial condition of the system is $\chi(t) = \begin{bmatrix} 2 & -4 & -2 \end{bmatrix}^T$, the state responses of the system are shown in Figure 7.5. We can see from the figure that all states asymptotically approach zero, indicating that the system is stable. At the same time, let $\lambda = s^\alpha$; then (i) $\det(\lambda E - A_0)$ is not identically zero, indicating that the system is regular, and (ii) $\deg(\det(\lambda E - A_0)) = \operatorname{rank}(E)$, indicating that the system is impulse-free. Therefore the system is admissible. The curve of $\operatorname{Sym}\{G(jw)\}(w \geq 0)$ is shown in Figure 7.6. We can see from the figure that the values of $\operatorname{Sym}\{G(jw)\}$ $(w \geq 0)$ are all greater than zero. Therefore the system satisfies admissibility and extended strict positive realness.

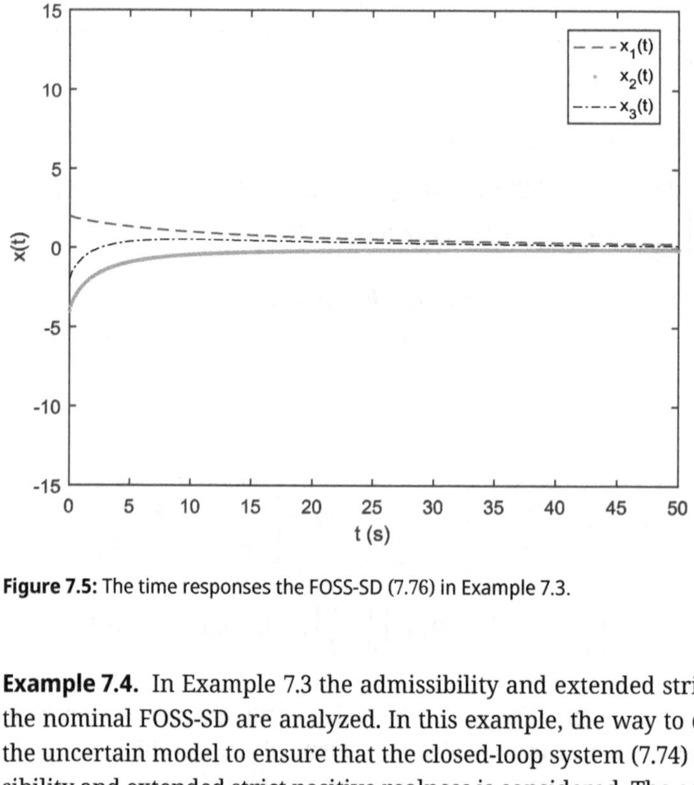

Figure 7.5: The time responses the FOSS-SD (7.76) in Example 7.3.

Example 7.4. In Example 7.3 the admissibility and extended strict positive realness of the nominal FOSS-SD are analyzed. In this example, the way to design a controller for the uncertain model to ensure that the closed-loop system (7.74) satisfies robust admissibility and extended strict positive realness is considered. The coefficient matrix of the control input $u(t)$ and the uncertain parameter matrix are given by

$$B_{10} = \begin{bmatrix} 8 \\ 8 \\ 8 \end{bmatrix}, \quad B_{20} = \begin{bmatrix} 1 \\ 1 \end{bmatrix}, \quad D_{B1} = \begin{bmatrix} 0.1 \\ 0 \\ 0.1 \end{bmatrix}, \quad G_{B1} = 0.1, \quad D_{B2} = \begin{bmatrix} 0.1 \\ 0.1 \end{bmatrix},$$

$$G_{A_h} = \begin{bmatrix} 0.1 \\ 0 \\ 0 \end{bmatrix}^T, \quad G_{B2} = 0.2, \quad D_A = \begin{bmatrix} 1.1 \\ 0.1 \\ 0 \end{bmatrix}, \quad G_A = \begin{bmatrix} 1 \\ 0.1 \\ 0 \end{bmatrix}^T, \quad D_B = \begin{bmatrix} 0.1 \\ 0.1 \\ 0.1 \end{bmatrix}, \tag{7.96}$$

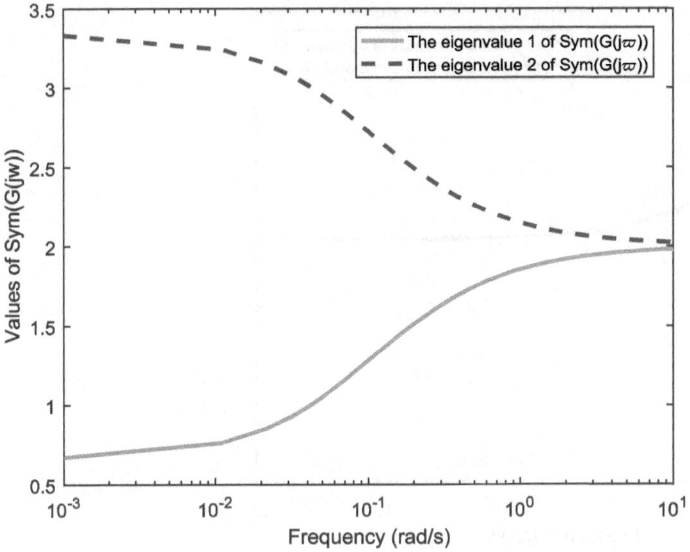

Figure 7.6: The eigenvalue curves of Sym$\{G(jw)\}$ $(w \geq 0)$ of the FOSS-SD (7.76) in Example 7.3.

$$G_B = \begin{bmatrix} 0.1 \\ 0.1 \end{bmatrix}^T, \quad D_C = \begin{bmatrix} 0.1 \\ 0.1 \end{bmatrix}, \quad G_C = \begin{bmatrix} 0.1 \\ 0.1 \\ 0.1 \end{bmatrix}, \quad D_D = \begin{bmatrix} 0.1 \\ 0.1 \end{bmatrix},$$

$$G_D = \begin{bmatrix} 0.1 \\ 0.1 \end{bmatrix}^T, \quad D_{A_h} = \begin{bmatrix} 0.1 \\ 0 \\ 0 \end{bmatrix}.$$

First, analyzing the admissibility and extended strict positive realness of the open-loop system, we can verify that when the control input $u(t) = 0$, the uncertain FOSS (4.1) cannot simultaneously satisfy admissibility and extended strict positive realness. For example, when the uncertain parameters F_Γ, $\Gamma \in \{A, A_h, B, C, D\}$, are all 1, the parameters A, A_h, B, C, D are as follows:

$$A = \begin{bmatrix} 0.1 & 0.11 & 0 \\ 0.1 & -1.99 & 0 \\ 1 & 1 & -1 \end{bmatrix}, \quad B = \begin{bmatrix} 1.01 & 0.01 \\ 0.01 & 1.01 \\ 0.01 & 0.01 \end{bmatrix}, \quad C = \begin{bmatrix} 0.01 & 1.01 & 0.01 \\ 1.01 & 0.01 & 0.01 \end{bmatrix},$$

$$D = \begin{bmatrix} 1.01 & 0.01 \\ 0.01 & 1.01 \end{bmatrix}, \quad A_h = \begin{bmatrix} -0.09 & 0 & 0 \\ 0 & -0.3 & 0 \\ 0 & 0 & 0 \end{bmatrix}.$$

In this case the curve of Sym$\{G(jw)\}$ $(w \geq 0)$ for the open-loop system is shown in Figure 7.7. We can see from the figure that some values of Sym$\{G(jw)\}$ $(w \geq 0)$ are less than 0, so the open-loop system does not satisfy extended strict positive realness.

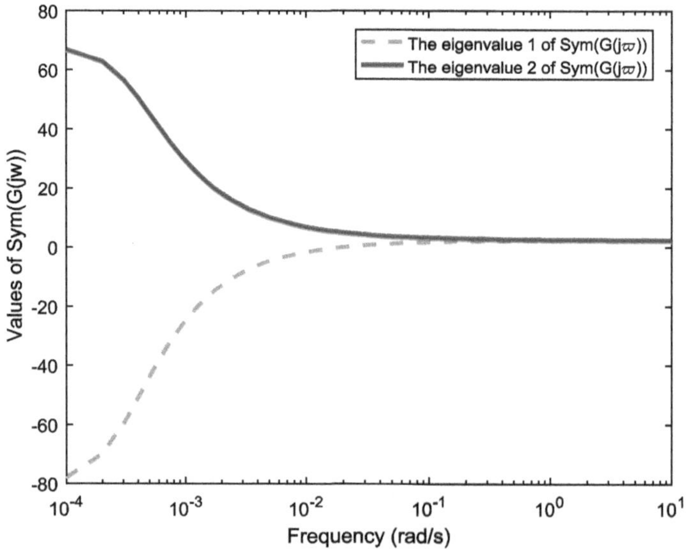

Figure 7.7: The eigenvalue curves of Sym$\{G(jw)\}$ ($w \geq 0$) of the open-loop FOSS-SD (7.72) in Example 7.4.

According to Theorem 7.9, we can find the following feasible solution:

$$P = \begin{bmatrix} 0.2583 & -0.158 - 0.0155i & 0 \\ -0.316 + 0.0311i & 0.5268 & 0 \\ 0.8499 & 0.4472 & 54.6759 \end{bmatrix},$$

$$Q = \begin{bmatrix} 16.8884 & 17.9041 & 24.167 \\ 17.9041 & 19.9962 & 24.8172 \\ 24.167 & 24.8172 & 69.5592 \end{bmatrix},$$

$$Y = \begin{bmatrix} -4.2974 & -4.6045 & 0 \end{bmatrix}.$$

Therefore the corresponding state feedback controller gain is $K = \begin{bmatrix} -22.3937 & -11.5265 & 0 \end{bmatrix}$.

To verify the effectiveness of the designed state feedback positive real controller, we take a set of parameters in the uncertain system and illustrate the results obtained by Theorem 7.9. For example, when $F_A, F_{A_h}, F_B, F_C, F_D, F_{B1}, F_{B2}$ are all 1, the system parameters are as follows:

$$\begin{aligned} &A = \begin{bmatrix} 0.1 & 0.11 & 0 \\ 0.1 & -1.99 & 0 \\ 1 & 1 & -1 \end{bmatrix}, \quad B = \begin{bmatrix} 1.01 & 0.01 \\ 0.01 & 1.01 \\ 0.01 & 0.01 \end{bmatrix}, \quad C = \begin{bmatrix} 0.01 & 1.01 \\ 1.01 & 0.01 \\ 0.01 & 0.01 \end{bmatrix}^T, \\ &A_h = \begin{bmatrix} -0.09 & 0 & 0 \\ 0 & -0.3 & 0 \\ 0 & 0 & 0 \end{bmatrix}, \quad B_1 = \begin{bmatrix} 8.01 \\ 8 \\ 8.01 \end{bmatrix}, \quad B_2 = \begin{bmatrix} 1.02 \\ 1.02 \end{bmatrix}, \quad D = \begin{bmatrix} 1.01 & 0.01 \\ 0.01 & 1.01 \end{bmatrix}. \end{aligned} \tag{7.97}$$

When the initial condition of the system is $\chi(t) = \begin{bmatrix}1 & -2 & 4.2812\end{bmatrix}^T$, under the control of the controller obtained in Theorem 7.9, the state responses of the closed-loop system are shown in Figure 7.8. We can see from the figure that all states asymptotically approach zero, indicating that the system is stable. At the same time, let $\lambda = s^\alpha$; then (i) $\det(\lambda E - A)$ is not identically zero, indicating that the system is regular, and (ii) $\deg(\det(\lambda E - A)) = \operatorname{rank}(E)$, indicating that the system is impulse-free. Therefore the system is robustly admissible. The curve of $\operatorname{Sym}\{G(jw)\}$ $(w \geq 0)$ for the closed-loop system is shown in Figure 7.9. We can see from the figure that all values of $\operatorname{Sym}\{G(jw)\}$ $(w \geq 0)$ are greater than 0, so the closed-loop system satisfies robust admissibility and extended strict positive realness.

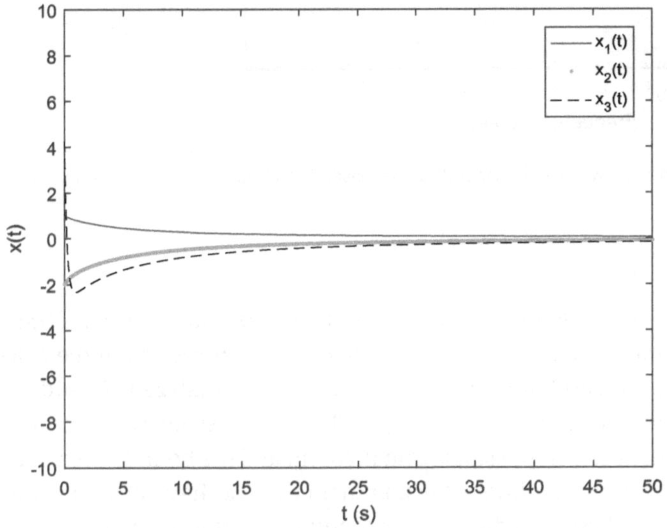

Figure 7.8: The time responses the closed-loop FOSS-SD (7.74) in Example 7.4.

Remark 7.7. In Example 7.3, it was verified that when the delay parameter $h = 0.1$, system (7.76) is admissible and satisfies extended strict positive realness. When the delay parameter h is taken arbitrarily as 5, 10, 15, 20, 25, 30, the system still satisfies regularity and impulse-freeness. The state response of the system under these conditions are shown in Figure 7.10. As we can see from the figure, when the delay parameter h is taken as 5, 10, 15, 20, 25, 30, the state curves of the system still tend to zero, indicating that the system remains admissible.

The eigenvalue curves of $\operatorname{Sym}\{G(jw)\}$ $(w \geq 0)$ for system (7.76) are shown in Figure 7.11. We can see from the figure that when the delay parameter h is taken as 5, 10, 15, 20, 25, 30, the eigenvalue curves of $\operatorname{Sym}\{G(jw)\}$ $(w \geq 0)$ for the system are all greater than zero, indicating that the system continues to satisfy extended strict positive realness. This verifies that the results of this chapter are delay-independent.

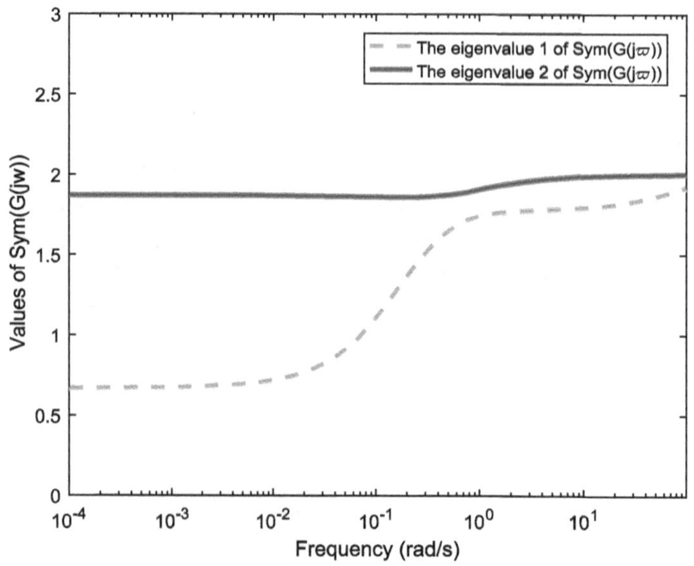

Figure 7.9: The eigenvalue curves of Sym$\{G(jw)\}$ ($w \geq 0$) of the closed-loop FOSS-SD (7.74) in Example 7.4.

7.5 Chapter summary

In this chapter, the robust stabilization, H_∞ control, and positive real control of FOSSs-SD with norm-bounded uncertainties have been studied. In the introduction the challenges encountered in the research on the FOSSs-SD have been analyzed. To address these challenges, the conditions for the regularity and nonimpulsiveness of nominal FOSSs-SD have been given. Based on the small-gain theorem and the bounded real lemmas for FOSSs, the admissibility conditions for nominal FOSSs-SD have been derived. Compared with the results in [98], the admissibility conditions in this paper are related to the system order. In contrast to the conclusions obtained using the Mittag-Leffler function analysis and norm inequality scaling methods in [42], the conclusions in this paper are presented in the form of LMIs, which can be more convenient to use for the controller design.

Based on the admissibility conditions for the FOSSs-SD, the bounded real lemmas (positive real lemmas) have been provided, giving the conditions under which the nominal systems simultaneously satisfy both the admissibility and H_∞ performance (extended strict positive realness). On this basis, the corresponding state feedback controller design methods have been presented for FOSSs-SD with norm-bounded uncertainties.

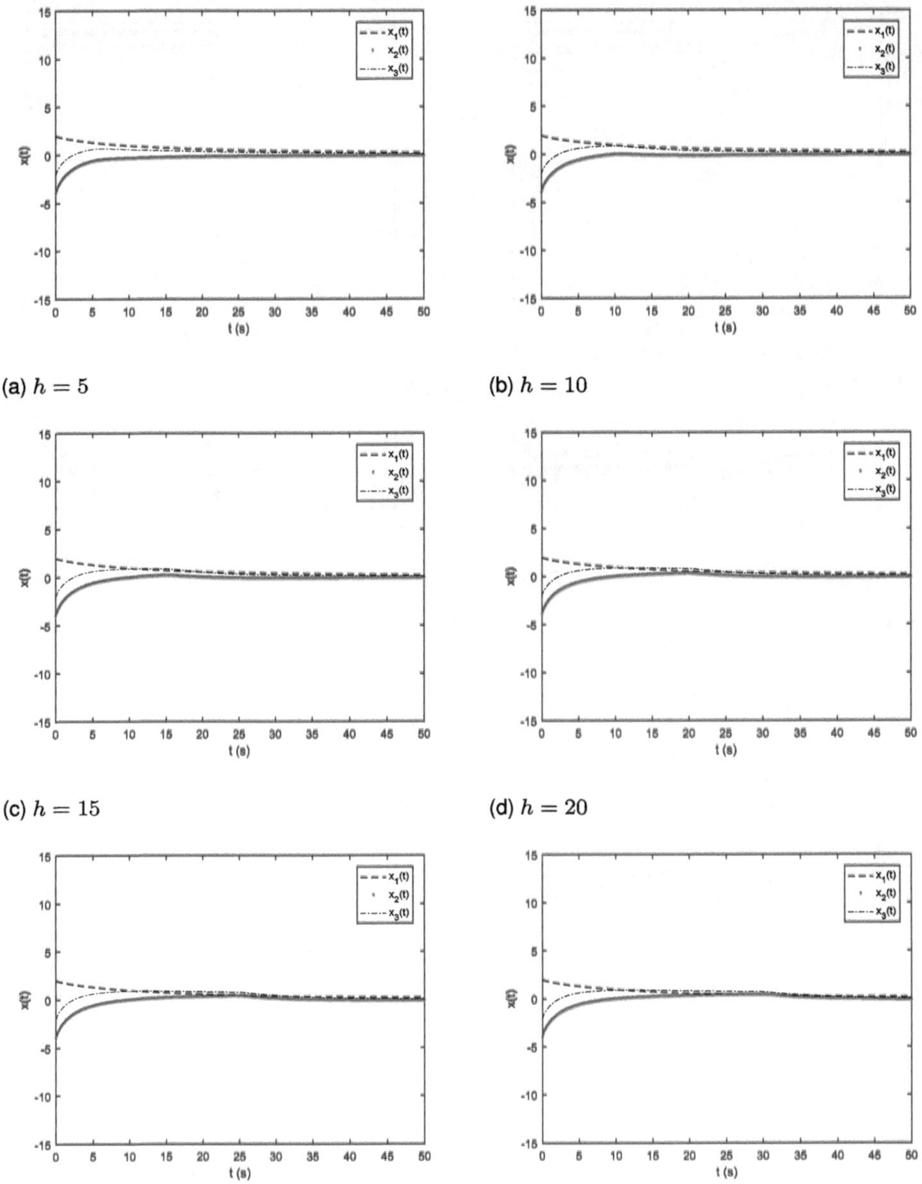

(a) $h = 5$ (b) $h = 10$

(c) $h = 15$ (d) $h = 20$

(e) $h = 25$ (f) $h = 30$

Figure 7.10: The time response of system (7.76) with various values of the delay parameter *h*.

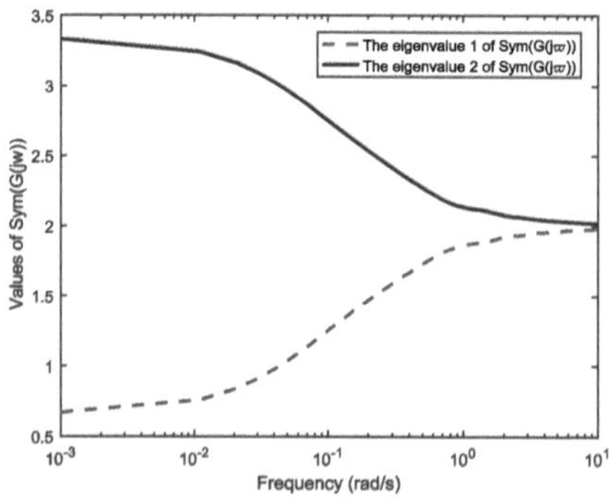

(a) $h = 5$

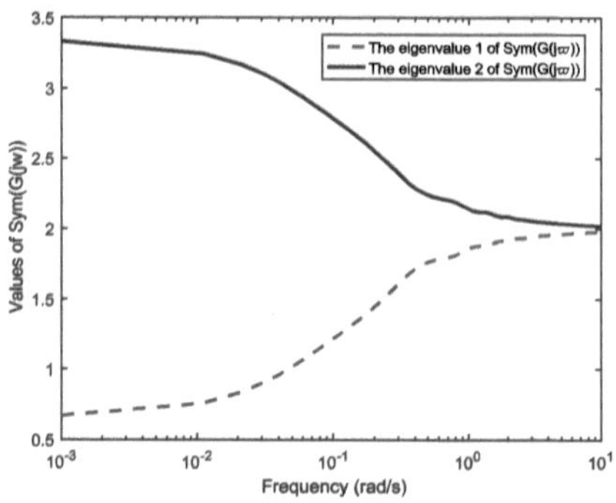

(b) $h = 10$

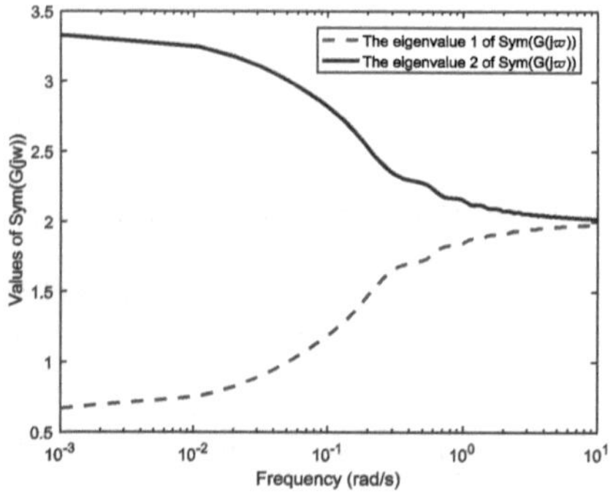

(c) $h = 15$

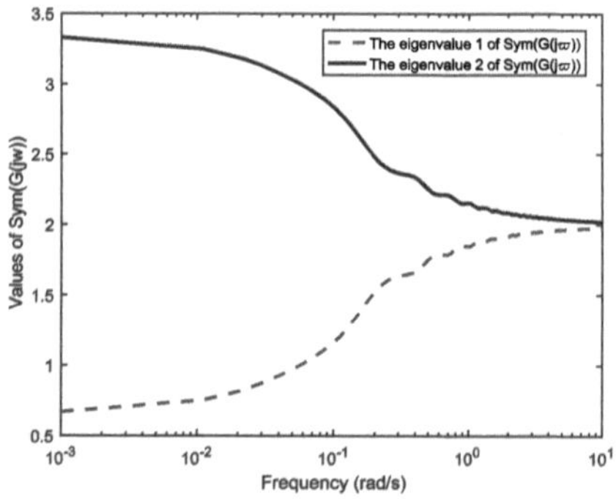

(d) $h = 20$

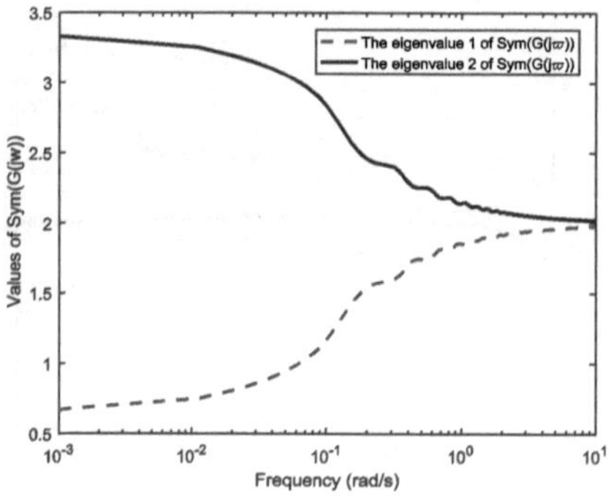

(e) $h = 25$

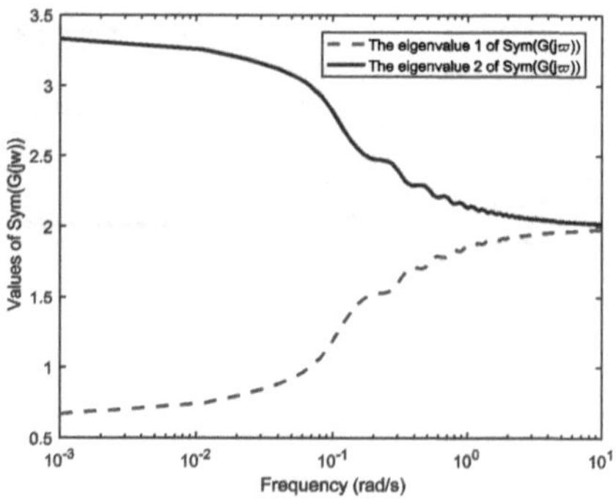

(f) $h = 30$

Figure 7.11: The eigenvalue curves of Sym$\{G(jw)\}$ ($w \geq 0$) of system (7.76) with various values of the delay parameter h.

8 Discrete systems

Abstract: In this chapter, we investigate the admissibility and robust stabilization of fractional-order singular discrete-time systems (FOSDTSs) with interval uncertainties. Firstly, based on the analysis of the regularity, causality, and stability, we derive several novel admissibility conditions for nominal FOSDTSs including a necessary and sufficient condition in terms of spectral radius and a sufficient condition in terms of nonstrict linear matrix inequalities (LMIs). To eliminate the coupling terms and propose strict LMI results, we obtain another novel admissibility condition, which is more tractable and reliable with the available LMI software solver and more suitable for the controller design compared with the existing results. Secondly, we address the state feedback controller synthesis for the FOSDTSs with interval uncertainties and design a state feedback controller. Finally, we demonstrate the efficiency of the proposed method by two numerical simulation examples.

8.1 Introduction

Singular systems, also called descriptor systems, implicit systems, generalized state-space systems (or generalized systems), or differential-algebraic systems, have been widely studied in the field of fractional-order control theory as they can depict control systems more directly. For fractional-order singular continuous-time systems (FOSCTSs), they are composed of fractional-order differential and algebraic equations (DAEs). The admissibility of FOSCTSs, including regularity, nonimpulsiveness, and stability, needs to be considered. A significant number of research results on different aspects of FOSCTSs have been proposed in monographs and papers; see, for example, [106, 68, 36, 116, 97, 63] and references therein. In some practical engineering applications, only discrete sampled data can be obtained, and thus researches focussing on FOSDTSs are necessary. FOSDTSs are composed of a set of fractional-order difference and algebraic equations. The admissibility of FOSDTSs also includes the regularity, causality, and stability. The regularity guarantees the existence and uniqueness of the solutions, whereas the causality ensures that at any time the state can be determined by the initial state condition and the inputs so far. Note that singular discrete systems are not always causal and the noncausality indeed exists in real systems [14]. For FOSDTSs, it is necessary to analyze the regularity and causality, which are as important as stability. In addition, FOSDTSs are inevitably influenced by uncertainties, for example, interval uncertainties [28]. It is necessary to solve the controller synthesis for FOSDTSs with interval uncertainties.

Inspired by the above discussions, in this chapter, we investigate the admissibility and robust stabilization of FOSDTSs with interval uncertainties. The main contributions are as follows:

(1) The regularity, causality, and stability conditions for nominal FOSDTSs are proposed by using the Kronecker equivalent standard form of singular systems and time do-

https://doi.org/10.1515/9783111628646-008

main solution analysis of FOSDTSs. The time domain methods in this chapter are more direct and intuitive than the frequency domain methods in [73, 74].

(2) Three admissibility conditions for nominal FOSDTSs are presented including a necessary and sufficient condition in terms of spectral radius, a sufficient condition in terms of nonstrict LMIs, and a sufficient condition in terms of strict LMIs. The third admissibility condition is more tractable and reliable with the available LMI software solver and eliminates the coupling terms. Compared with the existing works [40, 54], the results in this chapter simultaneously take regularity, causality, and stability into account and are more suitable for the controller design.

(3) For the FOSDTSs with interval uncertainties, the state feedback controller synthesis is solved, and the state feedback controllers are established such that the closed-loop FOSDTSs with interval uncertainties are robustly admissible.

8.2 Regularity, causality, and stability analysis

Consider the following FOSDTS with interval uncertainties:

$$E\Delta^{\alpha}x_{i+1} = A_U x_i + B_U u_i, \tag{8.1}$$

where $x_i \in \mathbb{R}^n$ and $u_i \in \mathbb{R}^p$ are the state and input vectors, respectively, $\alpha \in (0, 2)$ is the commensurate fractional order, $E \in \mathbb{R}^{n\times n}$ is a real constant matrix with $\operatorname{rank}(E) = m \le n$, and Δ^{α} represents the Grünwald–Letnikov fractional difference defined as

$$\Delta^{\alpha}x_i = \sum_{k=0}^{i} (-1)^k \binom{\alpha}{k} x_{i-k}, \tag{8.2}$$

which is the same as that in [89, 1],

$$\binom{\alpha}{k} = \begin{cases} 1, & k = 0, \\ \frac{\alpha(\alpha-1)\cdots(\alpha-k+1)}{k!}, & k = 1, 2, 3, \ldots, \end{cases}$$

and A_U and B_U are uncertain interval matrices of appropriate dimensions satisfying

$$\begin{aligned} A_U \in A_I &= [\underline{A}_U, \overline{A}_U] = \{[a_{ij}] : \underline{a}_{ij} \le a_{ij} \le \overline{a}_{ij}, 1 \le i, j \le n\}, \\ B_U \in B_I &= [\underline{B}_U, \overline{B}_U] = \{[b_{ij}] : \underline{b}_{ij} \le b_{ij} \le \overline{b}_{ij}, 1 \le i \le n, 1 \le j \le p\}, \end{aligned}$$

where $\underline{A}_U = [\underline{a}_{ij}]_{n\times n}$, $\overline{A}_U = [\overline{a}_{ij}]_{n\times n}$, $\underline{a}_{ij} \le \overline{a}_{ij}$, $\underline{B}_U = [\underline{b}_{ij}]_{n\times p}$, and $\overline{B}_U = [\overline{b}_{ij}]_{n\times n}$, $\underline{b}_{ij} \le \overline{b}_{ij}$. Denote

$$\begin{aligned} A &= 0.5(\underline{A}_U + \overline{A}_U), \quad \Delta A = 0.5(\overline{A}_U - \underline{A}_U) = \{\beta_{ij}\}_{n\times n}, \\ B &= 0.5(\underline{B}_U + \overline{B}_U), \quad \Delta B = 0.5(\overline{B}_U - \underline{B}_U) = \{e_{ij}\}_{n\times p}. \end{aligned}$$

Note that all elements of ΔA and ΔB are nonnegative. Denote

$$\begin{aligned}
D_A &= [\sqrt{\beta_{11}}e_1^n \quad \cdots \quad \sqrt{\beta_{1n}}e_1^n \quad \cdots \quad \sqrt{\beta_{n1}}e_n^n \quad \cdots \quad \sqrt{\beta_{nn}}e_n^n]_{n\times n^2}, \\
D_B &= [\sqrt{\epsilon_{11}}e_1^n \quad \cdots \quad \sqrt{\epsilon_{1p}}e_1^n \quad \cdots \quad \sqrt{\epsilon_{n1}}e_n^n \quad \cdots \quad \sqrt{\epsilon_{np}}e_n^n]_{n\times np}, \\
G_A &= [\sqrt{\beta_{11}}e_1^n \quad \cdots \quad \sqrt{\beta_{1n}}e_n^n \quad \cdots \quad \sqrt{\beta_{n1}}e_1^n \quad \cdots \quad \sqrt{\beta_{nn}}e_n^n]_{n^2\times n}^T, \\
G_B &= [\sqrt{\epsilon_{11}}e_1^p \quad \cdots \quad \sqrt{\epsilon_{1p}}e_p^p \quad \cdots \quad \sqrt{\epsilon_{n1}}e_1^p \quad \cdots \quad \sqrt{\epsilon_{np}}e_p^p]_{np\times p}^T,
\end{aligned}$$

where e_k^n ans e_k^p are all real column vectors with the kth element being 1 and all the other entries 0. Denote

$$\begin{aligned}
\Sigma_A &= \{\mathrm{diag}(\sigma_{11},\dots,\sigma_{1n},\dots,\sigma_{n1},\dots,\sigma_{nn}) \in \mathbb{R}^{n^2\times n^2},\ |\sigma_{ij}| \le 1, 1 \le i,j \le n\}, \\
\Sigma_B &= \{\mathrm{diag}(\tau_{11},\dots,\tau_{1p},\dots,\tau_{n1},\dots,\tau_{np}) \in \mathbb{R}^{np\times np},\ |\tau_{ij}| \le 1, 1 \le i \le n, 1 \le j \le p\}.
\end{aligned} \tag{8.3}$$

Substituting (8.2) into system (8.1), it can be transformed into

$$Ex_{i+1} = (A_U + E\alpha)x_i - E\sum_{k=2}^{i+1}(-1)^k\binom{\alpha}{k}x_{i-k+1} + B_U u_i. \tag{8.4}$$

The purposes of this chapter are as follows:

(1) To propose regularity, causality, and stability conditions for nominal FOSDTSs and then establish admissibility conditions for nominal FOSDTSs.
(2) To design state feedback controllers for FOSDTSs with interval uncertainties to make the systems robustly admissible.

To derive the main results, we introduce the following definitions and lemmas.

Definition 8.1 (Regularity [14]). Given proper initial conditions, if system (8.1) has a unique solution, then the system is said to be regular.

Definition 8.2 (Causality [14]). For system (8.1), if the states x_i $(i = 1,2,3,\dots)$ are determined completely by the initial state condition x_0 and inputs u_i $(i = 0,1,2,\dots)$, then the system is said to be causal.

Definition 8.3 (Stability [14, 40]). System (8.1) is said to be stable if the states x_i satisfy $\lim_{i\to\infty} x_i = 0$ for proper initial conditions.

Similarly to the definition of the admissibility of integer-order singular discrete-time systems (IOSDTSs) [14], the definition of the admissibility of FOSDTSs is as follows.

Definition 8.4 (Admissibility). The FOSDTS $E\Delta^\alpha x_{i+1} = Ax_i$ is said to be admissible if it is regular, causal, and stable.

Remark 8.1. The regularity, causality, and stability are three different properties of FOSDTSs, which are important components of admissibility. The fractional-order discrete-time systems (FODTSs) always have unique solutions, but the FOSDTSs may have unique solutions, infinite number of solutions, or no solutions. For example, consider the following FOSDTS:

$$\begin{bmatrix} 1 & 0 & 0 \\ 0 & 1 & 0 \\ 0 & 0 & 0 \end{bmatrix} \Delta^{\alpha} x_{i+1} = \begin{bmatrix} -1 & 0 & 0 \\ 0 & -1 & 0 \\ 0 & 0 & 0 \end{bmatrix} x_i,$$

where $x_i = \begin{bmatrix} x_i^{(1)T} & x_i^{(2)T} & x_i^{(3)T} \end{bmatrix}^T$. The substates $x_i^{(1)}$ and $x_i^{(2)}$ are constrained by fractional difference equations, but the substate $x_i^{(3)}$ has no constraints, which indicates that the above FOSDTS has a infinite number of solutions. To investigate the existence and uniqueness of solutions for FOSDTSs, the concept of regularity is introduced. The regular FOSDTSs have unique solutions. Since the fact of noncausality does indeed exist in some real systems, such as the Leontief dynamic models [14], the concept of causality is introduced. The causality ensures that the current states of FOSDTSs only depend on the initial state conditions and past inputs. The stability is the prerequisite for FOSDTSs to work properly and contains the stability of two subsystems, which will be analyzed in detail in Subsection 8.2.3.

Remark 8.2. The regularity is the premise of causality and stability. In other words, the causal or stable FOSDTSs must first satisfy the regularity conditions. The causality and stability of FOSDTSs are two different properties, which can be satisfied simultaneously or not simultaneously.

Lemma 8.1 ([55]). *Denote*

$$\begin{aligned} A_J &= \{A_U = A + D_A F_A G_A \mid F_A \in \Sigma_A\}, \\ B_J &= \{B_U = B + D_B F_B G_B \mid F_B \in \Sigma_B\}. \end{aligned}$$

Then we have $A_I = A_J$ and $B_I = B_J$, where Σ_A and Σ_B are defined in (8.3).

To derive the admissibility conditions for FOSDTSs with interval uncertainties, in this subsection, we propose the admissibility conditions for nominal FOSDTSs. Consider the following nominal FOSDTS:

$$E\Delta^{\alpha} x_{i+1} = Ax_i + Bu_i. \tag{8.5}$$

In this subsection, we derive novel conditions for the regularity, causality, stability, and admissibility of FOSDTSs.

8.2.1 Regularity analysis

Based on Definition 8.1, we propose the following theorem on the regularity of FOSDTSs.

Theorem 8.1. *System* (8.5) *is regular if and only if there exist nonsingular matrices* Ψ *and* Φ *satisfying*

$$\Phi A\Psi = \mathrm{diag}(A_1, I_{n-m}) \quad \text{and} \quad \Phi E\Psi = \mathrm{diag}(I_m, N),$$

where $A_1 \in \mathbb{R}^{m\times m}$, *and* $N \in \mathbb{R}^{(n-m)\times(n-m)}$ *is nilpotent.*

Proof. Similarly to the case of FOSCTSs in Chapter 2, the proof is based on the Kronecker equivalent standard form and is omitted here for brevity. □

8.2.2 Causality analysis

Based on Definition 8.2, we propose the following theorem on the causality of FOSDTSs.

Theorem 8.2. *The regular FOSDTS* (8.5) *is causal if and only if* $N = 0$, *where* N *is defined in Theorem* 8.1.

Proof. As the system is regular, there exist two nonsingular matrices Ψ and Φ such that system (8.5) can be expressed as

$$\Delta^\alpha x_{i+1}^{(1)} = A_1 x_i^{(1)} + B_1 u_i, \tag{8.6}$$

$$N\Delta^\alpha x_{i+1}^{(2)} = x_i^{(2)} + B_2 u_i, \tag{8.7}$$

with $x_{i+1} = \Psi\begin{bmatrix} x_{i+1}^{(1)} \\ x_{i+1}^{(2)} \end{bmatrix}$ and $\Phi B = \begin{bmatrix} B_1 \\ B_2 \end{bmatrix}$. Systems (8.6) and (8.7) are known as forward and backward subsystems, respectively.

Sufficiency. Suppose that $N = 0$. For subsystem (8.6), it is a normal FODTS, and the solution is given in [66] as

$$x_i^{(1)} = \Omega_i x_0^{(1)} + \sum_{k=0}^{i-1} \Omega_{i-k-1} B_1 u_k, \tag{8.8}$$

where Ω_i are determined by the equation

$$\Omega_{i+1} = \Omega_i (A_1 + I_m \alpha) + \sum_{k=2}^{i+1} (-1)^{k-1} \binom{\alpha}{k} \Omega_{i+1-k},$$

$$\Omega_1 = (A_1 + I_m\alpha)\Omega_0,$$

with the initial condition $\Omega_0 = I_m$. By equality (8.8) subsystem (8.6) is causal by Definition 8.2. For subsystem (8.7), due to the assumption that N is zero, we have

$$x_i^{(2)} = -B_2 u_i.$$

Therefore subsystem (8.7) is causal when N is zero. To summarize, the regular FOSDTS (8.5) is causal if $N = 0$.

Necessity. We prove by contradiction. Suppose that $N \neq 0$. Subsystem (8.7) is different from subsystem (8.6) as N is a nilpotent matrix. Let h be the nilpotent index of N. By applying the fractional difference definition in (8.2), subsystem (8.7) can be written as

$$N\left(x_{i+1}^{(2)} + \sum_{k=1}^{i+1}(-1)^k \binom{\alpha}{k} x_{i-k+1}^{(2)}\right) = x_i^{(2)} + B_2 u_i. \tag{8.9}$$

By repeatedly left multiplying by $N^0, N^1, N^2, \ldots, N^{h-1}$ both sides of equation (8.9), we have

$$N\left(x_{i+1}^{(2)} + \sum_{k=1}^{i+1}(-1)^k \binom{\alpha}{k} x_{i-k+1}^{(2)}\right) = x_i^{(2)} + B_2 u_i,$$

$$N^2\left(x_{i+2}^{(2)} + \sum_{k=1}^{i+2}(-1)^k \binom{\alpha}{k} x_{i-k+2}^{(2)}\right) = N(x_{i+1}^{(2)} + B_2 u_{i+1}),$$

$$N^3\left(x_{i+3}^{(2)} + \sum_{k=1}^{i+3}(-1)^k \binom{\alpha}{k} x_{i-k+3}^{(2)}\right) = N^2(x_{i+2}^{(2)} + B_2 u_{i+2}),$$

$$\vdots$$

$$N^h\left(x_{i+h}^{(2)} + \sum_{k=1}^{i+h}(-1)^k \binom{\alpha}{k} x_{i-k+h}^{(2)}\right) = N^{h-1}(x_{i+h-1}^{(2)} + B_2 u_{i+h-1}).$$

Add the above equations and notice that $N^h = 0$. Then

$$x_i^{(2)} = \sum_{j=1}^{h-1} N^j \sum_{k=1}^{i+j}(-1)^k \binom{\alpha}{k} x_{i-k+j}^{(2)} - \sum_{k=1}^{h-1} N^k B_2 u_{i+k} - B_2 u_i,$$

which indicates that the causality of $x_i^{(2)}$ is not satisfied by Definition 8.2 if $N \neq 0$. Therefore N is equal to zero if the regular FOSDTS (8.5) is causal. This completes the proof. □

Remark 8.3. Theorem 8.2 is the causality condition for FOSDTSs derived by using time domain methods. By comparison the causality conditions in [73, 74] are based on frequency domain methods, by which we cannot see that the state at any time is determined by the initial state condition and past inputs if FOSDTSs are causal. Therefore the time domain methods in this chapter are more direct than the frequency domain methods in [73, 74].

From Theorems 8.1 and 8.2, we can obtain the following theorem by a matrix transformation.

Theorem 8.3. *When the regularity of system* (8.5) *is not known, it is always possible to choose two nonsingular matrices M_1 and N_1 such that*

$$M_1EN_1 = \begin{bmatrix} I_m & 0 \\ 0 & 0 \end{bmatrix}, \quad M_1AN_1 = \begin{bmatrix} A_{11} & A_{12} \\ A_{21} & A_{22} \end{bmatrix}.$$

Then system (8.5) *is causal if* A_{22} *is nonsingular.*

Proof. Since A_{22} is nonsingular, denote

$$M_2 = \begin{bmatrix} I_m & -A_{12}A_{22}^{-1} \\ 0 & I_{n-m} \end{bmatrix}, \quad N_2 = \begin{bmatrix} I_m & 0 \\ -A_{22}^{-1}A_{21} & A_{22}^{-1} \end{bmatrix}.$$

Then

$$\begin{aligned} M_2M_1EN_1N_2 &= \begin{bmatrix} I_m & 0 \\ 0 & 0 \end{bmatrix}, \\ M_2M_1AN_1N_2 &= \begin{bmatrix} A_{11} - A_{12}A_{22}^{-1}A_{21} & 0 \\ 0 & I_{n-m} \end{bmatrix}. \end{aligned}$$

From Theorems 8.1 and 8.2, it follows that system (8.5) is causal. □

8.2.3 Stability analysis

Based on Definition 8.3, we propose the following theorem on the stability of FOSDTSs.

Theorem 8.4. *System* (8.5) *is asymptotically stable if and only if* $\rho(\bar{A}_1) < 1$, *where*

$$\bar{A}_1 = \begin{bmatrix} A_1 + \alpha I_m & -(-1)^2\binom{\alpha}{2} I_m & -(-1)^3\binom{\alpha}{3} I_m & \cdots \\ I_m & 0 & 0 & \cdots \\ 0 & I_m & 0 & \cdots \\ \vdots & \vdots & \vdots & \vdots \end{bmatrix},$$

and A_1 *is defined in Theorem* 8.1.

Proof. From the proof of Theorem 8.2 it follows that system (8.5) can be transformed into

$$\Delta^\alpha x_{i+1}^{(1)} = A_1x_i^{(1)} + B_1u_i, \tag{8.10}$$

$$N\Delta^\alpha x_{i+1}^{(2)} = x_i^{(2)} + B_2u_i. \tag{8.11}$$

By [40], subsystem (8.11) is asymptotically stable. Therefore the stability of system (8.5) is determined by the stability of subsystem (8.10). Subsystem (8.10) can be rewritten as

$$x_{i+1}^{(1)} + \sum_{k=1}^{i+1} (-1)^k \binom{\alpha}{k} x_{i-k+1}^{(1)} = A_1x_i^{(1)} + B_1u_i,$$

which is equivalent to

$$x_{i+1}^{(1)} = (A_1 + \alpha I_m)x_i^{(1)} - \sum_{k=2}^{i+1}(-1)^k \binom{\alpha}{k} x_{i-k+1}^{(1)} + B_1 u_i. \tag{8.12}$$

Then equality (8.12) can be rewritten as an infinite-dimensional system as $i \to \infty$:

$$\begin{bmatrix} x_{i+1}^{(1)} \\ x_i^{(1)} \\ x_{i-1}^{(1)} \\ \vdots \end{bmatrix} = \bar{A}_1 \begin{bmatrix} x_i^{(1)} \\ x_{i-1}^{(1)} \\ x_{i-2}^{(1)} \\ \vdots \end{bmatrix} + \bar{B}_1 u_i, \tag{8.13}$$

where

$$\bar{A}_1 = \begin{bmatrix} A_1 + \alpha I_m & -(-1)^2\binom{\alpha}{2} & -(-1)^3\binom{\alpha}{3} & \cdots \\ I_m & 0 & 0 & \cdots \\ 0 & I_m & 0 & \cdots \\ \vdots & \vdots & \vdots & \vdots \end{bmatrix}, \quad \bar{B}_1 = \begin{bmatrix} B_1 \\ 0 \\ 0 \\ \vdots \end{bmatrix}.$$

From [66, 19], it follows that system (8.13) is asymptotically stable if and only if $\rho(\bar{A}_1) < 1$. This completes the proof. □

Remark 8.4. Theorem 8.4 is the stability condition for FOSDTSs without conservatism, but the dimensions of $\bar{A}_1$ is infinite as $i \to \infty$. It is infeasible to apply Theorem 8.4 for the controller design purposes [89, 1]. As pointed out in [66, 19, 8], it is always supposed that the factor $(-1)^k\binom{\alpha}{k}$ in equality (8.12) is bounded by a certain $L \in \mathbb{Z}^+$ (called the memory length [19]) in practical problems.

Based on the above analysis, we propose another stability condition for FOSDTSs.

Theorem 8.5. *System* (8.5) *is asymptotically stable if*

$$\rho(A_1 + \alpha I_m) < r,$$

where

$$r = \begin{cases} 2 - \frac{\Gamma(L+1-\alpha)}{\Gamma(1-\alpha)\Gamma(L+1)} - \alpha, & \alpha \in [1, 2), \\ \frac{\Gamma(L+1-\alpha)}{\Gamma(1-\alpha)\Gamma(L+1)} + \alpha, & \alpha \in (0, 1), \end{cases}$$

$L \in \mathbb{Z}^+$ *is the memory length, and* A_1 *is defined in Theorem* 8.1

Proof. The stability of system (8.5) is determined by the stability of subsystem (8.10), which can be derived directly based on [66, 19]. The proof is omitted here for brevity. □

Note that it is not easy to apply Theorem 8.5 in robust analysis and synthesis of FOSDTSs. Thus we present a novel condition for the stability of FOSDTSs in the form of LMIs.

Theorem 8.6. *System* (8.5) *is asymptotically stable if there exists a Hermitian matrix* $P > 0$ *such that*

$$(A_1 + \alpha I_m)^T P(A_1 + \alpha I_m) - r^2 P < 0, \tag{8.14}$$

where

$$r = \begin{cases} 2 - \frac{\Gamma(L+1-\alpha)}{\Gamma(1-\alpha)\Gamma(L+1)} - \alpha, & \alpha \in [1,2), \\ \frac{\Gamma(L+1-\alpha)}{\Gamma(1-\alpha)\Gamma(L+1)} + \alpha, & \alpha \in (0,1), \end{cases}$$

$L \in \mathbb{Z}^+$ *is the memory length, and* A_1 *is defined in Theorem* 8.1.

Proof. Suppose that the ith eigenvalue of $A_1+\alpha I_m$ is λ_i and the corresponding eigenvector is v_i, i. e., $(A_1+\alpha I_m)v_i = \lambda_i v_i$. Left multiplying inequality (8.14) by v_i^* and right multiplying by v_i, we have

$$v_i^*(A_1 + \alpha I_m)^T P(A_1 + \alpha I_m)v_i - r^2 v_i^* P v_i < 0,$$

which is equivalent to

$$\lambda_i^* \lambda_i v_i^* P v_i - r^2 v_i^* P v_i < 0.$$

Then we have

$$\lambda_i^* \lambda_i - r^2 < 0. \tag{8.15}$$

By Theorem 8.5 system (8.5) is asymptotically stable. □

Remark 8.5. Theorem 8.5 can be applied to check the stability of FOSDTSs by calculating the spectral radius of $A_1 + \alpha I_m$. By using Theorem 8.5 and the properties of eigenvalues and eigenvectors, Theorem 8.6 is a novel condition for the stability of FOSDTSs in the form of LMIs, which can be conveniently settled by LMI software solver. By comparison Theorem 8.6 is more beneficial to the robust analysis and synthesis of FOSDTSs than Theorem 8.5. Therefore in the following section, we apply Theorem 8.6 to derive admissibility conditions for FOSDTSs and design robust controllers of FOSDTSs with interval uncertainties.

8.3 Admissibility and robust stabilization

Based on Definition 8.4, we propose the following theorem on the admissibility of FOSDTSs.

Theorem 8.7. *System* (8.5) *is admissible if and only if there exist invertible matrices* Ψ *and* Φ *such that*

$$\text{(i)}\quad \Phi A\Psi = \operatorname{diag}(A_1, I_{n-m}) \quad \textit{and} \quad \Phi E\Psi = \operatorname{diag}(I_m, 0), \tag{8.16}$$

$$\text{(ii)}\quad \rho(\bar{A}_1) < 1, \tag{8.17}$$

where $A_1 \in \mathbb{R}^{r\times r}$, *and*

$$\bar{A}_1 = \begin{bmatrix} A_1 + \alpha I_m & -(-1)^2\binom{\alpha}{2}I_m & -(-1)^3\binom{\alpha}{3}I_m & \cdots \\ I_m & 0 & 0 & \cdots \\ 0 & I_m & 0 & \cdots \\ \vdots & \vdots & \vdots & \vdots \end{bmatrix}.$$

Proof. By condition (i) system (8.5) is regular and causal by Theorems 8.1 and 8.2. By condition (ii) system (8.5) is asymptotically stable by Theorem 8.4. This completes the proof. □

Theorem 8.7 is the admissibility condition for FOSDTSs without conservatism, but it is not easy to apply in practice. Based on Theorem 8.6, we present a sufficient condition for the admissibility of FOSDTSs, which directly utilizes the system matrices E and A to judge the admissibility instead of utilizing the decomposition of these matrices.

Theorem 8.8. *System* (8.5) *is admissible if there exists a Hermitian matrix* P *such that*

$$E^T PE \geq 0, \tag{8.18}$$

$$(A + E\alpha)^T P(A + E\alpha) - r^2 E^T PE < 0, \tag{8.19}$$

where

$$r = \begin{cases} 2 - \frac{\Gamma(L+1-\alpha)}{\Gamma(1-\alpha)\Gamma(L+1)} - \alpha, & \alpha \in [1, 2), \\ \frac{\Gamma(L+1-\alpha)}{\Gamma(1-\alpha)\Gamma(L+1)} + \alpha, & \alpha \in (0, 1), \end{cases}$$

and $L \in \mathbb{Z}^+$ *is the memory length.*

Proof. When the regularity of system (8.5) is not known, there exist two invertible matrices M and N such that

$$E = M\begin{bmatrix} I_m & 0 \\ 0 & 0 \end{bmatrix}N, \quad A = M\begin{bmatrix} A_{11} & A_{12} \\ A_{21} & A_{22} \end{bmatrix}N. \tag{8.20}$$

Denote

$$M^T PM = \begin{bmatrix} P_1 & P_2 \\ P_3 & P_4 \end{bmatrix}. \tag{8.21}$$

Since P is a Hermitian matrix, we have $P_3 = P_2^T$ and $P_4 = P_4^T$. Inequality (8.18) yields

$$E^T PE = N^T \begin{bmatrix} P_1 & 0 \\ 0 & 0 \end{bmatrix} N \geq 0,$$

indicating that $P_1 \geq 0$. Substituting (8.20) and (8.21) into inequality (8.19) yields

$$N^T \begin{bmatrix} \Theta_1 & \Theta_2 \\ \Theta_3 & \Theta_4 \end{bmatrix} N < 0,$$

where

$$\begin{aligned}
\Theta_1 &= ((A_{11}^T + \alpha I_m)P_1 + A_{21}^T P_3)(A_{11} + \alpha I_m) + ((A_{11}^T + \alpha I_m)P_2 + A_{21}^T P_4)A_{21} - r^2 P_1, \\
\Theta_2 &= ((A_{11}^T + \alpha I_m)P_1 + A_{21}^T P_3)A_{12} + ((A_{11}^T + \alpha I_m)P_2 + A_{21}^T P_4)A_{22}, \\
\Theta_3 &= (A_{12}^T P_1 + A_{22}^T P_3)(A_{11} + \alpha I_m) + (A_{12}^T P_2 + A_{22}^T P_4)A_{21}, \\
\Theta_4 &= (A_{12}^T P_1 + A_{22}^T P_3)A_{12} + (A_{12}^T P_2 + A_{22}^T P_4)A_{22}.
\end{aligned}$$

Thus

$$\Theta_4 = A_{12}^T P_1 A_{12} + \mathrm{Sym}\left\{A_{12}^T P_2 A_{22} + \frac{1}{2} A_{22}^T P_4 A_{22}\right\} < 0,$$

which, together with $P_1 \geq 0$, implies

$$\mathrm{Sym}\left\{A_{12}^T P_2 A_{22} + \frac{1}{2} A_{22}^T P_4 A_{22}\right\} < 0.$$

From Lemma 2.10, it follows that

$$\begin{aligned}
\beta\left(\left(A_{12}^T P_2 + \frac{1}{2} A_{22}^T P_4\right)A_{22}\right) &\leq \mu\left(\left(A_{12}^T P_2 + \frac{1}{2} A_{22}^T P_4\right)A_{22}\right) \\
&= \frac{1}{2}\lambda_{\max}\left(\mathrm{Sym}\left\{\left(A_{12}^T P_2 + \frac{1}{2} A_{22}^T P_4\right)A_{22}\right\}\right) \\
&< 0,
\end{aligned}$$

which implies that A_{22} is nonsingular by Lemma 2.10. From Theorem 8.3 it follows that system (8.5) is causal and consequently regular.

Based on the causality of system (8.5), it follows from Theorems 8.1 and 8.2 that there exist two invertible matrices Ψ and Φ such that

$$A = \Phi^{-1} \begin{bmatrix} A_1 & 0 \\ 0 & I_{n-m} \end{bmatrix} \Psi^{-1}, \quad E = \Phi^{-1} \begin{bmatrix} I_m & 0 \\ 0 & 0 \end{bmatrix} \Psi^{-1}. \tag{8.22}$$

Denote

$$P = \Phi^T \begin{bmatrix} P_{11} & P_{12} \\ P_{21} & P_{22} \end{bmatrix} \Phi. \tag{8.23}$$

Since $E^T PE \geq 0$, $P_{11} \geq 0$. Substituting (8.22) and (8.23) into inequality (8.19), we have

$$(A_1 + aI_m)^T P_{11}(A_1 + aI_m) - r^2 P_{11} < 0,$$

where $P_{11} > 0$. It follows from Theorem 8.4 that system (8.5) is stable. Therefore system (8.5) is admissible. This completes the proof. □

Remark 8.6. The parameter L in Theorem 8.8 is the number of samples taken into account, called the memory length [19]. The larger the value of L, the weaker the conservatism of Theorem 8.8. As $L \to \infty$, Theorem 8.8 is a necessary and sufficient condition for the admissibility of FOSDTSs.

Theorem 8.8 is the admissibility condition for FOSDTSs. It should be pointed out that Theorem 8.8 has two shortcomings: (i) There exists a nonstrict LMI term, i. e., inequality (8.18); (ii) There exists a coupling term $(A + Ea)^T P(A + Ea)$ in inequality (8.19), which brings some trouble in controller design. To overcome these defects, in the following theorem, we propose a strict LMI condition for the admissibility of FOSDTSs, which also eliminates the coupling term.

Theorem 8.9. *System* (8.5) *is admissible if there exist matrices* $Y > 0$ *and* $X > 0$ *and a nonsingular matrix* G *such that*

$$\begin{bmatrix} \Sigma & (*) & (*) \\ (A + Ea)G - G^* & -G - G^* & (*) \\ 0 & G & -Y \end{bmatrix} < 0, \tag{8.24}$$

where $\Sigma = \operatorname{Sym}\{(A + Ea)G - \delta\Xi(A + Ea)G - \varepsilon r^2 EG\} + \delta^2 X + \varepsilon^2 r^2 Y$, $\delta \in \mathbb{R}$ *and* $\varepsilon \in \mathbb{R}$ *are any scalars,* $\Xi \in \mathbb{R}^{n\times n}$ *is any matrix with* $\operatorname{rank}(\Xi) = n - m$ *satisfying* $E^T \Xi^T = 0$,

$$r = \begin{cases} 2 - \frac{\Gamma(L+1-\alpha)}{\Gamma(1-\alpha)\Gamma(L+1)} - \alpha, & \alpha \in [1,2), \\ \frac{\Gamma(L+1-\alpha)}{\Gamma(1-\alpha)\Gamma(L+1)} + \alpha, & \alpha \in (0,1), \end{cases}$$

and $L \in \mathbb{Z}^+$ *is the memory length.*

Proof. Using the Schur complement lemma, inequality (8.24) yields

$$\begin{bmatrix} \Sigma & (*) \\ (A + Ea)G - G^* & G^* Y^{-1} G - G - G^* \end{bmatrix} < 0, \tag{8.25}$$

where $\Sigma = \operatorname{Sym}\{(A + Ea)G - \delta\Xi(A + Ea)G - \varepsilon r^2 EG\} + \delta^2 X + \varepsilon^2 r^2 Y$. Note that

$$\begin{aligned}
&\mathrm{Sym}\{-\delta\Xi(A+E\alpha)G\}+\delta^2X\\
&\quad=(G^*(A+E\alpha)^T\Xi^T-\delta X)X^{-1}(\Xi(A+E\alpha)G-\delta X)\\
&\qquad-G^*(A+E\alpha)^T\Xi^TX^{-1}\Xi(A+E\alpha)G\\
&\quad\geq -G^*(A+E\alpha)^T\Xi^TX^{-1}\Xi(A+E\alpha)G
\end{aligned}\tag{8.26}$$

and

$$\begin{aligned}
&\mathrm{Sym}\{-\varepsilon r^2EG\}+\varepsilon^2r^2Y\\
&\quad=r^2(G^*E^T-\varepsilon Y)Y^{-1}(EG-\varepsilon Y)-r^2G^*E^TY^{-1}EG\\
&\quad\geq -r^2G^*E^TY^{-1}EG.
\end{aligned}\tag{8.27}$$

Denote $\Sigma_2=\mathrm{Sym}\{(A+E\alpha)G\}-G^*(A+E\alpha)^T\Xi^TX^{-1}\Xi(A+E\alpha)G-r^2G^*E^TY^{-1}EG$, which implies that $\Sigma_2\leq\Sigma$. By inequality (8.25), we have

$$\begin{bmatrix}\Sigma_2 & (*)\\ (A+E\alpha)G-G^* & G^*Y^{-1}G-G-G^*\end{bmatrix}<0.\tag{8.28}$$

Pre- and postmultiplying inequality (8.28) by $\left[\begin{smallmatrix}G^{-*} & 0\\ 0 & G^{-*}\end{smallmatrix}\right]$ and $\left[\begin{smallmatrix}G^{-1} & 0\\ 0 & G^{-1}\end{smallmatrix}\right]$, we have

$$\begin{bmatrix}\Sigma_3 & (*)\\ G^{-*}(A+E\alpha)-G^{-1} & Y^{-1}-G^{-*}-G^{-1}\end{bmatrix}<0,\tag{8.29}$$

where $\Sigma_3=\mathrm{Sym}\{G^{-*}(A+E\alpha)\}-(A+E\alpha)^T\Xi^TX^{-1}\Xi(A+E\alpha)-r^2E^TY^{-1}E$. Denote $\hat{G}=G^{-1}$, $\hat{X}=X^{-1}$, and $\hat{Y}=Y^{-1}$. Then inequality (8.29) can be written as

$$\begin{bmatrix}\Sigma_4 & (*)\\ \hat{G}^*(A+E\alpha)-\hat{G} & \hat{Y}-\hat{G}^*-\hat{G}\end{bmatrix}<0,\tag{8.30}$$

where $\Sigma_4=\mathrm{Sym}\{\hat{G}^*(A+E\alpha)\}-(A+E\alpha)^T\Xi^T\hat{X}\Xi(A+E\alpha)-r^2E^T\hat{Y}E$. Pre- and postmultiplying inequality (8.30) by $[I \quad (A+E\alpha)^T]$ and $\left[\begin{smallmatrix}I\\ A+E\alpha\end{smallmatrix}\right]$, we have

$$(A+E\alpha)^T\hat{Y}(A+E\alpha)-(A+E\alpha)^T\Xi^T\hat{X}\Xi(A+E\alpha)-r^2E^T\hat{Y}E<0,\tag{8.31}$$

i. e.,

$$(A+E\alpha)^T(\hat{Y}-\Xi^T\hat{X}\Xi)(A+E\alpha)-r^2E^T\hat{Y}E<0.\tag{8.32}$$

Denote $P=\hat{Y}-\Xi^T\hat{X}\Xi$, where $P=P^*$. Then the following conditions hold:

$$\begin{aligned}
E^TPE&\geq 0,\\
(A+E\alpha)^TP(A+E\alpha)-r^2E^TPE&<0.
\end{aligned}\tag{8.33}$$

By Theorem 8.8 system (8.5) is admissible. This completes the proof. □

Remark 8.7. Theorem 8.9 is a novel admissibility condition for FOSDTSs in terms of strict LMI. Furthermore, there exist no coupling terms in Theorem 8.9. Compared with Theorem 8.8, Theorem 8.9 is more tractable and reliable, especially with the available LMI software solver. Compared with the existing works [40, 54], Theorem 8.9 simultaneously takes regularity, causality, and stability into account and is more suitable for controller design.

Based on the admissibility condition for FOSDTSs in Theorem 8.9, the state feedback controller synthesis can be addressed. To make the FOSDTS (8.1) with interval uncertainties satisfy the admissibility condition, design the state feedback controller

$$u_i = Kx_i, \tag{8.34}$$

by which system (8.1) can be written as

$$E\Delta^{\alpha}x_{i+1} = (A_U + B_U K)x_i. \tag{8.35}$$

The following theorem can be applied to solve the state feedback controller synthesis for FOSDTSs with interval uncertainties.

Theorem 8.10. *Given the uncertain system* (8.1)*, there exists a state feedback controller* (8.34) *such that the closed-loop system* (8.35) *is admissible if there exist matrices* $Y > 0$ *and* $X > 0$*, a nonsingular real matrix* G*, a real matrix* H*, and a real constant* $\gamma > 0$ *such that*

$$\begin{bmatrix} W_0 + \gamma \Upsilon_2 \Upsilon_2^T & \Upsilon_1^T \\ \Upsilon_1 & -\gamma I \end{bmatrix} < 0, \tag{8.36}$$

where

$$W_0 = \begin{bmatrix} \hat{\Sigma} & (*) & (*) \\ (A + E\alpha)G + BH - G^* & -G - G^* & (*) \\ 0 & G & -Y \end{bmatrix},$$

$$\hat{\Sigma} = \mathrm{Sym}\{(A + E\alpha)G + BH - \delta\Xi(A + E\alpha)G - \delta\Xi BH - \varepsilon r^2 EG\} + \delta^2 X + \varepsilon^2 r^2 Y,$$

$$\Upsilon_2 = \begin{bmatrix} (I - \delta\Xi)D_A & (I - \delta\Xi)D_B \\ D_A & D_B \\ 0 & 0 \end{bmatrix}, \quad \Upsilon_1 = \begin{bmatrix} G_A G & 0 & 0 \\ G_B H & 0 & 0 \end{bmatrix},$$

$\delta \in \mathbb{R}$ *and* $\varepsilon \in \mathbb{R}$ *are any given scalars,* $\Xi \in \mathbb{R}^{n\times n}$ *is any given matrix satisfying* $E^T \Xi^T = 0$ *with* $\mathrm{rank}(\Xi) = n - m$*,*

$$r = \begin{cases} 2 - \frac{\Gamma(L+1-\alpha)}{\Gamma(1-\alpha)\Gamma(L+1)} - \alpha, & \alpha \in [1, 2), \\ \frac{\Gamma(L+1-\alpha)}{\Gamma(1-\alpha)\Gamma(L+1)} + \alpha, & \alpha \in (0, 1), \end{cases}$$

and $L \in \mathbb{Z}^+$ *is the memory length. Then we can obtain the state feedback controller* $u_i = HG^{-1}x_i$.

Proof. The state feedback controller gain is $K = HG^{-1}$. We have

$$H = KG. \tag{8.37}$$

Using the Schur complement lemma, inequality (8.36) can be written as

$$W_0 + \gamma \Upsilon_2 \Upsilon_2^T + \gamma^{-1} \Upsilon_1^T \Upsilon_1 < 0. \tag{8.38}$$

Denote $F_{AB} = \left[\begin{smallmatrix} F_A & 0 \\ 0 & F_B \end{smallmatrix}\right]$. Then $F_{AB} F_{AB}^T \leq I$ by Lemma 8.1. Inequality (8.38), together with Lemma 4.6, becomes

$$W_0 + \mathrm{Sym}\{\Upsilon_2 F_{AB} \Upsilon_1\} < 0. \tag{8.39}$$

Substituting W_0, Υ_1, and Υ_2 into inequality (8.39) implies

$$\begin{bmatrix} \hat{\Sigma}_2 & (*) & (*) \\ (A_U + E\alpha)G + B_U H - G^* & -G - G^* & (*) \\ 0 & G & -Y \end{bmatrix} < 0, \tag{8.40}$$

where $\hat{\Sigma}_2 = \mathrm{Sym}\{(A_U + E\alpha)G + B_U H - \delta\Xi(A_U + E\alpha)G - \delta\Xi B_U H - \varepsilon r^2 EG\} + \delta^2 X + \varepsilon^2 r^2 Y$. Since $H = KG$, inequality (8.40) can be written as

$$\begin{bmatrix} \hat{\Sigma}_3 & (*) & (*) \\ (A_U + B_U K + E\alpha)G - G^* & -G - G^* & (*) \\ 0 & G & -Y \end{bmatrix} < 0, \tag{8.41}$$

where $\hat{\Sigma}_3 = \mathrm{Sym}\{(A_U + B_U K + E\alpha)G - \delta\Xi(A_U + B_U K + E\alpha)G - \varepsilon r^2 EG\} + \delta^2 X + \varepsilon^2 r^2 Y$. It follows from Theorem 8.9 that the closed-loop system (8.35) is admissible. This ends the proof. □

Remark 8.8. From Theorem 8.5 we can see that the results in this chapter can be applicable to FOSDTSs with $\alpha = 1$. When $\alpha = 1$, the stability radius r is 1 as $L \to \infty$. Therefore by Theorem 8.10 we can take $r = 1$ to design the state feedback controller when $\alpha = 1$.

Remark 8.9. In this chapter, the admissibility condition for nominal FOSDTSs in Theorem 8.9 is necessary but not sufficient for analyzing the controller synthesis for FOSDTSs with interval uncertainties. In the process of solving the controller synthesis, the properties of interval uncertainties, the Schur complement lemma, and the inequality scaling techniques are also necessary.

Remark 8.10. Compared with the existing results, the controller design methods in this chapter have the following advantages:

1. The controller design methods obtained in [1] are applicable to FODTSs. By comparison the controller design methods in this chapter are applicable to FOSDTSs, which are more general than those in [1].

2. The controller design methods obtained in [54] are in the form of matrix inequalities. By comparison the controller design methods in this chapter are in the form of LMIs, which can be more conveniently settled by the existing LMI software solver than those in [54].

The Grünwald–Letnikov operator is widely used for discretizing continuous models. Given that different fractional-order singular continuous models have been studied such as the nominal fractional-order singular continuous electrical circuits in [116] and Chapter 2, the FOSDTSs can be obtained by using the series connection of a zero-order holder (ZOH) and the sampler. For example, assuming that the ZOH and the sampler work at the same frequency $f_s = 1$ or, equivalently, with a sampling period $T = 1/f_s = 1$, a case of transforming a nominal fractional-order singular continuous electrical circuit to the FOSDTS is shown in Figure 8.1, in which the physical meanings of symbols in the electrical circuit can be found in [113]. The derivation process is similar to the case of transforming the fractional-order continuous-time systems (FOCTSs) to the FODTSs in [66] and is omitted here because this chapter focuses on the admissibility and robust stabilization of FOSDTSs instead of the modeling process of FOSDTSs. Therefore we present two numerical examples on the nominal FOSDTS (8.5) and FOSDTS (8.1) with interval uncertainties to verify the effectiveness of the results in this chapter.

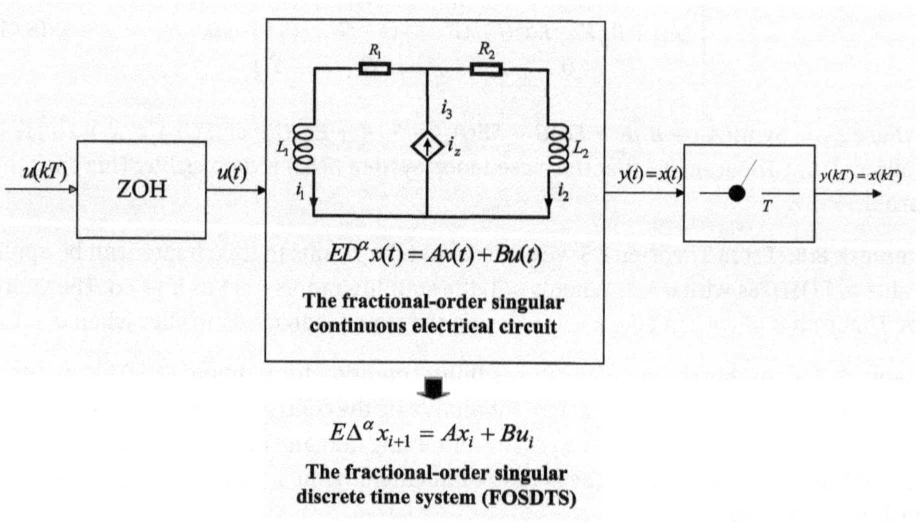

Figure 8.1: A case of transforming a nominal fractional-order singular continuous electrical circuit to the FOSDTS using a ZOH and a sampler.

Example 8.1. Consider the unforced nominal FOSDTS (8.5) with

$$E = \begin{bmatrix} 2 & 0 & 0 \\ 0 & 4 & 0 \\ 0 & 0 & 0 \end{bmatrix}, \quad A = \begin{bmatrix} -1 & 0.4 & 0 \\ 0.4 & -1.6 & 0 \\ 0 & 0 & 1 \end{bmatrix}, \quad \alpha = 0.5. \tag{8.42}$$

The purpose is to determine whether the unforced nominal FOSDTS (8.5) with these parameters is admissible or not.

Take $\delta = \varepsilon = 1$ and $\Xi = \left[\begin{smallmatrix} 0 & 0 & 0 \\ 0 & 0 & 0 \\ 0 & 0 & 10 \end{smallmatrix}\right]$. By applying Theorem 8.9 with memory length $L = 10$, we can obtain the following feasible solutions:

$$X = \begin{bmatrix} 373.0774 & -91.5279 & 0 \\ -91.5279 & 1284.0804 & 0 \\ 0 & 0 & 2759.0398 \end{bmatrix},$$

$$Y = \begin{bmatrix} 1967.7558 & 16.687966 & 0 \\ 16.687966 & 2366.1520 & 0 \\ 0 & 0 & 2421.6118 \end{bmatrix},$$

$$G = \begin{bmatrix} 1423.3729 & 523.2806 & 0 \\ 523.2806 & 1570.5153 & 0 \\ 0 & 0 & 360.1370 \end{bmatrix}.$$

By taking $L = 5, 15, 20, \dots$, we can also obtain feasible solutions. In fact, as long as one memory length $L \in \mathbb{Z}^+$ can be found such that Theorem 8.9 has feasible solutions, the unforced nominal FOSDTS (8.5) can be judged as admissible because Theorem 8.9 is a sufficient condition.

On the other hand, from Theorems 8.1 and 8.2, we choose two nonsingular matrices

$$\Phi = \begin{bmatrix} 0.5 & 0 & 0 \\ 0 & 0.25 & 0 \\ 0 & 0 & 1 \end{bmatrix} \quad \text{and} \quad \Psi = \begin{bmatrix} 1 & 0 & 0 \\ 0 & 1 & 0 \\ 0 & 0 & 1 \end{bmatrix}$$

such that

$$\Phi A \Psi = \operatorname{diag}(A_1, I_1) \quad \text{and} \quad \Phi E \Psi = \operatorname{diag}(I_2, 0),$$

where

$$A_1 = \begin{bmatrix} -0.5 & 0.2 \\ 0.1 & -0.4 \end{bmatrix}.$$

Thus the system is regular and causal. Given the initial condition $x_0 = \begin{bmatrix} 1 & -1 & 0 \end{bmatrix}^T$, the state responses of FOSDTS (8.5) with the parameters in (8.42) are shown in Figure 8.2. We can see that the state responses all tend to 0, which indicates that this system is stable. Thus, from the numerical point of view, the admissibility of FOSDTS (8.5) with parameters (8.42) can also be verified.

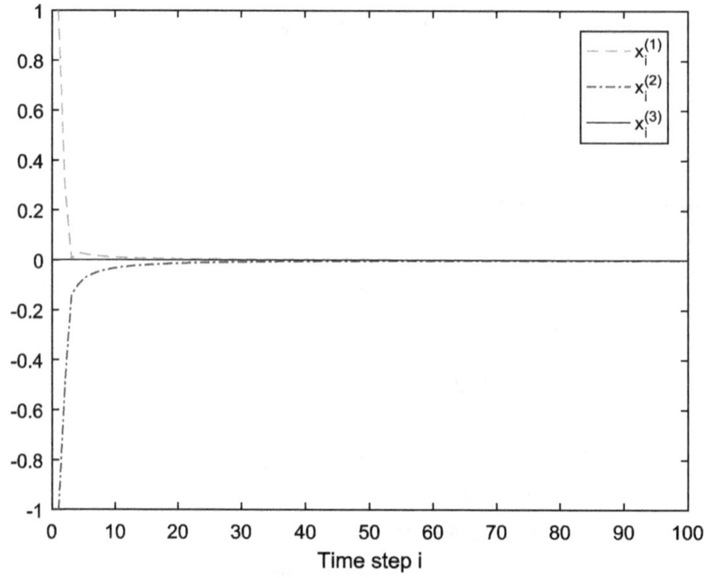

Figure 8.2: The state responses of FOSDTS (8.5) with parameters (8.42) in Example 8.1.

Example 8.2. Consider the FOSDTS (8.1) with interval uncertainties and with the following system parameters:

$$E = \begin{bmatrix} 8 & 0 & 0 \\ 0 & 5 & 0 \\ 0 & 0 & 0 \end{bmatrix}, \quad \overline{A} = \begin{bmatrix} -7.9 & 12.2 & 0.1 \\ 0.3 & -7 & 0.1 \\ 0.1 & 0.1 & 1.2 \end{bmatrix}, \quad \underline{A} = \begin{bmatrix} -8.1 & 11.8 & -0.1 \\ 0.1 & -8 & -0.1 \\ -0.1 & -0.1 & 0.8 \end{bmatrix},$$

$$\overline{B} = \begin{bmatrix} 1.1 \\ 1.2 \\ 1.5 \end{bmatrix}, \quad \underline{B} = \begin{bmatrix} 0.9 \\ 0.8 \\ 0.5 \end{bmatrix}, \quad \alpha = 1.2.$$

The purpose is to design the state feedback controller (8.34) such that the closed-loop FOSDTS (8.35) with these parameters is robustly admissible or not.

Take $\delta = \varepsilon = 1$ and $\Xi = \begin{bmatrix} 0 & 0 & 0 \\ 0 & 0 & 0 \\ 0 & 0 & 10 \end{bmatrix}$. By applying Theorem 8.10 with memory length $L = 10$, we can be obtain the following feasible solutions:

$$X = \begin{bmatrix} 33058.2850 & -3208.3086 & -3924.6433 \\ -3208.3086 & 1.0967 & 1100.6374 \\ -3924.6433 & 1100.6374 & 19407.497 \end{bmatrix},$$

$$Y = \begin{bmatrix} 57879.653 & -9203.6753 & 4480.3353 \\ -9203.6753 & 3342.8292 & -1150.7668 \\ 4480.3353 & -1150.7668 & 27808.975 \end{bmatrix},$$

$$G = \begin{bmatrix} 21823.262 & -2514.8172 & -3273.2053 \\ -136.54847 & 1104.1491 & 659.81126 \\ 4773.2034 & -664.9925 & 10453.932 \end{bmatrix},$$

$$H = [-4436.711 \quad 620.7317 \quad 2405.9553], \quad \gamma = 1968.1724.$$

By Theorem 8.10 the state feedback controller (8.34) can be determined as $u_i = HG^{-1}x_i = [-0.2351 \quad 0.1163 \quad 0.1492]\, x_i$. Taking $L = 5, 15, 20, \ldots$, we can also obtain feasible solutions. In fact, as long as one memory length $L \in \mathbb{Z}^+$ can be found such that Theorem 8.10 has feasible solutions, the state feedback controller (8.34) can be determined because Theorem 8.10 is a sufficient condition.

Take the randomly selected system parameters

$$E = \begin{bmatrix} 8 & 0 & 0 \\ 0 & 5 & 0 \\ 0 & 0 & 0 \end{bmatrix}, \quad A = \begin{bmatrix} -8 & 12 & 0 \\ 0.2 & -7.5 & 0 \\ 0 & 0 & 1 \end{bmatrix}, \quad B = \begin{bmatrix} 1 \\ 0.9 \\ 1.2 \end{bmatrix}, \quad \alpha = 1.2, \tag{8.43}$$

which implies that $A + BK = \begin{bmatrix} -8.2351 & 12.1163 & 0.1492 \\ -0.0116 & -7.3953 & 0.1343 \\ -0.2821 & 0.1396 & 1.1790 \end{bmatrix}$. From Theorems 8.1 and 8.2, we choose two nonsingular matrices

$$\Phi = \begin{bmatrix} 0.1250 & 0 & -0.0158 \\ 0 & 1.0000 & -0.1139 \\ 0 & 0 & 1.0000 \end{bmatrix} \quad \text{and} \quad \Psi = \begin{bmatrix} 1.0000 & 0 & 0 \\ 0 & 0.2000 & 0 \\ 0.2393 & -0.0237 & 0.8481 \end{bmatrix}$$

such that

$$\Phi(A + BK)\Psi = \operatorname{diag}(A_1, I_1) \quad \text{and} \quad \Phi E \Psi = \operatorname{diag}(I_2, 0),$$

where

$$A_1 = \begin{bmatrix} -1.0249 & 0.3025 \\ 0.0205 & -1.4822 \end{bmatrix}.$$

Given the initial condition $x_0 = [2 \quad -2 \quad 0.7154]^T$, the state responses of FOSDTS (8.35) with parameters (8.43) are shown in Figure 8.3, where the state responses all tend to 0, indicating that this system is stable. Thus, from the numerical point of view, the admissibility of FOSDTS (8.35) with parameters (8.43) can also be verified.

8.4 Chapter summary

In this chapter, we investigated the admissibility and robust stabilization of FOSDTSs with interval uncertainties. For nominal FOSDTSs, the regularity, causality, and stability conditions are derived based on the Kronecker equivalent standard form and time domain solution analysis methods. By using regularity, causality, and stability conditions,

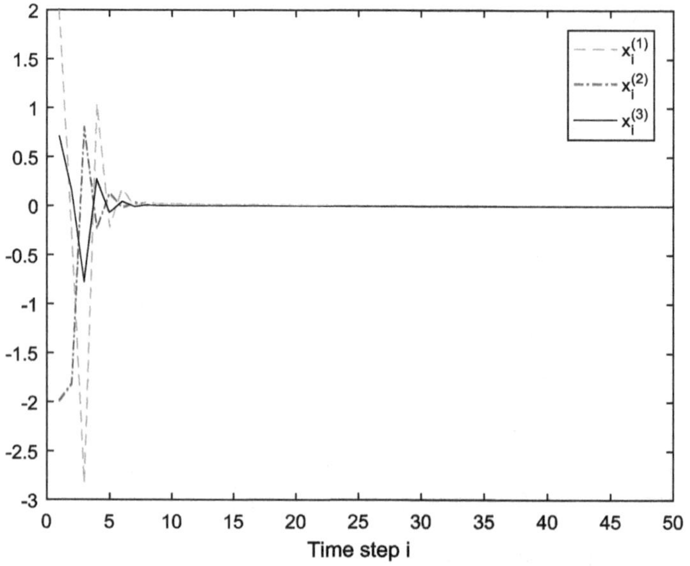

Figure 8.3: The state responses of FOSDTS (8.35) with parameters (8.43) in Example 8.2.

an admissibility condition for nominal FOSDTSs in terms of spectral radius without conservatism is given. Then another admissibility condition is given in terms of nonstrict LMIs. To convert the nonstrict LMIs into strict LMIs and to eliminate the coupling terms, a novel admissibility condition is established. Then, for FOSDTSs with interval uncertainties, the controller synthesis was addressed and the state feedback controller was designed. Finally, two numerical examples are provided to illustrate the proposed results.

Bibliography

[1] Andrea Alessandretti, Sérgio Pequito, George J. Pappas, and A. Pedro Aguiar. Finite-dimensional control of linear discrete-time fractional-order systems. *Automatica*, 115:108512:1–9, 2020.

[2] Vahid Badri and Mohammad Saleh Tavazoei. Stability analysis of fractional order time-delay systems: Constructing new Lyapunov functions from those of integer order counterparts. *IET Control Theory Appl.*, 13(15):2476–2481, 2019.

[3] Djillali Bouagada, Samuel Melchior, and Paul Van Dooren. Calculating the $\mathcal{H}_\infty$ norm of a fractional system given in state-space form. *Appl. Math. Lett.*, 79:51–57, 2018.

[4] Yassine Boukal, Mohamed Darouach, Michel Zasadzinski, and Nour-Eddine Radhy. Robust $\mathcal{H}_\infty$ observer-based control of fractional-order systems with gain parametrization. *IEEE Trans. Autom. Control*, 62(11):5710–5723, Nov. 2017.

[5] Yassine Boukal, Michel Zasadzinski, Mohamed Darouach, and Nour-Edine Radhy. Stability and stabilizability analysis of fractional-order time-varying delay systems via diffusive representation. In *2016 5th International Conference on Systems and Control (ICSC)*, pp. 262–266. IEEE, 2016.

[6] Otto Brune. *Synthesis of a finite two-terminal network whose driving-point impedance is a prescribed function of frequency*. PhD thesis, Massachusetts Institute of Technology, 1931.

[7] Pavol Brunovský, Aleš Černý, and Michael Winkler. A singular differential equation stemming from an optimal control problem in financial economics. *Appl. Math. Optim.*, 68(2):255–274, 2013.

[8] Mikolaj Busłowicz and Andrzej Ruszewski. Necessary and sufficient conditions for stability of fractional discrete-time linear state-space systems. *Bull. Pol. Acad. Sci., Tech. Sci.*, 61(4):779–786, 2013.

[9] Kai-Yuan Cai, Jie Chen, and Ron J. Patton. *Robust Model-Based Fault Diagnosis For Dynamic Systems*, vol. 3. Springer US, Boston, MA, 1999.

[10] M. Kanat Camlibel and Roberto Frasca. Extension of Kalman–Yakubovich–Popov lemma to descriptor systems. *Syst. Control Lett.*, 58(12):795–803, 2009.

[11] Liping Chen, Ranchao Wu, Yuan Cheng, and YangQuan Chen. Delay-dependent and order-dependent stability and stabilization of fractional-order linear systems with time-varying delay. *IEEE Trans. Circuits Syst. II, Express Briefs*, 67(6):1064–1068, 2020.

[12] Joaquin Collado, Rogelio Lozano, and Rolf Johansson. On Kalman–Yakubovich–Popov lemma for stabilizable systems. *IEEE Trans. Autom. Control*, 46(7):1089–1093, 2001.

[13] Víctor Costa da Silva Campos, Anh-Tu Nguyen, and Reinaldo Martínez Palhares. Adaptive gain-scheduling control for continuous-time systems with polytopic uncertainties: An LMI-based approach. *Automatica*, 133:109856, 2021.

[14] Liyi Dai. *Singular Control Systems*. Springer-Verlag, Berlin, Germany, 1989.

[15] Weihua Deng, Changpin Li, and Jinhu Lü. Stability analysis of linear fractional differential system with multiple time delays. *Nonlinear Dyn.*, 48(4):409–416, 2007.

[16] Kai Diethelm, Neville J. Ford, and Alan D. Freed. A predictor-corrector approach for the numerical solution of fractional differential equations. *Nonlinear Dyn.*, 29(1-4):3–22, 2002.

[17] John Doyle, Keith Glover, Pramod Khargonekar, and Bruce Francis. State-space solutions to standard H_2 and H_∞ control problems. In *Proceedings of American Control Conference (ACC)*, Atlanta, GA, USA, pp. 1691–1696. IEEE, 1988.

[18] Guang Ren Duan. *Analysis and Design of Descriptor Linear Systems*. Springer, New York, 2010.

[19] Andrzej Dzieliński and Dominik Sierociuk. Stability of discrete fractional order state-space systems. *J. Vib. Control*, 14(9-10):1543–1556, 2008.

[20] Christophe Farges, Lamine Fadiga, and Jocelyn Sabatier. $\mathcal{H}_\infty$ analysis and control of commensurate fractional order systems. *Mechatronics*, 23(7):772–780, 2013.

[21] Christophe Farges, Mathieu Moze, and Jocelyn Sabatier. Pseudo state feedback stabilization of commensurate fractional order systems. In *2009 European Control Conf. (ECC)*, pp. 3395–3400. IEEE, Aug. 2009.

https://doi.org/10.1515/9783111628646-009

[22] Christophe Farges, Mathieu Moze, and Jocelyn Sabatier. Pseudo-state feedback stabilization of commensurate fractional order systems. *Automatica*, 46(10):1730–1734, 2010.

[23] Yu Feng. Positive real control for discrete-time singular systems with affine parameter dependence. *J. Franklin Inst.*, 352(3):882–896, 2015.

[24] Augusto Ferrante, Alexander Lanzon, and Bernard Brogliato. A direct proof of the equivalence of side conditions for strictly positive real matrix transfer functions. *IEEE Trans. Autom. Control*, 65(1):450–452, 2020.

[25] Augusto Ferrante and Lorenzo Ntogramatzidis. Solvability conditions for the positive real lemma equations in the discrete time. *IET Control Theory Appl.*, 11(16):2916–2920, 2017.

[26] Pascal Gahinet and Pierre Apkarian. A linear matrix inequality approach to $\mathcal{H}_\infty$ control. *Int. J. Robust Nonlinear Control*, 4(4):421–448, 1994.

[27] Feliks Ruvimovich Gantmacher. *The Theory of Matrices*. Chelsea, New York, 1959.

[28] Zhe Gao. Robust stabilization criterion of fractional-order controllers for interval fractional-order plants. *Automatica*, 61:9–17, 2015.

[29] Panpan Gu, YangQuan Chen, and Senping Tian. Learnability of linear fractional-order ILC systems. *IEEE Trans. Circuits Syst. II, Express Briefs*, 68(3):963–967, 2021.

[30] Mojtaba Hakimi-Moghaddam and Augusto Ferrante. New results on the characterization of strictly positive real matrix transfer functions. *IEEE Trans. Autom. Control*, 66(1):335–339, 2021.

[31] Chang-Chun Hua, Dan Liu, and Xin-Ping Guan. Necessary and sufficient stability criteria for a class of fractional-order delayed systems. *IEEE Trans. Circuits Syst. II, Express Briefs*, 61(1):59–63, 2014.

[32] Salim Ibrir. New LMI conditions for stability and stabilizability of fractional-order systems with $\mathcal{H}_\infty$ performance. In *Proceedings of the 15th International Conference on Control & Automation (ICCA)*, Edinburgh, UK, pp. 952–957. IEEE, 2019.

[33] J. Y. Ishihara and M. H. Terra. Impulse controllability and observability of rectangular descriptor systems. *IEEE Trans. Autom. Control*, 45(6):991–994, 2001.

[34] Tetsuya Iwasaki and Robert E. Skelton. All controllers for the general $\mathcal{H}_\infty$ control problem: LMI existence conditions and state space formulas. *Automatica*, 30(8):1307–1317, 1994.

[35] Imad M. Jaimoukha, Zhenhai Li, and Vasilios Papakos. A matrix factorization solution to the H_-/H_∞ fault detection problem. *Automatica*, 42(11):1907–1912, 2006.

[36] Yude Ji and Jiqing Qiu. Stabilization of fractional-order singular uncertain systems. *ISA Trans.*, 56:53–64, 2015.

[37] Xiao-Chuang Jin and Jun-Guo Lu. Order-dependent LMI-based stability and stabilization conditions for fractional-order time-delay systems using small gain theorem. *Int. J. Robust Nonlinear Control*, 32(11):6484–6506, 2022.

[38] Tadeusz Kaczorek. Singular fractional linear systems and electrical circuits. *Int. J. Appl. Math. Comput. Sci.*, 21(2):379–384, 2011.

[39] Tadeusz Kaczorek. *Selected Problems of Fractional Systems Theory*. Springer-Verlag, Berlin, Germany, 2011.

[40] Tadeusz Kaczorek. Positivity and stability of a class of fractional descriptor discrete-time nonlinear systems. In *Theoretical Developments and Applications of Non-Integer Order Systems*, pp. 143–155. Springer, Switzerland, 2016.

[41] Tadeusz Kaczorek and Krzysztof Rogowski. *Fractional Linear Systems and Electrical Circuits*. Studies in Systems, Decision and Control, vol. 13. Springer International Publishing, Cham, Switzerland, 2015.

[42] Dazi Li, Liming Wei, Tianheng Song, and Qibing Jin. Study on asymptotic stability of fractional singular systems with time delay. *Int. J. Control. Autom. Syst.*, 18(4):1002–1011, 2020.

[43] He Li and Guang-Hong Yang. Fault detection for fractional-order linear systems in finite frequency domains. *IET Control Theory Appl.*, 13(15):2336–2345, 2019.

[44] Li Li, Qingling Zhang, and Baoyan Zhu. Fuzzy stochastic optimal guaranteed cost control of bio-economic singular Markovian jump systems. *IEEE Trans. Cybern.*, 45(11):2512–2521, 2015.

[45] Yuman Li, Yiheng Wei, Yuquan Chen, and Yong Wang. A universal framework of the generalized Kalman–Yakubovich–Popov lemma for singular fractional-order systems. *IEEE Trans. Syst. Man Cybern. Syst.*, 51(8):5209–5217, 2021.

[46] Shu Liang, Yi-Heng Wei, Jin-Wen Pan, Qing Gao, and Yong Wang. Bounded real lemmas for fractional order systems. *Int. J. Autom. Comput.*, 12(2):192–198, 2015.

[47] Fang Liao, Jian Liang Wang, and Guang-Hong Yang. Reliable robust flight tracking control: An LMI approach. *IEEE Trans. Control Syst. Technol.*, 10(1):76–89, 2002.

[48] Chong Lin, Bing Chen, Peng Shi, and Jin-Peng Yu. Necessary and sufficient conditions of observer-based stabilization for a class of fractional-order descriptor systems. *Syst. Control Lett.*, 112:31–35, 2018.

[49] Chong Lin, Bing Chen, and Qing-Guo Wang. Static output feedback stabilization for fractional-order systems in T-S fuzzy models. *Neurocomputing*, 218:354–358, 2016.

[50] Chong Lin, Jianliang Wang, and CB Soh. Maximum bounds for robust stability of linear uncertain descriptor systems with structured perturbations. *Int. J. Syst. Sci.*, 34(7):463–467, 2003.

[51] Jian Liu, Jian Liang Wang, and Guang-Hong Yang. An LMI approach to minimum sensitivity analysis with application to fault detection. *Automatica*, 41(11):1995–2004, 2005.

[52] L. Liu, S. Zhang, D. Xue, and Y. Chen. Robust stability analysis for fractional-order systems with time delay based on finite spectrum assignment. *Int. J. Robust Nonlinear Control*, 29:2283–2295, 2019.

[53] Song Liu, Xian-Feng Zhou, Xiaoyan Li, and Wei Jiang. Asymptotical stability of Riemann–Liouville fractional singular systems with multiple time-varying delays. *Appl. Math. Lett.*, 65:32–39, 2017.

[54] Xiang Liu, Peiguang Wang, and Douglas R. Anderson. On stability and feedback control of discrete fractional order singular systems with multiple time-varying delays. *Chaos Solitons Fractals*, 155:111740:1–7, 2022.

[55] Jun-Guo Lu and YangQuan Chen. Robust stability and stabilization of fractional-order interval systems with the fractional order α: The $0 < \alpha < 1$ case. *IEEE Trans. Autom. Control*, 55(1):152–158, Jan. 2010.

[56] Jun-Guo Lu, YangQuan Chen, and Weidong Chen. Robust asymptotical stability of fractional-order linear systems with structured perturbations. *Comput. Math. Appl.*, 66(5):873–882, 2013.

[57] David G. Luenberger. Dynamic equations in descriptor form. *IEEE Trans. Autom. Control*, 22(3):312–321, 1977.

[58] David G. Luenberger. Time-invariant descriptor systems. *Automatica*, 14(5):473–480, 1978.

[59] Siyou Luo and Jun-Guo Lu. Robust stability and stabilization of fractional-order systems with polytopic uncertainties via homogeneous polynomial parameter-dependent matrix forms. *Int. J. Gen. Syst.*, 50(8):891–914, 2021.

[60] Lei Ma, Zidong Wang, Chenxiao Cai, and Fuad E. Alsaadi. A dynamic event-triggered approach to $\mathcal{H}_\infty$ control for discrete-time singularly perturbed systems with time-delays and sensor saturations. *IEEE Trans. Syst. Man Cybern. Syst.*, 51(11):6614–6625, 2021.

[61] Benoit B. Mandelbrot. *The Fractal Geometry of Nature*. Academic Press, New York, 1982.

[62] Saliha Marir and Mohammed Chadli. Robust admissibility and stabilization of uncertain singular fractional-order linear time-invariant systems. *IEEE/CAA J. Autom. Sin.*, 6(3):685–692, May 2019.

[63] Saliha Marir, Mohammed Chadli, and Djillali Bouagada. New admissibility conditions for singular linear continuous-time fractional-order systems. *J. Franklin Inst.*, 354(2):752–766, 2017.

[64] Saliha Marir, Mohammed Chadli, and Djillali Bouagada. A novel approach of admissibility for singular linear continuous-time fractional-order systems. *Int. J. Control. Autom. Syst.*, 15(2):959–964, 2017.

[65] Denis Matignon. Stability results for fractional differential equations with applications to control processing. In *Computational Engineering in Systems Applications*, Lille, France, vol. 2, pp. 963–968, 1996.

[66] Concepción A. Monje, YangQuan Chen, Blas M. Vinagre, Dingyu Xue, and Vicente Feliu. *Fractional-Order Systems and Controls: Fundamentals and Applications*. Springer Science & Business Media, London, 2010.

[67] Mathieu Moze, Jocelyn Sabatier, and Alain Oustaloup. LMI tools for stability analysis of fractional systems. In *ASME 2005 International Design Engineering Technical Conf. and Computers and Information in Engineering Conf.*, pp. 1611–1619. ASME, Sep. 2005.

[68] Ibrahima N'Doye, Mohamed Darouach, Michel Zasadzinski, and Nour-Eddine Radhy. Robust stabilization of uncertain descriptor fractional-order systems. *Automatica*, 49(6):1907–1913, 2013.

[69] Ibrahima N'Doye, Michel Zasadzinski, Mohamed Darouach, and Nour-Eddine Radhy. Stabilization of singular fractional-order systems: An LMI approach. In *Proceedings of the 18th Mediterranean Conference on Control and Automation, MED'10*, Marrakech, Morocco, pp. 209–213. IEEE, 2010.

[70] Ibrahima N'Doye, Michel Zasadzinski, Mohamed Darouach, and Nour-Eddine Radhy. Regularization and robust stabilization of uncertain singular fractional-order systems. *IFAC Proc. Vol.*, 44(1):15031–15036, 2011.

[71] Ramine Nikoukhah, François Delebecque, and Laurent El Ghaoui. LMITOOL: A package for LMI optimization in Scilab. *INRIA Rocquencourt, France*, 1995.

[72] Zepeng Ning, Lixian Zhang, Ali Mesbah, and Patrizio Colaneri. Stability analysis and stabilization of discrete-time non-homogeneous semi-Markov jump linear systems: A polytopic approach. *Automatica*, 120:109080, 2020.

[73] Komeil Nosrati and Masoud Shafiee. Fractional-order singular logistic map: Stability, bifurcation and chaos analysis. *Chaos Solitons Fractals*, 115:224–238, 2018.

[74] Komeil Nosrati and Masoud Shafiee. Kalman filtering for discrete-time linear fractional-order singular systems. *IET Control Theory Appl.*, 12(9):1254–1266, 2018.

[75] Keith Oldham and Jerome Spanier. *The Fractional Calculus: Theory and Applications of Differentiation and Integration to Arbitrary Order*. Academic Press, New York, 1974.

[76] Manuel Duarte Ortigueira. Introduction to fractional linear systems. Part 1: Continuous-time case. *IEE Proc., Vis. Image Signal Process.*, 147(1):62–70, Feb. 2000.

[77] Mohammad Ali Pakzad, Sara Pakzad, and Mohammad Ali Nekoui. Exact method for the stability analysis of time delayed linear-time invariant fractional-order systems. *IET Control Theory Appl.*, 9(16):2357–2368, 2015.

[78] Huan Pan, Xinghuo Yu, Guohua Yang, and Li Xue. Robust consensus of fractional-order singular uncertain multi-agent systems. *Asian J. Control*, 22(6):2377–2387, 2020.

[79] Igor Podlubny. *Fractional Differential Equations*. Academic Press, New York, NY, USA, 1999.

[80] Vasile Mihai Popov and Radu Georgescu. *Hyperstability of Control Systems*. Springer, New York, USA, 1973.

[81] Fateme Pourdadashi Komachali and Masoud Shafiee. Sensor fault diagnosis in fractional-order singular systems using unknown input observer. *Int. J. Syst. Sci.*, 51(1):116–132, 2020.

[82] Timo Reis and Tatjana Stykel. Positive real and bounded real balancing for model reduction of descriptor systems. *Int. J. Control*, 83(1):74–88, 2010.

[83] Howard Harry Rosenbrock. Structural properties of linear dynamical systems. *Int. J. Control*, 20(2):191–202, 1974.

[84] Jocelyn Sabatier, Mathieu Moze, and Christophe Farges. LMI stability conditions for fractional order systems. *Comput. Math. Appl.*, 59(5):1594–1609, 2010.

[85] Maryam Sadati, Mohammad Hosseinzadeh, and Masoud Shafiee. Modelling and analysis of quadruped walking robot using singular system theory. In *Proceedings of the 21st Iranian Conference on Electrical Engineering (ICEE)*, Mashhad, Iran, pp. 1–6, IEEE, 2013.

[86] Robert E. Skelton, Tetsuya Iwasaki, and Dimitri E. Grigoriadis. *A Unified Algebraic Approach to Control Design*. Taylor Francis, New York, 1997.

[87] Robert E. Skelton, Tetsuya Iwasaki, and Karolos M. Grigoriadis. *A Unified Algebraic Approach to Linear Control Design*. Routledge, London, 2017.

[88] Xiaona Song, Leipo Liu, and Zhen Wang. Stabilization of singular fractional-order systems: A linear matrix inequality approach. In *2012 IEEE Int. Conf. Autom. Logist.*, pages 19–23. IEEE, Aug. 2012.

[89] Pantelis Sopasakis and Haralambos Sarimveis. Stabilising model predictive control for discrete-time fractional-order systems. *Automatica*, 75:24–31, 2017.
[90] Kenji Sugimoto, Xinyou Han, and Wataru Imahayashi. Stability of MIMO feedback error learning control under a strictly positive real condition. *IFAC-PapersOnLine*, 51(33):168–174, 2018.
[91] Weiqian Sun, Pramod P. Khargonekar, and Duksun Shim. Solution to the positive real control problem for linear time-invariant systems. *IEEE Trans. Autom. Control*, 39(10):2034–2046, 1994.
[92] Mohammad Tavazoei and Mohammad Hassan Asemani. On robust stability of incommensurate fractional-order systems. *Commun. Nonlinear Sci. Numer. Simul.*, 90:105344, 2020.
[93] Cuihong Wang, Huanhuan Li, and YangQuan Chen. $\mathcal{H}_\infty$ output feedback control of linear time-invariant fractional-order systems over finite frequency range. *IEEE/CAA J. Autom. Sin.*, 3(3):304–310, 2016.
[94] H. Wang and G.-H. Yang. A finite frequency domain approach to fault detection for linear discrete-time systems. *Int. J. Control*, 81(7):1162–1171, 2008.
[95] Zhenhua Wang, Peng Shi, and Cheng-Chew Lim. H_-/H_∞ fault detection observer in finite frequency domain for linear parameter-varying descriptor systems. *Automatica*, 86:38–45, 2017.
[96] Yiheng Wei, Peter W. Tse, Zhao Yao, and Yong Wang. The output feedback control synthesis for a class of singular fractional order systems. *ISA Trans.*, 69:1–9, 2017.
[97] Yiheng Wei, Jiachang Wang, Tianyu Liu, and Yong Wang. Sufficient and necessary conditions for stabilizing singular fractional order systems with partially measurable state. *J. Franklin Inst.*, 356(4):1975–1990, 2019.
[98] Qian Wu, Qiankun Song, Binxin Hu, Zhenjiang Zhao, Yurong Liu, and Fuad E. Alsaadi. Robust stability of uncertain fractional order singular systems with neutral and time-varying delays. *Neurocomputing*, 401:145–152, 2020.
[99] Lihua Xie and Yeng Chai Soh. Positive real control problem for uncertain linear time-invariant systems. *Syst. Control Lett.*, 24(4):265–271, 1995.
[100] Jie Xu. Research on some problems of fractional order system analysis and control. Master's thesis, Shanghai: Shanghai JiaoTong University, 2009.
[101] Shengyuan Xu and James Lam. *Robust Control and Filtering of Singular Systems*. Springer-Verlag, Berlin, Germany, 2006.
[102] Isaac Yaesh and Uri Shaked. Simplified adaptive control via improved robust positive real conditions. *Syst. Control Lett.*, 55(12):1033–1040, 2006.
[103] Lichang Yang, Wei Jiang, Jiale Sheng, Tingting Liu, and Musarrt Nawaz. Relative controllability of fractional neutral system with multiple delays. *Acta Math. Appl. Sin.*, 43(1):1–11, 2020.
[104] Le Hai Yen and Vu Ngoc Phat. Stability analysis of linear polytopic descriptor systems using a novel copositive matrix approach. *IEEE Trans. Autom. Control*, 64(11):4684–4690, 2019.
[105] Ma Yingdong. Research on some problems of robust control for uncertain fractional-order systems. Master's thesis, Shanghai: Shanghai JiaoTong University, 2014.
[106] Yao Yu, Zhuang Jiao, and Chang-Yin Sun. Sufficient and necessary condition of admissibility for fractional-order singular system. *Acta Autom. Sin.*, 39(12):2160–2164, 2013.
[107] George Zames. Feedback and optimal sensitivity: Model reference transformations, multiplicative seminorms, and approximate inverses. *IEEE Trans. Autom. Control*, 26(2):301–320, 1981.
[108] Ding Zhai, Li-Wei An, Jing-Hao Li, and Qing-Ling Zhang. Finite frequency fault detection for T-S fuzzy singular multiple timedelay systems. *Int. J. Control. Autom. Syst.*, 14(4):977–985, 2016.
[109] Ding Zhai, Qing-Ling Zhang, and Jing-Hao Li. Fault detection for singular multiple time-delay systems with application to electrical circuit. *J. Franklin Inst.*, 351(12):5411–5436, 2014.
[110] Fuzhen Zhang. *The Schur Complement and Its Applications*. Springer, New York, USA, 2005.
[111] Liqian Zhang, James Lam, and Shengyuan Xu. On positive realness of descriptor systems. *IEEE Trans. Circuits Syst. I, Regul. Pap.*, 49(3):401–407, 2002.
[112] Qing-Hao Zhang and Jun-Guo Lu. Robust stability of output feedback controlled fractional-order systems with structured uncertainties in all system coefficient matrices. *ISA Trans.*, 105:51–62, 2020.

[113] Qing-Hao Zhang, Jun-Guo Lu, Jie Xu, and YangQuan Chen. Solution analysis and novel admissibility conditions of SFOSs: The $1 < \alpha < 2$ case. *IEEE Trans. Syst. Man Cybern. Syst.*, 52(8):5056–5067, 2022.

[114] Shuo Zhang, Lu Liu, and Dingyu Xue. Nyquist-based stability analysis of non-commensurate fractional-order delay systems. *Appl. Math. Comput.*, 377:125111, 2020.

[115] Sunjie Zhang, Zidong Wang, Derui Ding, Huisheng Shu, Tasawar Hayat, and Abdullah M. Dobaie. On design of robust fault detection filter in finite-frequency domain with regional pole assignment. *IEEE Trans. Circuits Syst. II, Express Briefs*, 62(4):382–386, 2015.

[116] Xuefeng Zhang and YangQuan Chen. Admissibility and robust stabilization of continuous linear singular fractional order systems with the fractional order α: The $0 < \alpha < 1$ case. *ISA Trans.*, 82:42–50, 2018.

[117] Xuefeng Zhang and Kaijing Jin. Design of non-fragile controller for singular fractional order Takagi–Sugeno fuzzy systems. *Int. J. Fuzzy Syst.*, 22(4):1289–1298, 2020.

[118] Xuefeng Zhang and Zeli Zhao. Normalization and stabilization for rectangular singular fractional order T-S fuzzy systems. *Fuzzy Sets Syst.*, 381:140–153, 2020.

[119] Xuefeng Zhang, Zeli Zhao, and Li Li. An alternative admissibility theorem for singular fractional order systems. *IEEE Access*, 7:126005–126013, 2019.

[120] Xuefeng Zhang, Zeli Zhao, and Qingguo Wang. Static and dynamic output feedback stabilisation of descriptor fractional order systems. *IET Control Theory Appl.*, 14(2):324–333, 2020.

[121] Yingqi Zhang, Peng Shi, Ramesh K. Agarwal, and Yan Shi. Event-based dissipative analysis for discrete time-delay singular jump neural networks. *IEEE Trans. Neural Netw. Learn. Syst.*, 31(4):1232–1241, 2020.

[122] Meng Zhou, Zhengcai Cao, and Ye Wang. Robust fault detection and isolation based on finite-frequency H_-/H_∞ unknown input observers and zonotopic threshold analysis. *IEEE/CAA J. Autom. Sin.*, 6(3):750–759, 2019.

[123] Xi Zhou, Yiheng Wei, Shu Liang, and Yong Wang. Positive real lemmas for fractional order systems. In *Proceedings of the 53rd IEEE Conference on Decision and Control*, Los Angeles, pp. 6914–6919. IEEE, 2014.

Index

https://doi.org/10.1515/9783111628646-010

Fractional Calculus in Applied Sciences and Engineering

Volume 13
HongGuang Sun, Zhipeng Li, Shiqian Nie
Fractional and Fractal Derivative Models for Anomalous Sediment Transport
ISBN 978-3-11-134876-6, e-ISBN (PDF) 978-3-11-134883-4,
e-ISBN (EPUB) 978-3-11-134900-8

Volume 12
Mouffak Benchohra, Soufyane Bouriah, Abdelkrim Salim, Yong Zhou
Fractional Differential Equations. A Coincidence Degree Approach
ISBN 978-3-11-133434-9, e-ISBN (PDF) 978-3-11-133438-7,
e-ISBN (EPUB) 978-3-11-133446-2

Volume 11
Yong Zhou, Bashir Ahmad, Ahmed Alsaedi
Theory of Fractional Evolution Equations, 2022
ISBN 978-3-11-076918-0, e-ISBN (PDF) 978-3-11-076927-2,
e-ISBN (EPUB) 978-3-11-076936-4

Volume 10
Paweł Domański, Yang Quan Chen, Maciej Ławryńczuk (Eds.)
Outliers in Control Engineering, 2022
ISBN 978-3-11-072907-8, e-ISBN (PDF) 978-3-11-072912-2,
e-ISBN (EPUB) 978-3-11-072913-9

Volume 9
Ming Li
Theory of Fractional Engineering Vibrations, 2021
ISBN 978-3-11-072652-7, e-ISBN (PDF) 978-3-11-072615-2,
e-ISBN (EPUB) 978-3-11-072616-9

Volume 8
Vasily E. Tarasov, Valentina V. Tarasov
Economic Dynamics with Memory, 2021
ISBN 978-3-11-062460-1, e-ISBN (PDF) 978-3-11-062460-1,
e-ISBN (EPUB) 978-3-11-062481-6

Bei Fragen zur Produktsicherheit wenden Sie sich bitte an:
If you have any questions regarding product safety, please contact:

Walter de Gruyter
Genthiner Straße 13
10785 Berlin
productsafety@degruyterbrill.com